Plant Evolution and the Origin of Crop Species, 3rd Edition

_____

Plant Evolution and the Origin of Crop Species, 3rd Edition

James F. Hancock

Michigan State University, USA

www.cabi.org

CABI is a trading name of CAB International

CABI Head Office
Nosworthy Way
Wallingford
Oxfordshire OX10 8DE
UK

CABI North American Office
38 Chauncey St
Suite 1002
Boston, MA 02111
USA

Tel: +44 (0)1491 832111
Fax: +44 (0)1491 833508
Email: cabi@cabi.org
Web site: www.cabi.org

Tel: +1 617 395 4056
Fax: +1 617 354 6875
Email: cabi-nao@cabi.org

A catalogue record for this book is available from the British Library, London, UK.

Library of Congress Cataloging-in-Publication Data
Hancock, James F.
  Plant evolution and the origin of crop species / James F. Hancock. —3rd ed.
      p. cm.
  Includes bibliographical references and index.
  ISBN 978-1-84593-801-7 (alk.paper)
  1. Crops--Evolution. 2. Crops--Origin. 3. Plants--Evolution. I. Title.

SB106.O74H36 2012
633--dc23

2012001017

ISBN-13: 978-1-84593-801-7 (HB)
ISBN-13: 978-1-78064-477-6 (PB)

First published (HB) 2012
First paperback edition 2014

Commissioning editor: Meredith Carroll
Editorial assistants: Gwenan Spearing and Alexandra Lainsbury
Production editor: Tracy Head

Artwork provided by Marlene Cameron.
Typeset by SPi, Pondicherry, India.
Printed and bound in the UK by CPI Group (UK) Ltd, Croydon, CR0 4YY

Contents

Preface

_____

It has been almost 20 years since the first edition of this book was published, and the amount of information available on plant evolution and crop origins has skyrocketed, particularly in the molecular arena. Where the molecular technologies have had the greatest impact is in tracing crop origins – determining parental species, finding out where crops were domesticated and identifying the genes that made the domestications possible.

Molecular studies have also played a key role in testing long-standing evolutionary theories. Most of the hypotheses of Anderson, Grant, Heiser and Stebbins have proven robust, even as experimental technologies have dramatically shifted from morphological and cytogenetic comparisons to gene chips and genome sequencing. Hypotheses concerning hybrid speciation, polyploidy and the role of chromosome rearrangements have been tested, retested and expanded as each new technology has emerged. The molecular studies have generated some surprises, such as the amount of chromosome repatterning and gene expression alterations associated with interspecific hybridization, and the high level of gene duplication found in all sequenced genomes, but in general each new molecular technology has supported the evolutionary hypothesis formulated decades ago.

In this book I have tried to use, whenever possible, the historical evidence of evolutionary phenomena and crop origins. This may give the book an "old fashioned" feel, but I think that the information is still relevant and should not be lost in a cloud of new technologies. I have chosen to work from the past up to the present, rather than the reverse. It is likely that the growing wealth of molecular information will necessitate an alternative approach in any future editions, but so far I think that I can hold my ground.

Acknowledgments

_____

I would like to give special thanks to my wife Ann, who has so richly broadened my horizons with her passion and commitment to the living world around us. I would also like to thank the current crop of evolutionary biologists that I most admire: Norm Ellstrand, Loren Rieseberg, Tao Sang, Doug Schemske, Doug Soltis and Jonathan Wendel.

1

Chromosome Structure and Genetic Variability

_____

Introduction

Evolution is the force that shapes our living world. Countless different kinds of plants and animals pack the earth and each species is itself composed of a wide range of morphologies and adaptations. These species are continually being modified as they face the realities of their particular environments.

In its simplest sense, evolution can be defined as a change in gene frequency over time. Genetic variability is produced by mutation and then that variability is shuffled and sorted by the various evolutionary forces. The way organisms evolve is dependent on their genetic characteristics and the type of environment they must face.

A broad spectrum of evolutionary forces, including migration, selection and random chance, alter natural species. These same forces operated during the domestication of crops, as will be discussed in Chapter 6. In the next four chapters, we will describe these parameters and how they interact; but before we do this, we will begin by discussing the different types of genetic variability that are found in plants. The primary requirement for evolutionary change is genetic variability, and mutation generates these building blocks. A wide range of mutations can occur at all levels of genetic organization from nucleotide sequence to chromosome structure. In this chapter, we will discuss how plant genes are organized in chromosomes, and then we will discuss the kinds of genetic variability present and its measurement.

Gene and Chromosome Structure

Both gene and chromosome structure are complex. Genes are composed of coding regions called exons and non-coding regions called introns. Both the introns and the exons are transcribed, but the introns are removed from the final RNA product before translation (Fig. 1.1). There are also short DNA sequences, both near and far from the coding region, that regulate transcription but are not transcribed themselves. "Promotor" sequences are found immediately before the protein coding region and play a role in the initiation of transcription, while "enhancer" sequences are often located distant from the coding region and regulate levels of transcription.

Each chromosome contains not only genes and regulatory sequences, but also a large number of short, repetitive sequences. Some of these are concentrated near centromeres in the densely stained portions of chromosomes called heterochromatic regions, and may play a role in the homologous pairing of chromosomes and their separation. However, there are numerous other repeating units that are more freely dispersed over chromosomes and do not appear

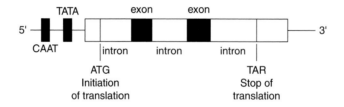

Fig. 1.1. Organization of a typical eukaryotic gene. A precursor RNA molecule is produced from which the introns are excised and the exons are spliced together before translation. The CAAT and TATA boxes play a role in transcription initiation and enhancement.

to have a functional role. These have been described by some as "selfish" or "parasitic", as their presence may stimulate further accumulation of similar sequences through transposition, a topic we will discuss more fully later.

The overall amount of DNA in nuclei can vary dramatically between taxonomic groups and even within species (Wendel *et al.*, 2002; Bennetzen *et al.*, 2005). The total DNA content of nuclei is commonly referred to as the genome. There is a 100-fold variation in genome size among all diploid angiosperms, and congeneric species vary commonly by threefold (Price *et al.*, 1986; Price, 1988). In some cases, genomic amplification occurs in a breeding population in response to environmental or developmental perturbations (Walbot and Cullis, 1985; Cullis, 1987). Most of these differences occur in the quantity of repetitive DNA and not in unique sequences.

Except in the very small genome of *Arabidopsis thaliana* (Barakat *et al.*, 1998), it appears that genes are generally found near the ends of chromosomes in clusters between various kinds of repeated sequences (Schmidt and Heslop-Harrison, 1998; Heslop-Harrison, 2000). The amount of interspersed repetitive DNA can be considerable, making the physical distances between similar loci highly variable across species. However, the gene clusters may be "hotspots" for recombination, making recombination-based genetic lengths much closer than physical distances (Mézard *et al.*, 2006).

Types of Mutation

There are four major types of mutation: (i) point mutations; (ii) chromosomal sequence alterations; (iii) chromosomal additions and deletions; and

(iv) chromosomal number changes. Point mutations arise when nucleotides are altered or substituted, for example, when the base sequence CTT becomes GTT. Chromosomal sequence alterations occur when the order of nucleotides is changed within a chromosome. Chromosomal duplications and deletions are produced when portions of chromosomes are added or subtracted. Chromosomal numerical changes arise when the number of chromosomes changes.

Point mutations

Nucleotide changes occur spontaneously due to errors in replication and repair at an average rate of 1×10^{-6} to 10^{-7}. These estimates have come largely from unicellular organisms such as bacteria and yeast, which are easy to manipulate and have tremendous population sizes, but good estimates have also been obtained in higher plants using enzymes (Kahler *et al.*, 1984) and a variety of seed traits (Table 1.1). Mutation rates can be increased by numerous environmental agents such as ionizing radiation, chemical mutagens and thermal shock.

Sequence alterations

Three types of DNA sequence alterations occur: translocations, inversions and transpositions. Small numbers of redundant nucleotide blocks may be involved or whole groups of genes. Translocations occur when nucleotide sequences are transferred from one chromosome to another. In homozygous individuals, nuclear translocations have no effect on fertility, but in heterozygous individuals only a portion of

Table 1.1. Spontaneous mutation rates of several endosperm genes in *Zea mays* (Stadler, 1942).

Gene	Character	No. of gametes tested	Mutation rate
R	Aleurone color	554,786	0.00049
I	Color inhibitor	265,391	0.00011
Pr	Purple color	647,102	0.000011
Su	Sugary endosperm	1,678,736	0.000002
C	Aleurone color	426,923	0.000002
Y	Yellow seeds	1,745,280	0.000002
Sh	Shrunken seeds	2,469,285	0.000001
Wx	Waxy starch	1,503,744	<0.000001

the gametes are viable due to duplications and deficiencies (Fig. 1.2). Translocations are widespread in a number of plant genera, including *Arachis, Brassica, Clarkia, Campanula, Capsicum, Crepis, Datura, Elymus, Galeopsis, Gossypium, Hordeum, Paeonia, Gilia, Layia, Madia, Nicotiana, Secale, Trillium* and *Triticum*.

Populations are generally fixed for one chromosomal type, but not all. Translocation heterozygotes are common in *Paeonia brownii* (Grant, 1975), *Chrysanthemum carinatum* (Rana and Jain, 1965), *Isotoma petraea* (James, 1965), and numerous species of *Clarkia* (Snow, 1960). Probably the most extreme example of translocation heterozygosity is in *Oenothera biennis* where all of its nuclear chromosomes contain translocations and a complete ring of chromosomes is formed at meiosis in heterozygous individuals (Cleland, 1972). In some cases, translocations have resulted in the fusion and fission of non-homologous chromosomes with short arms ("Robertsonian translocations").

Inversions result when blocks of nucleotides rotate 180°. Nuclear inversions are called *pericentric* when the rotation includes the centromere and *paracentric* when the centromeric region remains unaffected (Figs 1.3 and 1.4). Like translocations, individuals that are heterozygous produce numerous inviable gametes, but only if there is a crossover between chromatids; all the gametes of homozygotes are fertile regardless of crossovers.

Inversion polymorphisms have been described in a number of plant genera. One of the best documented cases of an inversion heterozygosity within a species is in *Paeonia californica*, where heterozygous plants are common throughout the northern range of the species (Walters, 1952). As we will describe more fully in the chapter on speciation, inversions on six chromosomes distinguish *Helianthus annuus, Helianthus petiolaris* and their hybrid derivative *Helianthus anomalus* (Rieseberg *et al.*, 1995). Tomato and potato vary by five inversions (Tanksley *et al.*, 1992), and pepper and tomato by 12 inversions (Livingstone *et al.*, 1999). The chloroplast genome of most angiosperm species has a large inverted repeat (Fig. 1.5), but its structure is highly conserved across families. Only a few species do not have the repeat and no intra-populational variation has been described (Palmer, 1985).

Transposition occurs when nucleotide blocks move from place to place in the genome (Bennetzen, 2000a; Fedoroff, 2000). Fragments from multiple chromosomal loci can even be fused to form new open reading frames (Jiang *et al.*, 2004; Hanada *et al.*, 2009). There are two major classes of transposons: DNA and RNA transposable elements. The RNA transposable elements (retroelements) amplify via RNA intermediates, while the DNA transposons rely on actual excision and reinsertment. Both classes of transpositions have been found in all plant species where detailed genetic analysis has been performed, and in many plant species, mobile elements actually make up the majority of the nuclear genome (SanMiguel and Bennetzen, 1998; Bennetzen, 2000a). Most of the transposons are inserted into non-coding regions, but sometimes they enter exons and when they do, they can have extreme effects on phenotype. The wrinkled-seed character described by Mendel is caused by a transpose-like insertion into the gene encoding a starch branching enzyme (Bhattacharyya *et al.*, 1990). Much of the flower color variation observed in the morning glory is due to the insertion and deletion of transposable elements (Clegg and Durbin, 2000; Durbin *et al.*, 2001).

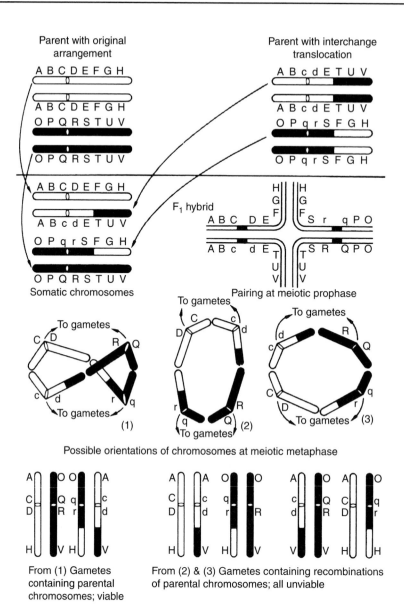

Fig. 1.2. Types of gametes produced by a plant heterozygous for a translocation. A ring of chromosomes is formed at meiosis and depending on how the chromosomes orient at metaphase and separate during anaphase, viable or inviable combinations of genes are produced. (Used with permission from T. Dobzhansky, © 1970, *Genetics of the Evolutionary Process*, Columbia University Press, New York.)

The DNA transposons range in size from a few hundred bases to 10 kb, and the most complex members are capable of autonomous excision, reattachment and alteration of gene expression. They all have short terminal inverted repeats (TIRs); the most complex ones encode an enzyme called transposase that recognizes the families' TIR and performs the excision and reattachment.

Retroelements (RNA transposons) are the most abundant class of transposons and they make up the majority of most large plant genomes. They transpose through reverse-transcription of RNA intermediates, and as a result they do not excise when they transpose, resulting in amplification. The most abundant class of retroelements in plants are the long

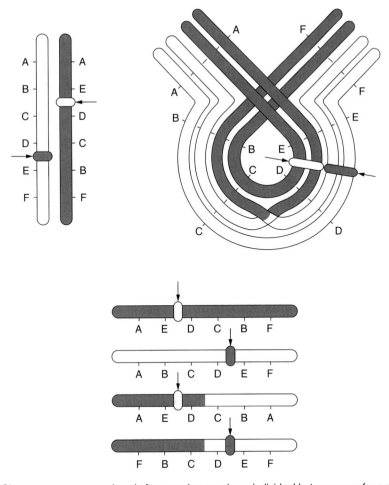

Fig. 1.3. Chromosome types produced after crossing over in an individual heterozygous for a pericentric inversion. Note the two abnormal chromatids, one with a duplication and the other with a deficiency. (Used with permission from T. Dobzhansky, © 1970, *Genetics of the Evolutionary Process*, Columbia University Press, New York.)

terminal repeat (LTR)-retrotransposons, varying in size from a few hundred to over 10,000 nucleotides.

Duplications and deficiencies

Chromosomal deficiencies occur when nucleotide blocks are lost from within a chromosome, while duplications arise when nucleotide sequences are multiplied. These are caused by unequal crossing-over at meiosis or translocation (Burnham, 1962). They also can occur when DNA strands mispair during replication of previously duplicated sequences (Levinson and Gutman, 1987). As previously mentioned, the genome is filled with high numbers of short, repeated sequences that vary greatly in length. So great, in fact, that some of them such as SSRs (simple sequence repeats) have proven valuable as molecular markers to distinguish species, populations and even individuals. Gene amplifications are so common that Wendel (2000) has suggested that "one generalization that has been confirmed and extended by the data emerging from the global thrust in genome sequencing and mapping is that most 'single-copy' genes belong to larger gene families, even in putatively diploid organisms". As more and more crop genomes are sequenced, it is becoming very clear that gene duplications are widespread. The fraction

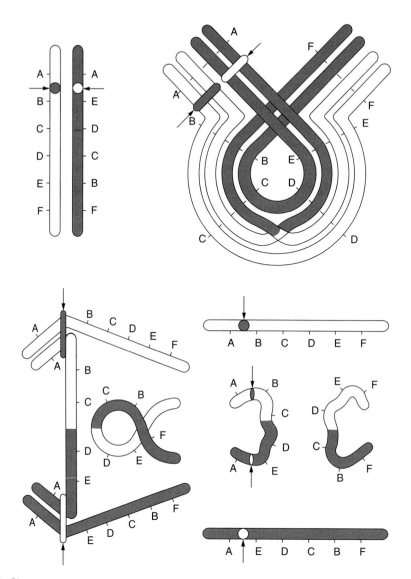

Fig. 1.4. Chromosome types produced after crossing over in an individual heterozygous for a paracentric inversion. Note the chromosomal bridge and the resulting chromatids with deletions. (Used with permission from T. Dobzhansky, © 1970, *Genetics of the Evolutionary Process*, Columbia University Press, New York.)

of the genome represented by duplications has been estimated to be 72% in maize (Ahn and Tanksley, 1993; Gaut and Doebley, 1997) and 60% in *Arabidopsis thaliana* (Blanc *et al.*, 2000).

Clusters of duplicated genes are often found scattered at multiple locations across the genome. When Blanc *et al.* (2000) compared the sequence of duplications on chromosomes 2 and 4 of *Arabidopsis thaliana*, they identified 151 pairs of genes, of which 59 (39%) showed highly similar nucleotide sequences. The order of these genes was generally maintained on the two chromosomes, except for a small duplication and an inversion. When they compared the sequence of these duplicated regions to the rest of the genome, they found 70% of the genes to be present elsewhere. The genes were duplicated in 18 large translocations and several smaller ones (Fig. 1.6).

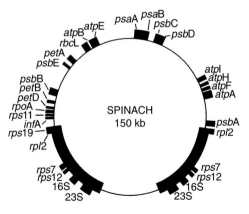

Fig. 1.5. The gene map of spinach chloroplast DNA. The two long thickenings in the lower half of the circle represent the inverted repeat. Gene designations: *rbc*L, the large subunit of ribulose bisphosphate carboxylase; *atp*A, *atp*B, *atp*E, *atp*H and *atp*I, the alpha, beta, epsilon, CF$_0$-I, CF$_0$-III, and CF$_0$-IV subunits of chloroplast coupling factor, respectively; *psa*A and *psa*B, the P700 chlorophyll-*a* apoproteins of photosystem I; *psb*A, *psb*B, *psb*C, *psb*D and *psb*E, the Q-beta (32 kilodaltons (kd), herbicide-binding), 51-kd chlorophyll-*a*-binding, 44-kd chlorophyll-*a*-binding, D2 and cytochrome-b-559 components of photosystem II; *pet*A, *pet*B and *pet*D, the genes for the cytochrome-f, cytochrome-b6 and subunit-4 components of the cytochrome-b6-f complex; *inf*A, initiation factor IF-1; *rpo*A, alpha subunit of RNA polymerase; *rpl*2, *rps*7, *rps*11, *rps*12 and *rps*19, the chloroplast ribosomal proteins homologous to *E. coli* ribosomal proteins L2, S7, S11, S12 and S19, respectively; 16S and 23S, the 16S and 23S ribosomal RNAs, respectively. (Used with permission from J.D. Palmer, © 1987, *American Naturalist* 130, S6–S29, University of Chicago Press, Chicago, Illinois.)

Chromosomal numerical changes

There are three primary types of numerical changes found in nuclear genomes: (i) aneuploidy; (ii) haploidy; and (iii) polyploidy. In aneuploidy, one or more chromosomes of the normal set are absent or present in excess. Haploids have half the normal chromosome set, while polyploids have more than two sets of homologous chromosomes.

Haploids are quite rare in nature, although they have been produced experimentally in many crops, including strawberries,

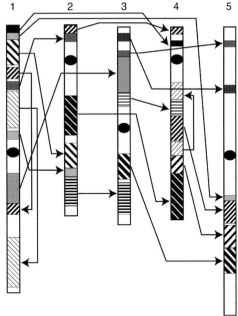

Fig. 1.6. Locations of duplications throughout the *Arabidopsis* genome. Similar blocks on different chromosomes are identified by similar colored regions and arrows. (Modified from Blanc *et al.*, 2000.)

lucerne and maize. They are generally inviable due to meiotic irregularities and low in vigor because lethal alleles are not buffered by heterozygosity.

Aneuploids are more common than haploids, although they are still relatively rare in natural populations. Most base numbers differ by a few chromosomes (*Citrus, Rubus, Poa, Nicotiana, Gossypium, Allium* and *Lycopersicon*), but extensive variations involving dozens of chromosomes occur in some groups (*Abelmoschus* and *Saccharum*). Aneuploids arise through the fusion and fission of chromosomes and when chromosomes migrate irregularly during meiosis. They are most common in polyploid species and hybrid populations resulting from transpecific crosses.

Polyploidy is quite prevalent in higher plants; between 35 and 50% of all angiosperm species are polyploid (Grant, 1981). The number of polyploid crop species is even higher (78%, Table 1.2) if we use the widely accepted assumption that chromosome numbers above $2n = 18$ represent polyploids. This is a conservative estimate, as there

Table 1.2. Chromosome numbers in selected crop species.

Diploid numbers	2n < 18	Polyploid numbers	2n > 18
Allium cepa (onion)	16	Abelmoschus esculentus (okra)	33–72
Avena sativa (oat)	14		
Beta vulgaris (sugarbeet)	18	Aleurites fordii (tung)	22
Beta oleracea (kale)	18	Ananas comosus (pineapple)	50
Carica papaya (papaya)	18	Avena abyssinica (oat)	28
Cicer arietinum (chickpea)	16	Avena sativa (oat)	42
Citrus spp. (citrus)	18	Avena nuda (oat)	42
Cucumis sativus (cucumber)	14	Arachis hypogaea (peanut)	40
Daucus carota (carrot)	18	Artocarpus altilis (breadfruit)	56, 84
Hordeum vulgare (barley)	14	Brassica campestris (turnip)	20
Lactuca sativa (lettuce)	18	Brassica napus (rape)	38
Lens culinaris (lentil)	14	Cajanus cajan (pigeon pea)	22
Medicago sativa (lucerne)	16	Camellia sinesis (tea)	30
Pennisetum americanum (bullrush millet)	14	Cannabis sativa (hemp)	20
		Capsicum spp. (peppers)	24
Pisum sativum (pea)	14	Carthamus tinctorius (safflower)	24
Prunus avium (sweet cherry)	16		
Prunus armeniaca (apricot)	16	Chenopodium quinoa (quinoa)	36
Prunus persica (peach)	16	Citrullus lanatus (watermelon)	22
Prunus amyglalus (almond)	16	Cocos nucifera (coconut)	32
Raphanus sativus (radish)	18	Coffee arabica (coffee)	44
Ribes nigrum (blackcurrants)	16	Colocasia esculenta (taro)	28
Ribes sativum (redcurrants)	16	Cucumis melo (muskmelon)	24
Secale cereale (rye)	14	Cucurbita spp. (squash)	20
Triticum monococcum (einkorn)	14	Dioscorea alata (yam)	30, 40, 50, 60, 70, 80
Vicia faba (bean)	12		
		Dioscorea esculenta (yam)	40
		Dioscorea rotundata (yam)	40
		Elaeis guineensis (oil palm)	32
		Elusine coracana (finger millet)	36
		Ficus carica (fig)	26
		Fragaria × ananassa (strawberry)	56
		Glycine max (soybean)	40
		Gossypium herbaceum (cotton)	26
		Gossypium arboreum (cotton)	26
		Gossypium hirsutum (cotton)	52
		Gossypium barbadense (cotton)	52
		Helianthus annuus (sunflower)	34
		Helianthus tuberosus (Jerusalem artichoke)	102
		Hevea brasiliensis (rubber)	36
		Impomoea batatas (sweet potato)	90
		Linum usitatissimum (linseed)	30, 32
		Lycopersicon esculentum (tomato)	24
		Malus × domestica (apple)	34, 51
		Mangifera indica (mango)	40
		Manihot esculenta (cassava)	36

Continued

Table 1.2. Continued.

Diploid numbers	2n <18	Polyploid numbers	2n >18
		Musa spp. (bananas)	22, 33, 44
		Nicotiana tabacum (tobacco)	48
		Olea europaea (olive)	46
		Oryza sativa (rice)	24
		Oryza globerrima (rice)	24
		Persea americana (avocado)	24
		Phaseolus vulgaris (common bean)	22
		Phaseolus coccineus (runner bean)	22
		Phaseolus acutifolius (tepary bean)	22
		Phaseolus lunatus (lima bean)	22
		Phoenix dactylifera (date)	36
		Pyrus spp. (pears)	34, 51
		Piper nigrum (pepper)	48, 52, 104, 128
		Prunus cerasus (sour cherry)	32
		Ricinus comminis (castor)	20
		Saccharum spp. (sugarcane)	60–205
		Sesamum indicum (sesame)	26
		Solanum melongena (aubergine)	24
		Solanum tuberosum (potato)	24, 36, 48, 60
		Sorghum bicolor (sorghum)	20
		Triticum timopheevi (wheat)	28
		Triticum turgidum (emmer)	28
		Triticum aestivum (bread wheat)	42
		Vigna unguiculata (cowpea)	22
		Vitus spp. (grape)	40
		Vaccinium corymbosum (blueberry)	48
		Zea mays (maize)	20

are known polyploid taxa, such as lucerne, that have haploid numbers lower than $x = 9$ (Grant, 1963). Most polyploids are thought to originate through the unification of unreduced gametes (Harlan and DeWet, 1975; Bretagnolle and Thompson, 1995). Only occasionally do polyploids arise through somatic doubling and the generation of polyploid meristems called "sports". Most commonly this is done artificially with the chemical colchicine.

Aneuploidy and polyploidy can have rather extreme effects on the physiology and morphology of individuals, since gene dosages are doubled. Cell sizes usually increase, developmental rates slow down, and fertility is often reduced. The whole of Chapter 4 is devoted to the physiological and evolutionary ramifications of polyploidy.

There are two major types of polyploids: autopolyploids and allopolyploids (amphiploids) (Fig. 1.7). Allopolyploids are derived from two different ancestral species, whose chromosome sets are unable to pair at meiosis. As a result, the chromosomes segregate and assort as in diploids, so inheritance of each individual duplicated loci of allopolyploids follows typical Mendelian patterns. In other words, allopolyploids display "disomic inheritance", where two alleles segregate at a locus. The chromosome sets of autopolyploids are derived from a single or closely related ancestral species and can pair at meiosis. The chromosomes associate together either as "multivalent" or random associations of bivalents. As a result, inheritance in autopolyploids does

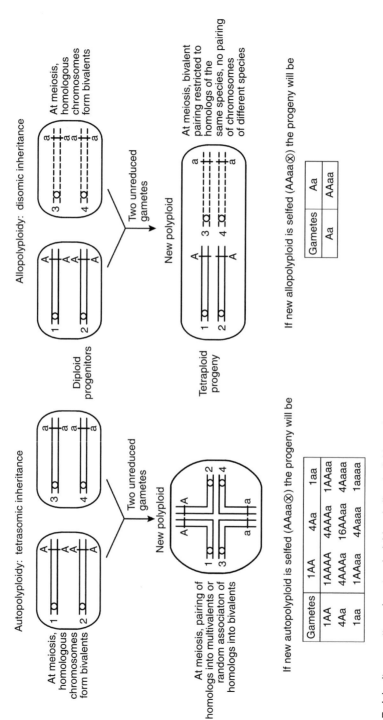

Fig. 1.7. Inheritance patterns in autopolyploid and allopolyploid species.

not follow typical Mendelian patterns, since the chromosomes carrying the duplicated loci can pair. Autopolyploids have more than two alleles at a locus and display what is called "polysomic inheritance".

The two major types of polyploids can be further divided into the following groups (Stebbins, 1947; Grant, 1981):

1. Autopolyploids
   (i) Unipartite autopolyploids (one progenitor, polysomic inheritance)
   (ii) Bipartite autopolyploids (closely related progenitors, polysomic inheritance)
2. Amphiploids
   (i) Segmental allopolyploids (partially divergent progenitors, mixed inheritance)
   (ii) Genomic allopolyploids (divergent progenitors, disomic inheritance)
   (iii) Autoallopolyploids (complex hybrid of related and divergent progenitors)

Unipartite autopolyploids are based on the doubling of one individual, while bipartite autopolyploids arise after the hybridization of distinct individuals within the same or closely related species. They form multivalents at meiosis or there is a random association of homologs into bivalents resulting in polysomic inheritance. The simplest case is tetrasomic inheritance (four alleles at a locus), which is found in autotetraploids. In some cases, polysomic inheritance is observed in polyploids derived from what have been classified as separate species (Qu et al., 1998). Because of their chromosomal behavior, these should be considered bipartite autopolyploids.

Genomic allopolyploids originate from separate species with well differentiated genomes, form mostly bivalents at meiosis and display disomic inheritance. The progenitors of segmental allopolyploids are differentiated structurally to an intermediate degree and have varying levels of chromosomal associations resulting in mixed inheritance. Autoallopolyploids have gone through multiple phases of doubling involving similar and dissimilar species.

Nuclear autopolyploids have been traditionally considered rarer than allopolyploids in wild and cultivated species, but the number of autopolyploid species may have been greatly underestimated (Soltis et al., 2007). Many of the previous classifications were based solely on morphological and cytogenetic data without information on inheritance patterns. Inheritance data are usually the only equivocal means of distinguishing between auto- and allopolyploidy (Hutchinson et al., 1983). Looking at metaphase I for bivalent pairing is insufficient, since chromosomal associations in autopolyploids can occur either as multivalents or as bivalents as described above. In fact, Ramsey and Schemske (1998) discovered in a survey of the published literature that autopolyploids average fewer mutivalents than expected due to random chiasmata formation, and allopolyploids average more. Tetrasomic inheritance has been documented in several bivalent pairing polyploids including lucerne (Quiros, 1982), potato (Quiros and McHale, 1985), Haplopappus (Hauber, 1986), Tolmiea (Soltis and Soltis, 1988), Heuchera (Soltis and Soltis, 1989b) and Vaccinium (Krebs and Hancock, 1989; Qu et al., 1998).

All plant organelle genomes are highly autopolyploid. There are 20–500 plastids per leaf cell and within each plastid are hundreds of identical plastid genomes (plastomes) depending on species, light levels and stage of development (Scott and Possingham, 1981; Boffey and Leech, 1982; Baumgartner et al., 1989). There are also high numbers of mitochondria per cell, but there are few estimates of genome copy number per mitochondria. Lamppa and Bendich (1984) reported 260 copies per leaf in mature pea leaves and 200–300 copies in etiolated hypocotyls of watermelon, zucchini and muskmelon.

Measurement of Variability

Plant evolutionists typically assess genetic variability at several different organizational levels from the actual DNA base sequence to quantitative morphological traits. In spite of the low mutation rates in plants, large amounts of genetic variation have accumulated at all levels of organization, and recent molecular technologies have uncovered an astonishing degree of genetic polymorphism in natural and domesticated plant populations.

Morphological variation has been described for both single and multiple gene systems by countless investigators. The first modern genetic analysis of Mendel (1866) was based on allelic variation at several independent loci in cultivated peas. He looked at discrete loci controlling such things as pod color, seed surface and leaf position. Since these seminal studies, the genetics of numerous monogenic traits have been described. A few examples are listed in Table 1.3.

In the early 1900s, geneticists began to wonder whether more continuous traits such as plant height and seed weight were inherited according to Mendelian laws. Johannsen (1903) showed that such variation was indeed influenced by genes, but that the environment also played a role – he was the first to distinguish between genotype and phenotype. Yule (1902) hypothesized that quantitative variation could be caused by several genes having small effects, and Nilsson-Ehle (1909) and East (1916) confirmed this suspicion using wheat and tobacco.

The greater the number of gene loci that determine a trait, the more continuous the variation will be (Fig. 1.8). The genes that have cumulative effects on variation in quantitative traits are called polygenes or quantitative trait loci (QTL). As Johannsen originally described, the expression of quantitative traits is confounded by the environment, so that variation patterns are generally a combination of both genetic and environmental influences.

Several statistical techniques have been developed to partition the total variability within a population into its genetic and environmental components (for excellent reviews see Mather and Jinks, 1977; Falconer and Mackay, 1996). The overall relationship can be written as:

$$V^P = V^G + V^E + V^{GE} \qquad (1.1)$$

where V^P represents the phenotypic or total variation within a population, V^G the genetic variation and V^E the environmental variation. V^{GE} represents the interaction between the environmental and genetic variance where the performance of the individuals is dependent on the particular environment in which they are placed.

Table 1.3. Commonly studied traits regulated by one gene (sources: Hilu, 1983; Gottleib, 1984).

Structure	Trait	Example
Flower	Corolla shape	*Primula sinensis* (primrose)
	Gender	*Cucumis sativus* (cucumber)
	Male sterility	*Solanum tuberosum* (potato)
	Petal number	*Ipomoea nil* (morning glory)
	Pistil length	*Eschscholzia californica* (poppy)
	Self-incompatibility	*Brassica oleracea* (cabbage)
	White versus colored	*Viola tricolor* (violet)
	2- or 6-rowed inflorescences	*Hordeum vulgare* (wild barley)
Leaf	Angle	*Collinsia heterophylla* (innocence)
	Chlorophyll deficiency	*Hordeum sativum* (barley)
	Cyanogenic glucosides	*Lotus corniculatus* (bird's foot trefoil)
	Leaflets versus tendrils	*Pisum sativum* (garden pea)
	Margin	*Lactuca serriola* (lettuce)
	Rust resistance	*Triticum aestivum* (wheat)
Seeds and fruit	Fruit location	*Phaseolus vulgaris* (common bean)
	Fruit pubescence	*Prunus persica* (peach)
	Fruit shape	*Capsicum annuum* (pepper)
	Fruit spiny	*Cucumis sativus* (cucumber)
	Fruit surface	*Spinachia oleracea* (spinach)
	Pod clockwise versus anticlockwise	*Medicago truncatula* (wild lucerne)
	Rachis persistence	Cereal grasses
	Seeds winged versus wingless	*Coreopsis tinctoria* (tickseed)
Physiological	Annual versus biennial	*Meliotus alba* (clover)
	Determinate versus indeterminate growth	*Lycopersicon esculentum* (tomato)
	Flowering photoperioid	*Fragaria × ananassa* (strawberry)
	Tall versus short stature	*Oryza sativa* (rice)

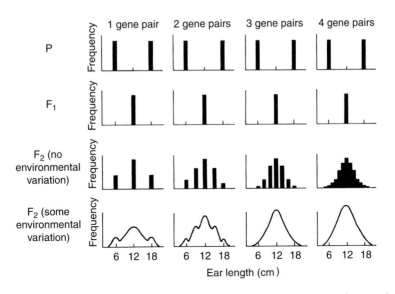

Fig. 1.8. Distribution of progeny from crosses involving plants that differ at one, two, three or six gene loci. The F_2 populations are shown without any environmental variation (row 3) and with 25% environmental variation (row 4). (Used with permission from Francisco Ayala, © 1982, *Population and Evolutionary Genetics: A Primer*, Benjamin/Cummings Publishing Company, Menlo Park, California.)

The genetic variance can be further broken down into additive (V^A), dominance (V^D) and epistatic or interaction variation (V^I). With additive genes, the substitution of a single allele at a locus results in a regular increase or decrease in a phenotypic value, e.g. aa = 4, aA = 5, AA = 6. Dominance effects occur when the heterozygote has the phenotype of one of the homozygotes, e.g. aa = 4, aA = 6, AA = 6. Epistatic effects occur when the influence of a gene at one locus is dependent on the genes at *another* locus, e.g. aA = 5 in the presence of Bb, but aA = 6 in presence of BB. Thus, genotypic variation can be written as:

$$V^G = V^A + V^D + V^I \tag{1.2}$$

and phenotypic variation can be written as:

$$V^P = V^A + V^D + V^I + V^E + V^{GE} \tag{1.3}$$

The most commonly employed measurement of quantitative variation is called heritability (h^2). Heritability is expressed in a "broad" or "narrow" sense, depending on which component of genetic variation is considered. Broad sense heritability is the ratio of the total genetic variance to the total phenotypic variance ($h^2 = V^G/V^P$). Heritability in the narrow sense is the ratio of just the additive genetic variation to the phenotypic variation – the effects of dominance and epistatic interactions are statistically removed ($h^2 = V^A/V^P$).

High heritabilities have been found in a plentitude of traits in both natural and cultivated populations (Table 1.4). Most measurement traits determining dimension, height and weight are quantitative, but numerous exceptions have been reported (Gottlieb, 1984). As we will discuss later, a large range in the contribution of quantitative trait loci (QTL) is often found. It is not unusual to find one gene that has a "major" effect on a trait and a number that "modify" its effects only slightly.

Over the last several decades, plant evolutionists have used several classes of biochemical compounds to further assess levels of genetic variability and calculate evolutionary relationships. Alkaloids and flavonoids achieved some popularity (Harborne, 1982), but gel electrophoresis of enzymes became the predominant mode of measuring genetic variability until the last two decades. This technique takes advantage of the fact that proteins with different amino acid sequences often have different charges and physical conformations, and they migrate

Table 1.4. Broad sense heritability estimates for several traits in representative plant species.

Species	Trait	Heritability	Reference
Solanum tuberosum (potato)	Tuber number	0.25	Tai (1976)
	Tuber weight	0.87	
	Yield	0.26	
Avena sativa (oat)	Days to heading	0.86	Sampson and Tarumoto (1976)
	Plant height	0.90	
	Stem diameter	0.76	
	Yield	0.70	
Fragaria × *ananassa* (strawberry)	Fruit size	0.20	Hansche *et al.* (1968)
	Firmness	0.46	
	Appearance	0.02	
	Yield	0.48	
Holeus lanatus (grass)	Tiller weight	0.24	Billington *et al.* (1988)
	Stolon number	0.17	
	Stolon weight	0.15	
	Leaf width	0.17	
	Flowering time	0.10	
	Inflorescence number	0.14	

at different rates through a charged gel matrix (Fig. 1.9). The major advantage of this type of analysis is that many loci and individuals can be measured simultaneously and most alleles are codominant so that heterozygotes can be identified (Fig. 1.10).

Electrophoretic variability in plant and animal populations is described in a number of different ways. Different molecular forms of an enzyme that catalyze the same reaction are called *isozymes* if they are coded by more than one locus, and *allozymes* if they are produced by different alleles of the same locus. An individual with more than one allozyme at a locus is referred to as *heterozygous*. A population or species with more than one allozyme at a locus is called *polymorphic*. Populations are usually represented by their proportion of polymorphic loci (P) or the average frequency of heterozygous individuals per locus (H).

Striking levels of polymorphism have been observed within most of the plant species examined for electrophoretic variation. An average of 2–3 alleles at a locus is the norm in natural and cultivated populations (Table 1.5), and plant breeders have even found sufficient variability in cultivated varieties to use isozymes in varietal patent applications (Bailey, 1983). Populations of the same species (conspecific) generally have

genetic identity values in the range of 0.95–1.00, while identities between species range from 0.28 in *Clarkia* to 0.99 in *Gaura*, with an average 0.67 ± 0.04 (Gottleib, 1984). The genetic identity of crops and their wild progenitors is generally above 0.90 (Doebley, 1989).

While the use of electrophoresis uncovers more variability than morphological and physiological analyses, it still misses a substantial amount of the variability in the total DNA sequence. The DNA code is redundant, so mutations in many base pairs do not result in amino acid substitutions and not all amino acid substitutions result in large enzyme conformational or activity changes. Only about 28% of the nucleotide substitutions cause amino acid replacements that change electrophoretic mobility (Powell, 1975).

A number of DNA marker systems have been developed that more fully represent the molecular diversity in plants (Schlötterer, 2004). These molecular markers measure variability directly at the DNA sequence level and thus uncover more polymorphisms than isozymes, and they are frequently more numerous, allowing for examination of a greater proportion of the genome. There are five major marker systems that have emerged: (i) restriction fragment length polymorphisms (RFLPs), where the

(a)

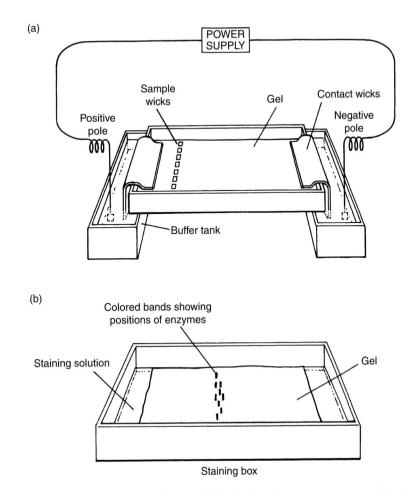

(b)

Fig. 1.9. Technique of starch gel electrophoresis. (a) Crude tissue homogenates are extracted in a buffer, loaded into paper wicks and placed in a gel, which is subjected to an electric current. (b) The enzymes migrate at different rates in the gel for several hours and their position is visualized by removing the gel from the electrophoresis unit and putting it in a box with protein-specific chemicals. The genotype of each individual can be determined from the spots (bands) that develop on each gel. (Used with permission from Francisco Ayala, © 1982, *Population and Evolutionary Genetics: A Primer*, Benjamin/Cummings Publishing Company, Menlo Park, California.)

DNA is cut into fragments using enzymes called restriction endonucleases and specific sequences are identified after electrophoresis by hybridizing them with known, labeled probes; (ii) randomly amplified polymorphic DNA (RAPDs), where a process called polymerase chain reaction (PCR) is used to amplify random sequences; (iii) simple sequence repeats (SSRs) or microsatellites, where PCR is used to amplify known repeated sequences; (iv) amplified fragment length polymorphisms (AFLPs), where the DNA is cut with restriction enzymes and the resulting fragments are amplified with PCR; and (v) single nucleotide polymorphisms (SNPs), where single nucleotide differences are examined in DNA sequences amplified by PCR (Rafalski, 2002; Hamblin *et al.*, 2007).

In the RFLP analysis, DNA is digested by specific restriction enzymes, which recognize 4, 5 or 6-base sequences. For example, the enzyme Hind II recognizes and cleaves the nucleotide sequence CCGG. The enzyme cuts the DNA

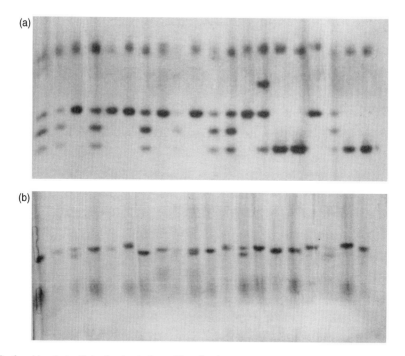

Fig. 1.10. A gel loaded with leaf extracts from 20 red oak trees and stained for either (a) leucine aminopeptidase (LAP) or (b) phosphoglucoisomerase (PGI). LAP is a monomeric enzyme (one subunit) and plants with one band are homozygous and those with two bands are heterozygotes. PGI is a dimeric enzyme (two subunits) and plants with only one band are also homozygous, but heterozygous individuals have three bands. (Picture courtesy of S. Hokanson.)

wherever this recognition site occurs in the molecule. The samples are then electrophoresed in agarose or acrylamide gels and the different fragments migrate at different rates due to their size differences (Fig. 1.11). In cases where the fragments are quite prevalent they can be seen directly under ultraviolet light after staining with ethidium bromide (Fig. 1.12). Because of the low complexity of the chloroplast genome and its high copy number, such restriction site analyses were widely used by molecular systematists to construct phylogenies. However, when the fragments are uncommon, as is usually the case with nuclear sequences, they are "blotted" from the gel on to a nitrocellulose filter, denatured into single-stranded DNA, and known pieces of labeled DNA are hybridized with the test samples (called Southern analysis). The fragments that "light up" are the RFLPs. These markers are codominant, meaning that both alleles can be recognized in heterozygous individuals.

In the RAPD analysis involving PCR, a reaction solution is set up, which contains DNA, short DNA primers (usually ten oligonucleotides), the four nucleotide triphosphates found in DNA (dNTPs) and a special heat stable enzyme, Taq polymerase. The DNA strands are separated by heating and then cooled, allowing the primers to hybridize (anneal) to complementary sequences on the DNA (Fig. 1.13). The polymerase enzyme then synthesizes the DNA strand between the primers. The solution is then heated and cooled in numerous cycles, and the DNA is amplified over and over again by the same set of events. The resulting fragments are then separated by electrophoresis, as in the RFLP analysis, and visualized by staining them with ethidium bromide. RAPD polymorphisms originate from DNA sequence variation at primer binding sites (whether the primers anneal or not) and DNA length differences between primer binding sites due to insertion or deletions of nucleotide sequences. RAPD fragments are dominant, being either present (dominant) or absent (recessive) and, as a result, heterozygotes cannot be distinguished from homozygous dominant individuals (Fig. 1.14).

Table 1.5. Genetic variability uncovered by electrophoresis in selected plant species (sources: Gottleib, 1981; Doebley, 1989).

Species	Number		Average no. of alleles	Proportion of loci polymorphic
	Populations	Loci		
Capsicum annuum var. annuum[a]	–	26	1.4	0.31
Chenopodium pratericola	26	12	2.2	0.33
Clarkia rubicunda	4	13	2.3	0.46
Cucurbita pepo var. ovifera[a]	17	12	1.7	0.32
Gaura longifolia	3	18	3.2	0.33
Glycine max[a]	109	23	2.1	0.36
Lens culinaris subsp. culinaris[a]	31	15	1.6	0.33
Limnathes alba subsp. alba	6	15	2.8	0.80
Lycopersicon cheesmanii	54	14	2.6	0.57
Lycopersicon pimpinellifolium	43	11	4.5	1.00
Oenothera argillicola	10	20	2.2	0.20
Phlox cuspidata	43	20	2.4	0.25
Phlox drummondii	73	20	2.3	0.30
Picea abies	4	11	3.9	0.86
Pinus ponderosa	10	23	2.2	0.61
Raphanus sativus[a]	24	8	2.6	0.49
Stephenomeria exigua subsp. carotifera	11	14	4.8	0.57
Tragapogon dubius	6	21	2.5	0.24
Xanthium strumarium subsp. chinense	2	12	2.5	0.17
Zea mays subsp. mays	94	23	7.1	0.50

[a]Crop species.

In AFLP analysis, the DNA from plants of interest is first cut with restriction enzymes. Specific primer sequences of DNA are then attached (ligated) to the resulting fragments and these are then amplified using PCR. The resulting fragments are separated by electrophoresis and visualized with DNA specific stains. AFLP polymorphisms originate from DNA sequence variation (where the cut sites are located) and DNA length differences between primer binding sites. They are dominant markers like RAPDs, based on the presence or absence of a band (Fig. 1.15).

The analysis of SSRs or microsatellites is similar to RAPDs, except that primers are used that flank known rather than unknown sequences. The recognized sequences are highly repeated units of 2–5 bases (for example ... GCAGCAGCA...). These repeat units are quite common in plants and can consist of hundreds of copies, which vary greatly in number across individuals. They are identified by cloning DNA fragments of individuals into bacterial carriers or vectors (bacterial artificial chromosomes, BACs), determining which BACs carry the repeat units by Southern hybridization, and then sequencing

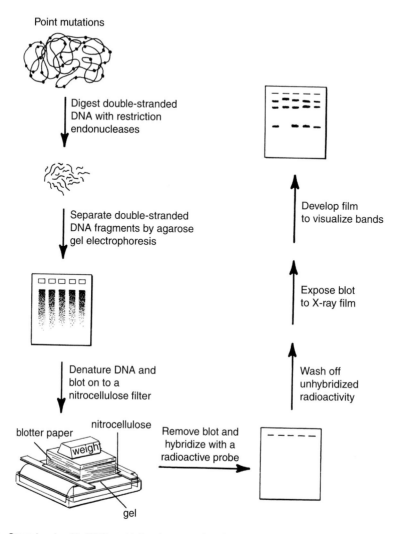

Fig. 1.11. Steps involved in DNA restriction fragment length analysis. DNA is cut into pieces using restriction endonucleases; the dots depicted above on the DNA strands represent cut sites. The different-sized fragments are separated by electrophoresis in a gel made of agarose and then are transferred by blotting on to a nitrocellulose filter. The fragments are then denatured into single-stranded DNA and known pieces of radioactively labeled DNA are hybridized with them. The filters are then placed on X-ray film and bands "light up" where successful hybridizations have occurred. (Drawing courtesy of M. Khairallah.)

the DNA flanking the microsatellite regions. Primers are then developed from the flanking sequences that will amplify the microsatellites using PCR. Polymorphisms in SSRs are observed when there are different numbers of tandem repeats in different individuals. The SSRs are codominant (heterozygotes can be recognized) (Fig. 1.16).

In a SNP analysis, previous knowledge about the presence of nucleotide variation in the genome is used to develop PCR primers to amplify DNA fragments containing single nucleotide polymorphisms. The amplified segments are then examined to determine which SNP individuals carry. This is commonly done by sequencing, although other techniques can be

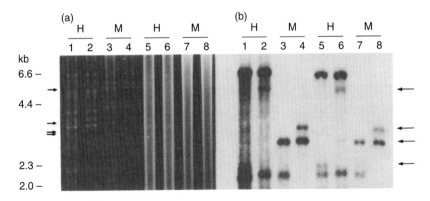

Fig. 1.12. Ethidium bromide stained gel (a) and Southern blot (b) of two genotypes of lucerne. Lanes 1–4 represent purified chloroplast DNA and lanes 5–8 contain total cell DNA. The DNA was digested with two enzymes Hind II (H) and Msp I (M), and hybridized with a piece of tomato plastid DNA in the Southern analysis. The total cell DNA is blurred in the ethidium bromide stained gel because it represents a much larger genome than the plastid DNA and has many more cut sites, which produce a continual array of fragment lengths (Schumann and Hancock, 1990).

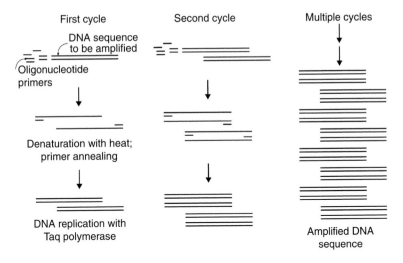

Fig. 1.13. Schematic drawing of the polymerase chain reaction (PCR). See text for details.

utilized (Schlötterer, 2004; Ganal *et al.*, 2009). SNP markers are codominant.

In comparing the various marker systems (Table 1.6), RFLPs have the advantage of being codominant and often represent known DNA sequences, but they are sometimes avoided because radioactivity is used in the process and the evaluation of segregating populations takes the greatest investment of time. RAPDs are probably the cheapest markers, but suffer from problems with reproducibility and the fact that they

are dominant. SSRs produce extremely high levels of allelic variability, but they are very expensive to generate. AFLPs produce the highest amount of variability per gel run of any of the marker systems, but their dominance remains limiting. SNPs can be found at many more loci than the other types of markers, but they are generally limited to two alleles at a locus. Their cost of development is very high.

At present, SSRs are the most commonly used markers in plant evolutionary studies.

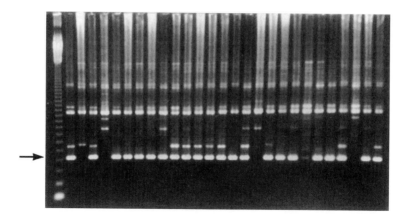

Fig. 1.14. A gel showing randomly amplified polymorphic DNA (RAPD) fragments of blueberries. (Photo courtesy of Luping Qu.)

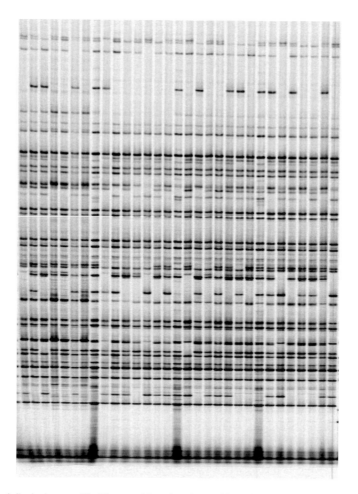

Fig. 1.15. A gel displaying amplified fragment length polymorphisms (AFLPs) of sugarbeets. (Picture courtesy of Daniele Trebbi.)

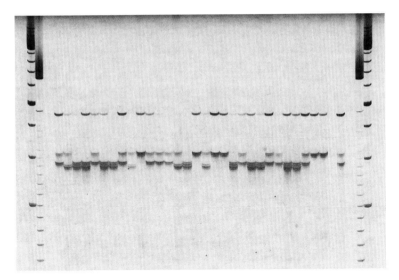

Fig. 1.16. A gel showing simple sequence repeats (SSRs) of sour cherries. From left: lanes 1 and 2, size markers; 3, blank; 4, 'NY 54'; 5 'Emperor Francis'; 6–35, first 30 individuals in an F1 population between 'NY54' and 'Emperor Francis'; 36, blank; 37 and 38, size markers. The marker was generated from a genomic library of almond. (Picture courtesy of Jim Olmstead.)

Table 1.6. A comparison of the various molecular marker systems.

Characteristic	Marker system				
	AFLPs	RAPDs	RFLPs	SSRs	SNPs
Cost of development	Low	Low	Intermediate	High	Very high
Genotyping cost	High	Low	High	Low	High
Cost per marker	Low	Low	High	Intermediate	High
Number of polymorphic loci	High	Intermediate	Intermediate	Intermediate	Very high
Number of alleles per locus	Low	Low	Intermediate	High	Low
Nature of gene action	Dominant	Dominant	Codominant	Codominant	Codominant

AFLPs are probably the leading choice when other molecular markers are not available in a crop or high numbers of new markers are desired for systematic or mapping projects. RFLPs remain important in those crops where they have already been developed and their utility has been established. RAPDs are only used where low cost and rapid development are deemed most important. SNPs appear to be the marker choice of the future, as genomic sequence data become more available. Because a high proportion of loci contain them, they are particularly useful where broad chromosomal coverage is needed.

Biochemical and DNA markers have been successfully used to estimate genetic distance in a wide range of native and crop species. For the codominant marker data (isozymes, RFLPs, SSRs and SNPs), genetic distances are often estimated using the equation of Nei and Li (1979). For dominant markers (RAPD and AFLPs), a simple matching coefficient is generally employed to measure genetic distance (Jaccard, 1908).

High levels of genetic variation have also been elucidated by the sequencing of a number of known plant genes. The mainstay of molecular phylogenetic studies has been rbcL (large subunit of ribulose 1,5-bisphosphate carboxylase/oxygenase) and the 18S ribosomal DNA. These genes have proven most useful for inferring higher relationships, as their sequences are evolutionarily conserved. Several other chloroplast genes have also been commonly used in phylogenetic studies, including rDNA internal transcribed spacer (ITS), atpB (a subunit of ATP synthase), ndhF (a subunit of NADH dehydrogenase), matK (a maturase involved in splicing introns) and the atpB–rbcL intergenic region. Rates of evolution in matK, ndhF and the atpB–rbcL intergenic region appear to be high enough to resolve intergeneric and interspecific relationships.

While plant molecular systematists have relied heavily on genes from the chloroplast, the use of nuclear gene sequences is also common. Most efforts have involved ribosomal DNA, although several other genes have gained favor, including adh1 and adh2, G3pdh (glyceraldehyde 3-phosphate dehydrogenase) and Pgi (phosphoglucose isomerase). The rDNA gene is composed of several genes including the 5S, 5.8S, 18S and 26S rDNAs, and ITS. The 18S and 26S sequences are highly conserved and have been used to resolve higher order relationships, while the ITS regions are much more variable and have proven useful in distinguishing species. The 5S and 5.8S have been little used.

Variation in Gene Expression Patterns

Over the last decade, a growing number of studies has been conducted to assess levels of variability in gene expression patterns. These have been utilized most heavily in plant evolutionary studies to determine the effects of hybrid speciation and polyploidy across the whole genome (see Chapters 4 and 5). The most common method of evaluation has been a variant of the membrane-based Southern hybridization approach described in the section on RFLP analysis, where small silica chips are covered by thousands of short nucleotides of known sequence. This approach, called gene chip microarray analysis, allows for changes in gene expression to be sensitively measured for very large numbers of genes. In short, complementary DNAs

(cDNAs) are made from RNA isolated from the plants of interest. These cDNAs are labeled and allowed to hybridize with the spots of DNA on the gene chip. The intensity of signal of each probe is proportional to the amount of cDNA bound to it. Most recently, the analysis of relative gene expression is also being done in what is called "real time quantitative PCR", where a targeted DNA molecule is amplified and simultaneously quantified.

Tracing the Evolutionary and Geographical Origins of Crops

Molecular marker frequency data have been used in a number of instances to trace the evolutionary and geographical origins of crops (e.g. Heun et al., 1997; Badr et al., 2000; Kwak et al., 2009). This is done by collecting a representative sample of wild and domesticated genotypes from across their geographical range, analyzing each genotype for a wide array of molecular markers and then calculating genetic distances or identity values between each of the genotypes using individual marker frequency data. After the genetic identities are calculated, a phenetic analysis is conducted to generate a visual representation of the genetic relationships among the sampled genotypes. These can be in the form of a tree-based diagram or principal component plot.

The trees are commonly built using a nearest-neighbor joining analysis, where the genotypes are placed on the tree by their degree of similarity. The branch lengths on the tree represent the genetic distance between the branches or nodes.

One example of this type of analysis can be found in the work of Özkan et al. (2002), where they evaluated AFLP marker data at 204 loci from 43 domesticated hexaploid and 99 wild tetraploid populations of wheat from a broad area of the Middle East. They found that wild samples from southeastern Turkey were most closely related to the crop lines and they speculated that this was the region of domestication (Fig. 1.17).

In a principal component analysis, allelic frequency data from a group of individuals are placed in a multi-dimensional matrix, and a "best-fitting line" (first principal component) is calculated that minimizes the sum of squares

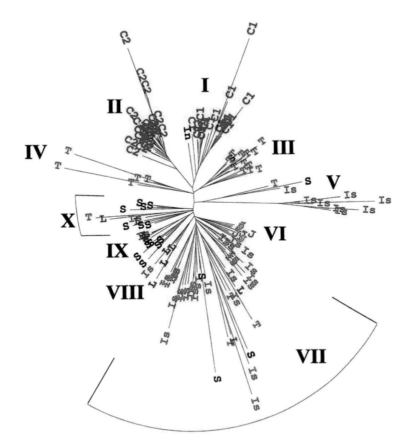

Fig. 1.17. Amplified fragment length polymorphism (AFLP) phylogeny of 99 wild emmer lines from southeastern Turkey (T; 22), Israel (Is; 37), Jordan (J; 8), Lebanon (L; 13), Syria (S; 18), Iran (In; 1) and 43 cultivated tetraploid wheats from across the world (C1 = 19 hulled landraces of emmer wheat; C2 = 24 free-threshing hard wheat varieties). Cultivated emmer lines were from Turkey, Romania, Iran, India, Germany and Italy, the hard wheat lines were from France, Palestine, Italy, Tunisia, Jordan, Cyprus, Spain, Syria, Greece, Ukraine, Tajikistan and Mexico. The neighbor-joining tree (Saitou and Nei, 1987) of Dice (1945) genetic distances is shown. Note that the wild lines clustering the closest to the cultivated ones (C1 and C2) are from southeastern Turkey (T). (Redrawn from Özkan *et al.* (2002) AFLP analysis of a collection of tetraploid wheats indicates the origin of emmer and hard wheat domestication in Southeast Turkey. *Molecular Biology and Evolution* 19, 1797–1801.)

of the distances of each data point from the line itself (Cavalli-Sforza and Bodmer, 1971). A second principal component is then generated at right angles (orthogonal) to the first and a third axis can be constructed perpendicular to the first two. The amount of information provided by each axis is estimated as the proportion of the total variance. An example of this approach can be found in Salamini *et al.* (2004), where they reanalyzed the data of Özkan *et al.* (2002) using a principal component analysis and came to the same conclusions (Fig. 1.18).

Sequence data have also been used to trace crop ancestries. In this approach, individual genes are sequenced in a wide array of germplasm, including representatives of the domesticated species and putative wild relatives. This information is often depicted as "gene trees" that represent the inferred relationships among the alleles found in the germplasm being evaluated (Schaal and Olsen, 2000). Alleles are placed on trees based on the minimal number of mutational steps between them. In these diagrams, observed alleles are generally represented by letters or

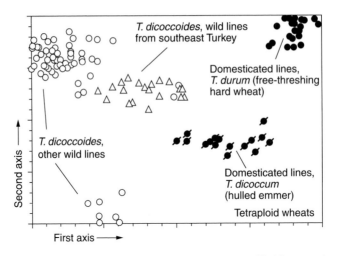

Fig. 1.18. Results of the principal coordinate analysis (PCA) of the amplified fragment length polymorphism (AFLP) data sets used for the phylogenetic reconstructions of relationships between domesticated and wild lines of wheat using the data base of Özkan *et al.* (2002). Nineteen domesticated lines of hulled emmer, 23 of domesticated free-threshing hard wheat, and 95 wild emmer lines originating from several locations in the Fertile Crescent were included. The analyses were carried out using the computer package NTSYSPc (Rohlf, 1998). The pairwise genetic distances were calculated according to the JACCARD algorithm (Jaccard, 1908). Note that the wild lines from southeast Turkey (Karakadag) clustered closest to the domesticated lines. (Redrawn from Salamini *et al.* (2004) Comment on "AFLP data and the origins of domestic crops". *Genome* 47, 615–620.)

symbols, while putative intermediate alleles that were not sampled are represented by dots. As an example, Olsen and Schaal (1999) were interested in determining where *Manihot esculenta* (cassava) was domesticated and whether its close relative *Manihot pruinosa* was part of the domestication process. They extracted DNA from 212 individuals of cultivated cassava, wild *M. esculenta* subsp. *flabellifolia* and its close relative *M. pruinosa* from several Brazilian states and sequenced *G3pdh*. They identified 28 different sequences (called haplotypes) among the individuals and found that only six are present in cultivated cassava (Fig. 1.19). These six haplotypes were only found in wild *M. esculenta* subsp. *flabellifolia* in the southern Amazon border region of Brazil, which indicates that *M. pruinosa* probably played little role in the origin of cassava and that domestication may have occurred in a relatively limited region.

It is important to note that several authors have addressed the potential pitfalls associated with using DNA fingerprinting and phylogenetic analysis to determine where crops originated, particularly in making arguments about whether crop domestications occurred once (monophyletic) or multiple times (polyphyletic) (Abbo *et al.*, 2001; Allaby *et al.*, 2008; Gross and Olsen, 2010). First, the analysis is only as good as the breadth of sampling, as too limited sampling could exclude additional areas of domestication. Secondly, a species could have been domesticated at multiple locations but evidence of only one line now persists in contemporary germplasm if all the others were replaced by a single superior line. It is also possible that more than one domesticated lineage may have emerged, but the evidence of all but one was lost through hybridization, repeated backcrossing and selective breeding. In this case, single genes may trace back to multiple domestications, but the majority of the remaining genes would trace to a single progenitor (Nichols, 2001).

Construction of Genetic Maps and Genome Evolution

Molecular markers have been widely used to construct genetic maps of species. These maps have greatly enhanced our knowledge

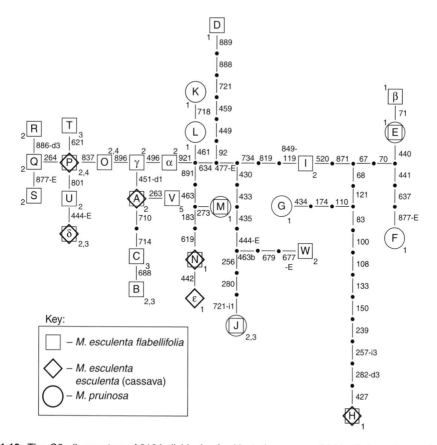

Fig. 1.19. The *G3pdh* gene tree of 212 individuals of cultivated cassava, wild *Manihot esculenta* subsp. *flabellifolia* and its close relative *Manihot pruinosa* from several Brazilian states. Shapes around letters indicate the taxa in which haplotypes were found. Each line between haplotypes represents a mutational step, with numbers on lines indicating the variable base pair position. Note that only six haplotypes were found in cultivated cassava (diamonds around symbols) and these were only present in wild *M. esculenta* subsp. *flabellifolia* and not in *M. pruinosa*. This indicates that *M. pruinosa* was probably not a progenitor of cultivated cassava. (Used with permission from Olsen, K.M. and Schaal, B.A. (1999) Evidence on the origin of cassava: Phylogeography of *Manihot esculenta. Proceedings of the National Academy of Sciences USA* 96, 5586–5591.)

of genome structure and evolution, although comparisons of whole genome sequences will provide even more information in the very near future as more and more crops are sequenced. By comparing maps of related species, we can evaluate the kinds of genomic changes that accompany speciation. Genetic maps are constructed by observing the frequency of recombination between markers in large segregating populations. It is assumed that the frequency of recombination is associated with the degree of linkage or the physical distance across chromosomes. A single map

unit is called a centimorgan (cM) and is equal to 1% recombination.

The degree of linkage conservation or "genome evolution" appears to vary greatly across plant species (Bennetzen, 2000b; Paterson *et al.*, 2000). Gene order is highly conserved among the *Fabaceae* lentil and pea (Weeden *et al.*, 1992), mung bean and cowpea (Menacio-Hautea *et al.*, 1993), and the solanaceous potato and tomato (Tanksley *et al.*, 1992). Linkage arrangements in the *Poaceae* species maize, rice and sorghum are so highly conserved that a circular map can be constructed that contains representatives of all three species (Devos and Gale, 2000). However, the *Gramineae* rye and wheat differ by at least 13 chromosomal rearrangements (Devos *et al.*, 1992a,b), and in the *Solanaceae*, pepper is quite divergent from its relative tomato (Livingstone *et al.*, 1999). Among the highest levels of divergence are found among the diploid and amphiploid *Brassica* species (Paterson *et al.*, 2000).

In spite of these genomic reshufflings, linkage relationships within gene blocks are often sufficiently conserved that numerous homologous regions can be identified across species, even in groups as widely diverged as *Arabidopsis* and tomato (Ku *et al.*, 2000), and *Arabidopsis*, common bean, cowpea and soybean (Lee *et al.*, 2001). The location of gene clusters is greatly shuffled across species, but the order of the genes within the clusters is highly conserved.

Molecular markers have also provided a means to dissect the genetic basis of the complex traits that are regulated by many individual QTL (Tanksley, 1993; Paterson *et al.*, 1998). The analysis is done by developing a genetic map of molecular markers, and searching for those markers whose presence is significantly associated with a phenotypic difference in a large segregating population, such as large versus small seeds. A significant association indicates that the marker is genetically linked to one or more of the QTL. The number of genes regulating a particular trait can be determined through this process along with an estimate of their relative importance. Plant breeders now regularly employ QTL analyses to identify the key genes regulating agronomic traits, and evolutionary biologists have found the technique to be valuable in identifying the genetics of important adaptive traits and reconstructing the speciation process (Rieseberg, 2000). This technique has been employed to identify key loci associated with the domestication of maize, bean and rice (and others), a topic that will be covered more fully in the chapter on plant domestication and speciation.

Summary

Most species have accumulated high levels of genetic variability through the constant slow pressure of mutation. Widespread variation has been observed in nucleotide sequence (point mutations), gene order on chromosomes (translocations, inversions and deficiencies), gene number (deficiencies and duplications) and genome number (polyploidy). Polyploidy is particularly important in plants, as up to half of all species probably originated in this manner. Much recent effort has concentrated on measuring molecular variation in populations and species. Levels of variation in most species are strikingly high, and a wide array of marker systems have now been successfully used to estimate genetic distance in numerous plant species, and trace where crop domestications first occurred. Genetic maps have also been used to compare rates of genome evolution. The degree of genome evolution appears to vary greatly across plant species, although blocks of genes can be found whose linkage relationships are conserved across many plant groups.

2

Assortment of Genetic Variability

_____

Introduction

Evolution is the process by which the genetic constitution of populations is changed through time. Evolutionary change is a two-step process. First, mutation produces hereditary variation and then that variability is shuffled and sorted from one generation to another. The major forces of evolution are *migration, selection* and *genetic drift*. Migration is a movement of genes within and between populations via pollen or seeds. Selection is a *directed* change in a population's gene frequency due to differential survival and reproduction. Genetic drift is a non-directed change in a population's gene frequency due to random events.

This chapter is dedicated to the processes by which gene frequencies change. We will first describe how populations behave with completely random mating and no extenuating circumstances, and then we will discuss how gene frequencies are influenced by migration, drift and selection. The various evolutionary forces will be described individually, as if the others were not acting, and then in concert.

Random Mating and Hardy–Weinberg Equilibrium

In 1908, G.H. Hardy and W. Weinberg formulated what has become known as the "Hardy–Weinberg Principle". What these two men discovered mathematically is that genotype frequencies will reach equilibrium in one generation of random mating in the absence of any other evolutionary force (Hardy, 1908; Weinberg, 1908). The frequencies of different genotypes will then depend only upon the allele frequencies of the previous generation. If gene frequencies do not accurately predict genotype frequencies, then plants are crossing in a non-random fashion or some other evolutionary force is operating.

The general relationship between gene and genotype frequencies can be described in algebraic terms. If p is the frequency of one allele (A) in a population and q the frequency of another (a), then $p + q = 1$ when there are no other alleles. The equilibrium frequencies of the genotypes are given by the expansion of the binomial:

$$(p+q)^2 = p^2 + 2pq + q^2 \qquad (2.1)$$

Suppose you have a population with the following composition:

    55 AA   40 Aa   5 aa

Because each individual is diploid and carries two alleles at each locus, total allele frequencies are:

$$pA = 2(55) + 40/2(55 + 40 + 5) = 0.75$$

$$qA = 40 + 2(5)/2(55 + 40 + 5) = 0.25$$

Under Hardy–Weinberg assumptions the expected genotype *frequencies* would be the binomial expansion:

$AA = p^2 = (0.75)^2 = 0.562$

$Aa = 2pq = 2(0.75)(0.25) = 0.375$

$aa = q^2 = (0.25)^2 = 0.063$

The expected *number* of individuals of each genotype would be expected gene frequency × number of individuals in population:

$AA = (0.562)(100) = 56.2$

$Aa = (0.375)(100) = 37.5$

$aa = (0.063)(100) = 6.3$

A chi-square (X^2) test is most commonly used to determine if a population varies significantly from Hardy–Weinberg expectations, although other comparisons such as the G test are sometimes used (Sokal and Rohlf, 1995). In the chi-square test,

$$X^2 = \Sigma(\text{obs.} - \text{exp.})^2/\text{exp.}$$

where obs. represents the observed number in a genotype class, exp. refers to the expected number in that same genotype class, and Σ indicates that the values are to be summed over all genotypic classes. A statistical table or graph is consulted to determine the probability (p) that a sampling error could have produced the deviation between the observed and expected values (Fig. 2.1). The degrees of freedom (df) are equal to the number of genotype classes minus one. When the deviations are small, high probability values are produced, which indicate that the population is in Hardy–Weinberg equilibrium and the differences observed are probably due to minor sampling errors. Low probability values reflect deviations too great to be explained solely by chance and, therefore, some other evolutionary force could be operating on the populations. Usually, a population is considered in Hardy–Weinberg equilibrium unless its probability value is less than or equal to 0.05. Using the data generated in the above example:

A X^2 value of 0.47 with two degrees of freedom has a probability value of 0.80. Therefore, our population is very close to Hardy–Weinberg expectations and one of the evolutionary forces is probably not influencing the population, unless drift, selection and migration are acting in opposition and balancing each other out (Workman, 1969).

If more than two alleles exist in the populations, the Hardy–Weinberg formula takes the form of the polynomial square. For three alleles whose frequencies are p, q and r, then $p+q+r=1$ and the equilibrium frequency is given by the trinomial square $(p+q+r)^2 = p^2 + 2pq + 2pr + q^2 + 2qr + r^2$.

While the Hardy–Weinberg formula is very useful in describing the completely randomized situation, it is a rare natural population that is in fact in Hardy–Weinberg equilibrium. More commonly, mating is not random and the populations are subjected to other evolutionary forces such as migration, genetic drift and natural selection. The next section describes how differential levels of gene migration can influence variation patterns; later sections discuss how drift and natural selection affect gene frequencies. Linkage can also have a profound effect on gene frequencies and will be discussed in the next chapter.

Migration

Inter-populational

The influx of genes from another population can increase variability in the recipient population if the immigrants are unique. The factors limiting the relative importance of gene flow are the difference in gene frequency between populations and the rate of migration. The rate of migration (m) varies from 0 to 1;

	AA	Aa	aa	Total
Observed	55.0	40	5	100
Expected	56.2	37.5	6.3	100
Obs. – exp.	1.2	2.5	1.3	
Above number squared	1.44	6.25	1.69	
Above divided by expected	0.03	0.17	0.27	
	$\chi^2 = 0.03 + 0.17 + 0.27 = 0.47$, df = 2			

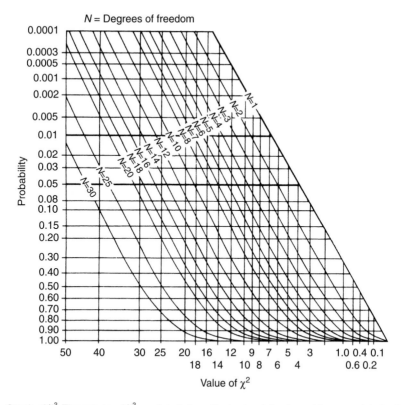

Fig. 2.1. Graph of χ^2. The values of χ^2 are listed along the horizontal axis and the associated values of p are listed along the vertical axis. N = degrees of freedom (Crow, 1945).

$m=0$ signifies no immigrant alleles and $m=1$ represents a complete replacement by immigrants. Mathematically:

m = number of immigrants per generation/ total

If the frequency of an allele among the natives is represented by q_0 and among the immigrants by q_m, then the allele frequency in the mixed population will be:

$q = m(q_m - q_o) + q_o$

The change in gene frequency per generation (q) due to migration is:

$\Delta q = m(q_m - q_o)$

For example, if the native population originally contains $q_0 = 0.3$, and the immigrant group is $q_m = 0.5$, and the migration rate is $m = 0.001$, then the change in gene frequency due to one generation of immigration is:

$\Delta q = 0.001 (0.5 - 0.3) = 0.0002$

Obviously, two populations must have different gene frequencies to be greatly affected by gene flow (unless an unequal sample of genes is transmitted), and the greater the input of novel genes, the greater the overall change.

Numerous methods of measuring migration or gene flow among plant populations have been developed (Moore, 1976; Handel, 1983a,b; Slatkin, 1985). In some, pollen and seeds are simply captured in "traps" and identified, while in others the movement of pollinators and dispersers is closely monitored. Pollen and seeds have also been tagged with dyes and other chemical markers. The most popular method is to find unique alleles such as allozymes or molecular markers in parents, and trace their movement into progeny (Ellstrand and Marshall, 1985; Ouborg et al., 2005).

Small amounts of geographic separation have been found to greatly minimize levels of migration. The bulk of pollen from both wind- and insect-pollinated species travels 5–10 m at most (Fig. 2.2; Colwell, 1951; Levin and Kerster, 1969; Hokanson *et al.*, 1997a). Seed dispersal is also generally leptokurtic and limited to several meters (Levin and Kerster, 1969). These observations suggest that most populations are isolated against high influxes of foreign genes.

This is not to say that long-distance dispersal does not occur at all. Wind-carried pollen has been found hundreds of kilometers from its source (Ehrlich and Raven, 1969; Moore, 1976) and insects have been shown to carry viable pollen for several kilometers (Ellstrand and Marshall, 1985; Devlin and Ellstrand, 1990). Many seeds have long distance dispersal mechanisms such as wings to catch air patterns, "stickers" that attach to animal fur, or hard seed coats that survive trips through the digestive tracts of migrating animals. All these mechanisms ensure at least some long-range dispersal of genes, although only a small percentage travels more than a few meters from a plant.

Human beings have played an important role in plant evolution by planting cultigens next to wild species, and by providing disrupted sites for population expansion (Anderson, 1949). Numerous crop species such as barley, carrots, wheat, oats, sorghum and rye have been shown to hybridize with their wild progenitors (Harlan, 1965; Wijnheijmer *et al.*, 1989). This sexual transfer of genes to weedy relatives has brought on concerns about the escape of engineered genes into the natural environment (Ellstrand and Hoffman, 1990). The long-term effects of interspecies hybridizations will be discussed in depth later (Chapter 5).

Not only has migration been important in the development of plant species, but it may also have played an important role in the spread of agriculture among human beings. There has long been a debate over whether agriculture spread from the Middle East to Europe via "word of mouth" or actual migration of agricultural peoples. Recent analyses of blood proteins in extant human populations indicate that agricultural peoples slowly spread from their Middle East origin and hybridized along the way with local peoples (Sokal *et al.*, 1991). This led to a slow diffusion of genes from east to west over a period of 4000 years (see Chapter 7).

Intra-populational

The degree of gene movement affects variation patterns not only between populations, but also within populations. Limited gene flow results in a highly substructured population in which only adjacent plants have a high likelihood of mating. Several terms have been developed to describe the substructuring of plant populations due to limited gene flow. A deme or panmictic unit is the group within which random mating occurs (Wright, 1943). A neighborhood is defined as the number of individuals in a circle with a radius twice the standard deviation of the

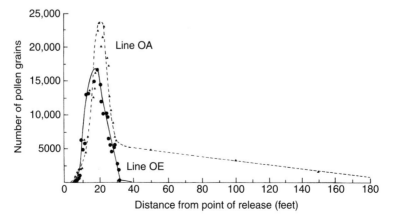

Fig. 2.2. Distance pine pollen travels after release. Line OA is directly downwind from a tree, while line OE is 45° from line OA (Colwell, 1951).

migration distance or the number of individuals with which a parent has a 99% certainty of crossing (Wright, 1943). Neighborhood sizes have been estimated to vary from four individuals in *Lithospermum caroliniense* (Kerster and Levin, 1968) to 282 in *Phlox pilosa* (Levin and Kerster, 1967).

The size of a neighborhood has a strong effect on the variation patterns observed in polymorphic populations. In substructured populations with many small neighborhoods, genes will move very slowly across the total population resulting in patchy distributions, while in large less structured populations patterns will appear much more regular (Fig. 2.3). Expected levels of heterozygosity in the population as a whole can also be reduced by small neighborhood size if the gene frequencies within individual subpopulations vary significantly from the populational mean (Wahlund, 1928). This occurs because

the expected number of heterozygotes calculated from average gene frequencies is not always the same as the average of the subpopulations. For example, in the substructured population shown in Table 2.1, the expected number of heterozygotes calculated using the average gene frequency across all populations is H = 2pq = 2(0.5)(0.5) = 0.5, but the actual average across neighborhoods is H = 0.40. Thus, subpopulations can be in Hardy–Weinberg equilibrium even though the population as a whole is not.

Another factor that has a strong influence on variation patterns is the breeding system of the species. Largely self-pollinated species tend to be much more homozygous than outcrossed ones, since heterozygote percentages are reduced by 50% with each generation of selfing. For instance, if we start with a population of only heterozygotes (Aa) and self-pollinate the progeny each year, the ratios shown in

(a)

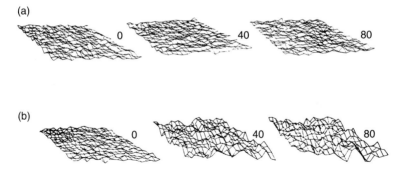

(b)

Fig. 2.3. A computer simulation of the influence of neighborhood size on gene frequency variation. Random changes in gene frequency are simulated for a group of 10,000 evenly spaced individuals. In series (a), all individuals have an equal probability of mating, while in series (b), each individual mates within a neighborhood of nine individuals. Local gene frequency differences are represented by the heights of peaks above the plane after 0, 40 and 80 generations. (Used with permission from F.J. Rohlf and G.D. Schnell, © 1971, *American Naturalist*, University of Chicago Press, Chicago, Illinois.)

Table 2.1. Expected frequency of heterozygosity in a substructured population.

Neighborhood	Gene frequency		Expected frequency of heterozygotes (2pq)
	p	*q*	
A	0.2	0.8	0.32
B	0.5	0.5	0.50
C	0.8	0.2	0.32
D	0.3	0.7	0.42
E	0.7	0.3	0.42
Average	0.5	0.5	0.40

Table 2.2 will be produced over time. The reduction in heterozygosity occurs because half of each heterozygote's progeny will be homozygous each generation due to Mendelian segregation, while the homozygotes continue to produce only homozygotes.

Any level of inbreeding will reduce the number of heterozygotes expected from random mating. The *inbreeding coefficient F* is commonly used to measure levels of inbreeding and is defined as the probability that two genes in a zygote are identical by descent (Wright, 1922). Mathematically, the chance of a newly arisen homozygote with identical alleles is $1/2N$ for any generation, where N is the number of breeding diploid individuals. The probability that the remaining zygotes $1-(1/2N)$ have identical genes is the previous generation's inbreeding coefficient. Thus for succeeding generations:

$$F_o = 0$$

$$F_1 = 1/2N$$

$$F_2 = 1/2N + (1 - 1/2N) F_1$$

$$F_3 = 1/2N + (1 - 1/2N) F_2$$

The inbreeding coefficient for any generation n is:

$$F_n = 1/2N + (1 - 1/2N) F_{n-1}$$

The increase in the inbreeding coefficient F with different types of mating is depicted in Fig. 2.4.

Plant species range widely in their breeding systems from obligate outcrossers to completely selfed, but most fall somewhere in the middle (Table 2.3). It is not unusual to find some degree of cross-fertilization (1–5%) in even some of the strongest self-pollinating crops such as barley, oats and broadbean (Allard and Kahler, 1971; Martin and Adams, 1987a,b). Rick and Fobes (1975) found outcrossing rates in some populations of native self-pollinated tomatoes to range from 0 to 40% across South America.

When selfing does occur, it is generally the result of proximity – anthers and pistils are often very close to each other in hermaphroditic flowers and this close alignment encourages self-pollination. The presence of multiple flowers on the same plant also encourages self-pollination. Probably the most extreme mechanism insuring selfing is the rare phenomenon called cleistogamy, where flowers never open at all (groundnuts and violets). There are also

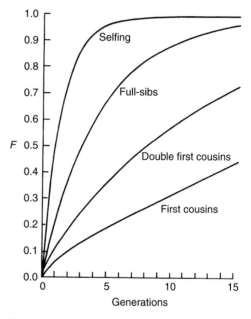

Fig. 2.4. Effect of selfing on the inbreeding coefficient F in random mating and inbred populations. (Used with permission from F.J. Ayala, © 1982, *Population and Evolutionary Genetics: A Primer*, Benjamin/Cummings Publishing Company, Menlo Park, California.)

Table 2.2. Heterozygotes and self-pollination.

Generation	Genotypes			Proportion of heterozygotes relative to those in the initial population
	AA	Aa	aa	
0	0	1	0	1
1	1/4	1/2	1/4	1/2
2	3/8	1/4	3/8	1/4
3	7/16	2/16	7/16	1/8
4	15/32	2/32	15/32	1/16
∞	1/2	0	1/2	0

Table 2.3. Reproductive systems of selected crops (sources: Allard, 1960; Fryxell, 1957).

Predominantly self-pollinated	Predominantly cross-pollinated	
Apricot	Almond	Papaya
Aubergine	Apple	Pear
Barley	Asparagus	Pecan
Broadbean	Banana[a]	Pistachio
Chickpea	Beet	Plum
Citrus[a]	Broccoli	Radish
Common bean	Brussels sprout	Raspberry
Cotton	Blueberry	Rhubarb
Cowpea	Cabbage	Rye
Date	Carrot	Ryegrass
Fig[a]	Cauliflower	Safflower
Flax	Celery	Spinach
Groundnut	Cherry	Squash
Hemp	Clover	Strawberry
Lettuce	Cucumber	Sunflower
Lima bean	Date	Sweet potato
Mung bean	Fig	Turnip
Papaya	Filbert	Walnut
Pea	Grape	Watermelon
Peach	Hemp	
Rice	Lucerne	
Sorghum	Maize	
Soybean	Mango	
Spinach	Muskmelon	
Tomato	Olive	
Wheat	Onion	

[a]Some parthenocarpic types.

"apomictic" species, which do not undergo any sexual reproduction.

Several mechanisms encourage outcrossing, including dioecy, dichogamy and incompatibility systems. Dioecy is found in a small percentage of species and is represented by separate pistillate (female) and staminate (male) plants. Dioecy has originated independently in many families, but there are no common underlying mechanisms (Ainsworth, 2000). In a few rare cases, even sex chromosomes have evolved. Some important crops with this system are hops, date palm, asparagus, spinach, strawberry and hemp. Dichogamy is also quite rare and occurs when the pollen of hermaphroditic plants is shed at times when the stigmata are not receptive. Representative species with this mating system are maples, oaks and sugarcane.

Self-incompatibility (SI) systems are very common and fall into two broad classes:

gametophytic and sporophytic (Fig. 2.5). In gametophytic systems, pollen germination and tube growth is dependent on the genotype of the parent; while in sporophytic systems, pollen performance is based on the genotype of the male parent. SI evolved independently in numerous lineages of plants; gametophytic systems are the most common and have been found in 60 families of angiosperms (Lewis, 1979; deNettancourt, 2001). Modes of SI differ widely across dicots, but they all are regulated by tightly linked multigene complexes referred to as haplotypes. Information on the genetics of self-incompatibility in monocot grasses is still emerging, but it is known that at least two independent loci are involved in self recognition (Yang et al., 2008, 2009).

SI can act both before and after fertilization. Prezygotic SI results when there is an interaction between maternal tissue and the male gametophyte, which prevents self pollen tube development (Lewis, 1979; Williams et al., 1994). Postzygotic reductions in self-fertility can arise due to "late acting" or ovarian self-incompatibility (OSI), where there is synchronous embryo failure (Seavy and Bawa, 1986; Sage et al., 1994, 1999).

The modes of action of SI vary widely (Takayama and Isogai, 2005; Franklin-Tong, 2008). In the *Brassicaceae* (crucifers) with sporophytic incompatibility, self pollen germination is inhibited through the interaction of two pollen proteins (SP11/SCR), a stylar receptor kinase (SKR), a glycoprotein (SLG) and several other proteins. In the *Papaveraceae* (poppy) with gametophytic SI, self pollen tube growth is rejected at the stigmatic surface by the interaction of a stigmatic S protein, a pollen S receptor (SBP), a calcium-dependent protein kinase (CDPK) and a glycoprotein. Pollen tube growth is arrested by a cascade of events associated with an increase in cytosolic calcium and a disruption of the cytoskeleton. In the gametophytic *Solanaceae* (tobacco, tomato, petunia and potato) and *Rosaceae* (apple, cherry and pear), self pollen tube growth is inhibited within styles by a pistil-released RNase (S-RNAase) that selectively destroys the RNA of self pollen tubes.

Self-sterility in some species is also caused by "early acting inbreeding depression", where embryos abort as deleterious alleles are expressed during seed development or the outcrossed

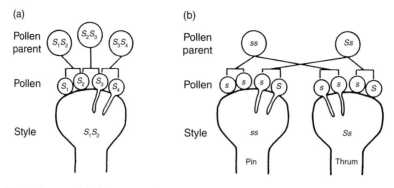

Fig. 2.5. Pollen tube growth in (a) gametophytic and (b) sporophytic incompatibility. The letters (*S*) denote incompatibility genes. Note that in the gametophytic system the genotype of the pollen *itself* determines whether it germinates or not, while in the sporophytic system it is the genotype of the pollen *parent* that is important. In gametophytic incompatibility, fertilization only occurs from pollen with a different allele from that of the female; thus, an S_1S_2 stigma cannot be fertilized by S_1 or S_2 pollen and only S_3 and S_4 pollen can germinate. In sporophytic incompatibility, the pollen can germinate if it contains the same allele as the stigmata, as long as its parent had a different genotype; thus, *s* pollen can grow on an *ss* stigma if it comes from an *Ss* parent. (Used with permission from K. Mather, © 1973, *Genetic Structure of Populations*, Chapman & Hall, London.)

progeny display a hybrid vigor that enables them to outcompete the selfed ones (Charlesworth and Charlesworth, 1987). This system has been shown to operate in a wide array of species, particularly long-lived, outcrossing ones (Busbice, 1968; Crowe, 1971; Krebs and Hancock, 1991; Seavey and Carter, 1994; Carr and Dudash, 1996). Such species are often misclassified as self-incompatible due to poor selfed seed set, but they are distinct from SI in that they commonly display: (i) a range in self-fertility among different genotypes; (ii) a significant positive correlation between self and outcross fertility; (iii) a significant correlation between the percent aborted ovules and the inbreeding coefficient; and (iv) embryos that abort at different stages of development (Hokanson and Hancock, 2000).

Selection

Selection can be defined as a change in a population's gene frequency due to differential survival and reproduction. Many factors influence the persistence of a gene, including germination rate, seedling survival, adult mortality, fertility and fecundity.

The *fitness* of a genotype is mathematically defined as the mean number of offspring left by that genotype relative to the mean number of progeny from other, competing genotypes. It is commonly designated by the letter *W*. Since it is a relative measure, the genotype with the highest fitness (most fit) is assigned a value of 1. Other fitness values are decimal fractions ranging between 1 and 0. The *selection coefficient* is the proportional reduction in each genotype's fitness due to selection.

Suppose you had the following genotype frequencies before and after selection:

	AA	Aa	aa
Frequency before selection	0.25	0.50	0.25
Frequency after selection	0.35	0.48	0.17

The relative reproductive contribution of each genotype would be:

	Reproductive contribution
AA	0.35/0.25 = 1.40
Aa	0.48/0.50 = 0.96
aa	0.17/0.25 = 0.68

To assign the most fit a value of 1, we must divide each genotype by the reproductive contribution of the most successful genotype:

	Fitness (W)
AA	1.4/1.4 = 1
Aa	0.96/1.4 = 0.7
aa	0.68/1.4 = 0.5

The selection coefficient is $1 - W$; so:

AA $1.0 - 1.0 = 0$

Aa $1.0 - 0.7 = 0.3$

aa $1.0 - 0.5 = 0.5$

Jain and Bradshaw (1966) found selection coefficients in nature to range from 0.001 to 0.5. Humans probably used even more extreme values in the domestication of crop species (Ladizinsky, 1985).

The response of quantitative traits to selection can be predicted by the equation $R = h^2 S$, where h^2 is the estimate of heritability and S is the selective differential or the difference between the mean of the selected parents (μ_s) and the mean of all individuals in the parental population (μ) (Falconer and Mackay, 1996). The truncation point (T) represents the point at which individuals are either selected for or against (Fig. 2.6). For example, if we are interested in a trait such as pod length with a high heritability of $h^2 = 0.60$ and our total parental population has a mean of 12.3 cm and our selected population has a mean of 15.2 cm, then $R = 0.6 (15.2 - 12.3) = 1.8$ cm. This means that our progeny population will have pods 1.8 cm longer than the original population.

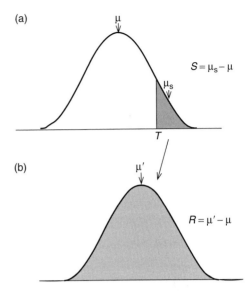

Fig. 2.6. (a) Quantitative distribution of phenotypes in a parental population with the mean μ. Only those individuals with phenotypes above the truncation point (T) produced progeny in the next generation. The selected parents are denoted by shading and their mean phenotype by μ_s. (b) Distribution of phenotypes in the next generation from offspring derived from the selected parents. The mean phenotype is denoted μ'. S represents the selection differential, and R is called the response to selection. (Used with permission from D.L. Hartl, © 1980, *Principles of Population Genetics*, Sinauer Associates, Inc., Sunderland, Massachusetts.)

Types of Selection

There are three primary types of selection: (i) directional; (ii) stabilizing; and (iii) disruptive or diversifying (Fig. 2.7). In directional selection, one side of a distribution is selected against, resulting in a directional change. Under stabilizing selection, both extremes are selected against and the intermediate type becomes more prevalent. In disruptive selection, the intermediate types are selected against, resulting in the increase of divergent types. The amount of genetic variability present in a population and the strength of the selection coefficient determine how fast and how much a population will change.

Numerous examples of directional selection have been presented in the evolutionary literature. Perhaps the most striking and repeatable is the evolution of heavy metal tolerance in those plant species that come in contact with mine borders (Jain and Bradshaw, 1966; Antonovics, 1971). In less than 200 years, pasture species such as *Anthoxanthum odoratum* and *Agrostis tenuis* evolved distinct morphologies and means of coping with high levels of copper, lead and zinc in the face of substantial gene flow. As is shown in Fig. 2.8, a distinct change in metal tolerance can be observed across only a few meters.

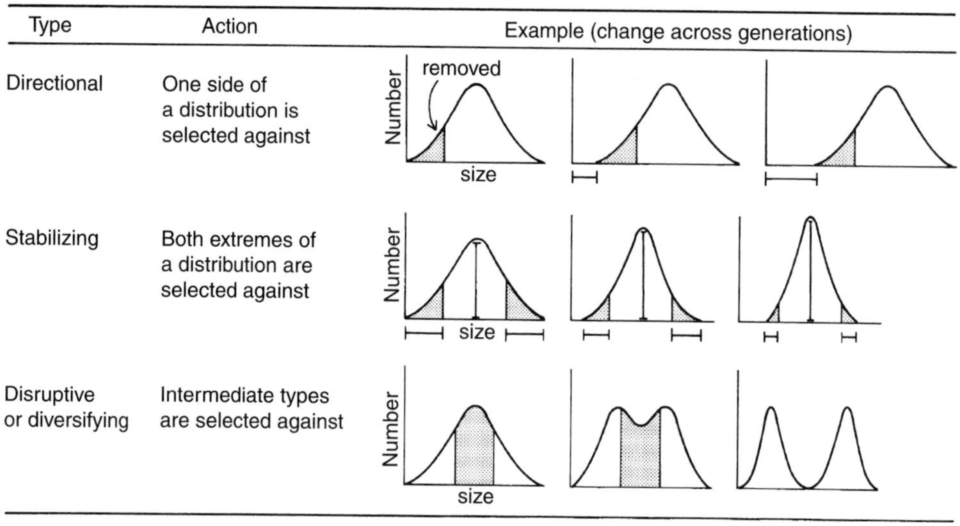

Type	Action	Example (change across generations)
Directional	One side of a distribution is selected against	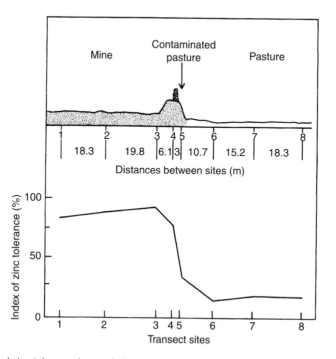
Stabilizing	Both extremes of a distribution are selected against	
Disruptive or diversifying	Intermediate types are selected against	

Fig. 2.7. Comparison of the three major types of natural selection. Each population is depicted initially as a normal distribution with the cross-hatched areas representing the less fit individuals. The rate of change is dependent on the degree of genetic variability present and the selection coefficient (see text). The bars represent how the proportions change each generation.

Fig. 2.8. Lead and zinc tolerance in populations of *Anthoxanthum odoratum* located at different distances from a mine border. (Redrawn with permission from T. McNeilly and J. Antonovics, © 1968, *Heredity* 23, 205–218 and J. Antonovics and A.D. Bradshaw, © 1970, *Heredity* 27, 349–362.)

The domestication of maize also offers a dramatic example of directional selection. Under human guidance, the size of the ear was increased between 7000 and 3500 BP (before present) from less than 2.5 cm to well over 15 cm (Fig. 2.9).

The fact that we recognize most species as definable entities argues that stabilizing selection is common in nature, or else species lines would be much more blurred than they are. Huether (1969) provided some of the strongest evidence of stabilizing selection when he showed that *Linanthus androsaceus* almost always have five lobes per flower even though their environments vary greatly and there is genetic variation for the trait at every site. Plant taxonomists often find floral traits to be the most dependable species markers, because they are generally less variable than other traits.

Documentation of disruptive selection has been done most frequently by correlating gene frequency with environmental variation – one of the most commonly cited examples is the work of Allard and co-workers (Allard and Kahler, 1971; Allard *et al.*, 1972), where they found reproducible correlations between allozyme frequencies in the oat *Avena barbata* and the relative moisture content of soils in California. Distinct allozyme frequencies are found on the wettest (mesic) and driest (xeric) sites (Table 2.4). These

allozymes may not have been the specific traits selected, but they were at least associated with the selected loci through genetic linkage and/or partial selfing (Hedrick, 1980; Allard, 1988).

Many unique races of crops were developed by human beings through disruptive selection as they gathered and grew populations from distinct regions and habitats (see Chapter 7). Perhaps the most distinct alteration came in *Brassica oleracea*, where several distinct crops were developed including kale, broccoli, cabbage, kohlrabi and Brussels sprouts (Fig. 11.3). Diverse races of rice, chickpea and chili peppers also emerged in response to isolation and differential selection pressure from humans.

Plant breeders have on numerous occasions employed diversifying selection to produce differences in harvest dates and quality factors. For example, dramatic changes in the protein content of maize were produced over 60 generations by selecting repeatedly for high and low values (Fig. 2.10). In a long-term study selecting for high and low oil levels in kernel oil content,

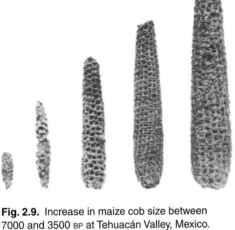

Fig. 2.9. Increase in maize cob size between 7000 and 3500 BP at Tehuacán Valley, Mexico. The cob on the left was about 2.5 cm long. (Used with permission from D.S. Byers (ed.), © 1967, *The Prehistory of the Tehuacan Valley*, Andover Foundation for Archaeological Research.)

Table 2.4. Genotype frequencies in two populations of oats (*Avena barbata*) found in California. Three loci of esterase (E_1, E_4 and E_{10}) and one each of phosphatase (P_5) and anodal peroxidase (APX_5) are represented (Marshall and Allard, 1970).

Locus	Genotype	Mesic population	Xeric population
		Frequency	
E_1	11	0.76	0.00
	12	0.07	0.00
	22	0.17	1.00
E_4	11	1.00	0.30
	12	0.00	0.11
	22	0.00	0.59
E_{10}	11	1.00	0.46
	12	0.00	0.13
	22	0.00	0.41
P_5	11	0.00	0.40
	12	0.00	0.15
	22	1.00	0.45
APX_5	11	0.86	0.48
	22	0.00	0.41
	33	0.09	0.00
	12	0.00	0.11
	13	0.05	0.00
	23	0.00	0.00

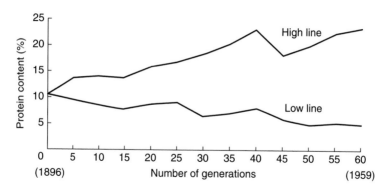

Fig. 2.10. Change in protein content of maize during 60 generations of artificial selection by E. Leng and Associates at the Illinois State Experiment Station. (Used with permission from V. Grant, © 1977, *Organismic Evolution*, W.H. Freeman and Company, San Francisco, California.)

a 20-fold difference was obtained between high and low lines, affecting more than 50 QTL (Laurie *et al.*, 2004).

While directional selection can eliminate variability in a population, stabilizing and disruptive selection often act to maintain it. Dobzhansky (1970) called this maintenance of genetic polymorphism *balancing selection*. Disruptive selection maintains polymorphism when different genotypes are "favored" in different environments. Stabilizing selection acts to maintain polymorphisms if heterozygotes produce the favored phenotype. Numerous examples have been provided where superior vegetative or reproductive vigor has been associated with heterozygosity (Mitton and Grant, 1984; Strauss and Libby, 1987).

Disruptive selection occurs not only across spatially heterogeneous environments, but also across time. Polymorphisms can be maintained across temporarily varying environments, different life stages and variant sexes. There are also examples of polymorphisms being maintained by differing population frequencies and genetic backgrounds. Some alleles, such as the *S* alleles associated with self-incompatibility, are even favored when they are in low frequency, but disfavored at high.

Factors Limiting the Effect of Selection

Several important factors limit the influence of selection on gene frequencies: (i) amount of genetic variability present; (ii) generation time; (iii) strength of the selective coefficient; (iv) degree of dominance; (v) initial frequency of the advantageous allele; and (vi) intra-genomic interactions.

The only way a population can change is if there is genetic variability present. Many a breeding program has stalled or a species has gone extinct when the genetic variability for a particular trait was extinguished. This is true for both single gene traits and for quantitative traits where more than one locus is involved. In 1930, Fisher outlined his fundamental theorem of natural selection, which stated "the rate of increase in fitness of any organism at any time is equal to its genetic variance at that time".

The mating behavior of a species has important long-term evolutionary ramifications. Outcrossed species often have high levels of heterozygosity, and gene combinations are continually shuffled during reproduction through crossing over and independent assortment. Selfing species are frequently very homozygous and specific gene combinations are rarely disrupted by sexual recombination. An outcrossed, heterozygous species has a considerable amount of genic flexibility with which to meet environmental change, but the selfing, homozygous ones may be better adapted to specific, unchanging conditions. An intermediate strategy with variable outcrossing rates may yield the greatest long term success in an unpredictable world; this may be why few extreme selfers or outcrossers are found (Allard, 1988). Gornall (1983) has suggested that "those colonizing species which had extraordinary flexible recombination systems, which allowed them both to store and release large amounts of variability, were in a sense pre-adapted to successful cultivation".

The strength of selection will critically influence the rate of change in a population as the greater the proportion of individuals being selected each year, the more rapid the change. Likewise, the generation time of an organism can have a strong effect on the rate of evolution simply because the longer the period between reproductive episodes, the longer the separation between meiotic reassortments of genes and seedling selection. Generation times in higher plants vary from a few weeks in *Arabidopsis* to many years in some tree species.

The degree of dominance has an influence on rates of change because recessive alleles in the heterozygous state are masked from selection. Frequencies of a newly arisen recessive allele will therefore change very slowly in a population until its frequency is quite high (Fig. 2.11). Frequencies of codominant alleles or those with an intermediate effect in a heterozygote will change more rapidly than recessives, since they have a partial influence on the phenotype of heterozygotes. Frequencies of dominants will initially change more abruptly than either recessives or intermediates, but the rate of change will eventually slow as the frequency of recessives become so low that most are "hidden" in heterozygotes. In all cases, populations with intermediate gene frequencies change more rapidly than those with low frequencies of deleterious or advantageous alleles (Fig. 2.11).

The number of selective deaths involved in one complete allelic substitution has been called the *genetic cost* (Haldane, 1957, 1960). By similar logic, the number of individuals that will die each generation due to selection has been called the *genetic load* (Wallace, 1970). The cost factor is high when the favored allele is at a low frequency, and decreases as the initial frequency increases. Obviously, a population can tolerate only a limited number of genetic deaths each generation if it is to avoid extinction.

The adaptiveness of an allele is not only influenced by the external environment, but also by the other genes in the genome. In most of our discussion so far, we have treated genes as if they are independent entities; however, the genome is in reality a tightly interacting

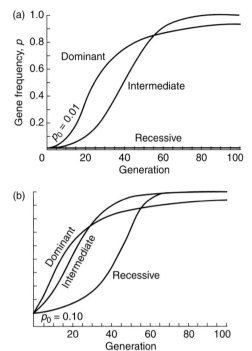

Fig. 2.11. Increase of advantageous alleles with different degrees of dominance (recessive, dominant and intermediate): (a) new mutation in a very low frequency ($p = 0.01$), (b) existing polymorphism at $p = 0.10$. The degree of dominance has an influence on rates of change because recessive alleles in the heterozygous state are masked from selection. (Used with permission from D.J. Futuyma, © 1979, *Evolutionary Biology*, Sinauer Associates, Sunderland, Massachusetts.)

unit. This concept will be discussed at length in the next chapter on the complexity of plant genomes.

Finding Loci Subject to Strong Selection

Genes subject to selection can be found using "top-down" or "bottom-up" approaches (Wright *et al.*, 2005; Chapman *et al.*, 2008). In top-down approaches, investigators observe phenotypic variation that appears correlated with an environmental gradient or the process of domestication, use a QTL analysis to identify the chromosomal regions containing the gene regulating the trait or search for candidate genes that have already been shown to influence similar traits in other species, and then sequence the chromosomal area carrying the gene. This approach has been used to identify several important genes associated with domestication, including *tb1*, which controls the transition from the long branches of teosinte tipped by tassels versus the short branches of maize tipped by ears, *fw2.2*, which controls a large portion of the difference in fruit mass between wild and domesticated tomatoes, *tga1*, which controls the formation of the casing around kernels of teosinte (the ancestor of maize), and *Q*, which controls spike shattering in wheat as well as several other traits (Doebley *et al.*, 2006).

In bottom-up approaches, levels of DNA sequence variation are evaluated across the genome to find examples of sequences that are highly conserved across genotypes and signal strong selection pressure or variation patterns that are strongly associated with a trait of interest. For example, Chapman *et al.* (2008) examined the sequences of 492 loci in wild versus cultivated sunflower and found 36 genes probably involved in the domestication of the sunflower. Zhao *et al.* (2008) investigated sequence variation in 72 candidate genes in maize and found 17 that were likely targets of selection during domestication. Mariac *et al.* (2010) found the gene *PgMADS11* to be strongly associated with flowering time variation and annual rainfall levels in varieties of pearl millet. Jones *et al.* (2008) found polymorphisms in the gene *Photoperiod-HI* that were associated with variation in flowering time in wild and cultivated barley, and they identified wild accessions carrying a nonresponsive haplotype that may have contributed to the development of European cultivars.

Coevolution

Most of this chapter has concerned selection within a single species. However, there are numerous examples of what is called coevolution, where species have evolved together rather than independently. The most commonly cited examples concern mutualism,

character displacement and host–pathogen evolution. In mutualism, there is a symbiotic reaction in which two species benefit by their interaction. Two classic examples in plants are soil mycorrhiza and nitrogen-fixing bacteria. In both cases, microbes have become intimately associated with plant roots such that the plant gains mineral nutrients and the fungi or bacteria receive a carbon source. An even more dramatic example is found between the tree *Acacia* and the ant *Pseudomyrmex*, where the ant receives nourishment from the tree's nectaries and the ants defend the tree against herbivores and competing vegetation (Janzen, 1966).

When two species have overlapping ranges, one species is either selectively eliminated or the two undergo character displacements that minimize competition (Roughgarden, 1976). For example, where the ranges of *Phlox pilosa* and *Phlox glaberrima* do not overlap, they both have pink flowers; but in areas of overlap, *P. pilosa* frequently has white flowers. Interspecific hybridizations are minimized because insects have a tendency to carry pollen from white to white flowers rather than from pink to white or vice versa (Levin and Kerster, 1967). Ecological races of many species also have different flowering times, which reduce the chance of hybridization. In some cases, these displacements are thought to be an early step in speciation (Chapter 5).

There are also numerous examples of host–parasite evolution, where there is widespread matching of host alleles conferring resistance to specific parasite strains (Flor, 1954; Vanderplank, 1978; Allard, 1990). Most of the best studied examples come from agricultural systems where constant battles are fought to find new sources of resistance to rapidly evolving pest populations (Wahl and Segal, 1986; Allard, 1990). For example, dominant alleles have been identified at more than five loci in wheat that confer resistance to the hessian fly, but alleles exist for each locus in the fly that confers counter resistance (Hatchett and Gallum, 1970). Dozens of examples of "gene-for-gene" relations have been described in pathogen/host systems where there is a gene for susceptibility or resistance in the crop to match each gene for virulence or avirulence in the pathogen (Table 2.5).

Table 2.5. The reaction of 16 genotypes of wheat to three Canadian races of *Puccinia graminis tritici* (Vanderplank, 1978).

Resistance genes	Race[a]		
	C10	C33	C35
Sr 5	S	S	S
Sr 6	R	R	S
Sr 7a	R	S	S
Sr 8	R	S	S
Sr 9a	S	R	S
Sr 9b	S	R	S
Sr 9d	S	S	R
Sr 9e	S	S	R
Sr 10	S	S	R
Sr 11	S	S	R
Sr 13	S	R	R
Sr 14	S	S	S
Sr 15	S	R	S
Sr 17	S	R	R
Sr 22	R	R	R
Sr T2	S	R	R

[a]S, susceptible; R, resistant.

Genetic Drift

Many people stress the importance of selection in shaping natural populations, but non-directional forces such as genetic drift can also play an important role. The common inclination is to assume that nature is ordered and that all is optimized. The simple truth is that variation patterns are regulated to a large extent by luck. A highly adapted genotype will not predominate in a population if most of its seeds fall on a rocky outcropping where they cannot germinate or there is no pollinator activity due to cool conditions when its pollen is dehisced. To proliferate, a genotype must be both well adapted and its progeny must reach maturity before some inadvertent accident eliminates it.

While it is easy to document selection by observing directional changes over time or genotype/environment correlations, drift is difficult to measure because it is non-directional and often masked by other forces (Schemske and Bierzychudek, 2001). Most demonstrations of drift have come by the process of elimination, i.e. no apparent associations were observed between any environmental parameter and an allele's frequency. For example, on a hillside of *Liatris cylindracea* in Illinois, Schaal (1975) found a substantial amount of allozyme variation in several loci that she could not associate with any environmental parameter, including moisture, soil pH, nutrient content or percentage organic matter (Fig. 2.12).

Sewell Wright (1931, 1969) provided numerous models that show how drift might operate. Drift will ultimately result in the fixation of one allele in a population like directional selection, but the approach will be much more variable (Fig. 2.13). Two primary factors influence the effects of drift: allele frequency and population size. The more frequent an allele, the greater its chances of being fixed, and the smaller the population, the faster it will stumble towards fixation. The probability that an allele will be fixed is equal to its frequency in a population.

In the 1980s, there was considerable debate over the relative importance of drift and selection in natural populations, particularly as they related to allozyme variation (Hedrick, 1983; Nei, 1987). The "selectionists" felt that most variability was maintained by selection, while the "neutralists" felt that drift was the most important force. While this debate continues to go on, a reasonable compromise is to assume that both drift and selection are important and that the relative strengths of the two forces are dictated by each individual set of circumstances. The critical point to realize is that forces other than selection can shape variation patterns. This is particularly true in plant populations where limited migration results in most populations being effectively small.

Extreme cases of random genetic drift occur when a new population is initiated by only a few individuals – this is called the founder effect (Mayr, 1942). Gene frequencies may be quite different in a few colonizers compared to the population from which they originated because of sampling errors. Chance variations in allelic frequencies can also occur when populations undergo drastic reductions in numbers ("bottlenecks") after environmental catastrophes. The early stages of most crop domestications were probably influenced by founder effects, as the bulk of our domesticated strains are based on relatively few genotypes (Ladizinsky, 1985). In general, crop plants carry only a fraction of their wild relatives' genes (Tanksley and McCouch, 1997).

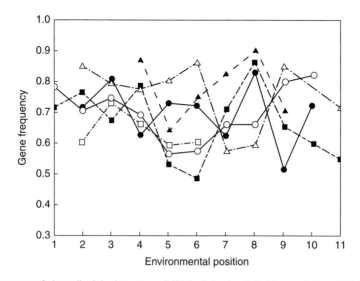

Fig. 2.12. Freqency of six malic dehydrogenase (MDH) alleles in a hillside population of *Liatris cylindracea*. The lines represent a transect down the hill. Point 1 is at the top and point 11 is at the bottom. (Used with permission from B.A. Schaal, © 1975, *American Naturalist* 109, 511–528, University of Chicago Press, Chicago, Illinois.)

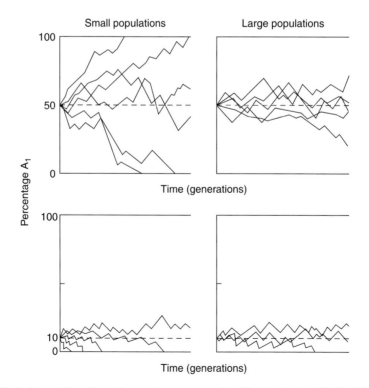

Fig. 2.13. Effect of population size and gene frequency on rate of fixation due to drift. Each line represents a different population.

Evolution in Organelles

Drift may also play an important role in the evolution of plastid and mitochondrial genomes. Plastids and mitochondria replicate independently from the nucleus and their assortment into daughter cells is generally random (Michaelis, 1954; Birky, 1983). As we have already discussed, these organelles are polyploid and exist in multiple copies in mature leaf cells, but in meristematic cells their numbers are greatly reduced (Butterfass, 1979). Both organelles are also inherited maternally due to the unequal contribution of egg cells at fertilization, with only a few exceptions (Sears, 1980; Schumann and Hancock, 1990). Any mutations that arise in an organelle genome can become fixed in cells during cell division due to random sorting and ultimately establish sexual tissues, which produce gametes with a unique cytoplasm (Fig. 2.14).

As with nuclear genes, the ultimate evolutionary consequence of organelle replacement is dependent on its physiological ramifications. Many changes have little effect on the overall phenotype and are completely subject to drift, while others are dramatic and are greatly affected by selection. In the non-photosynthetic parasitic angiosperm *Epifogus virginiana*, de Pamphilis and Palmer (1990) found the photosynthetic genes to be largely deleted, while those involved in gene expression were still present. This indicates that once photosynthesis was unnecessary, the plastid genes associated with it were selectively eliminated, but the plastid "housekeeping" genes necessary for replication were maintained.

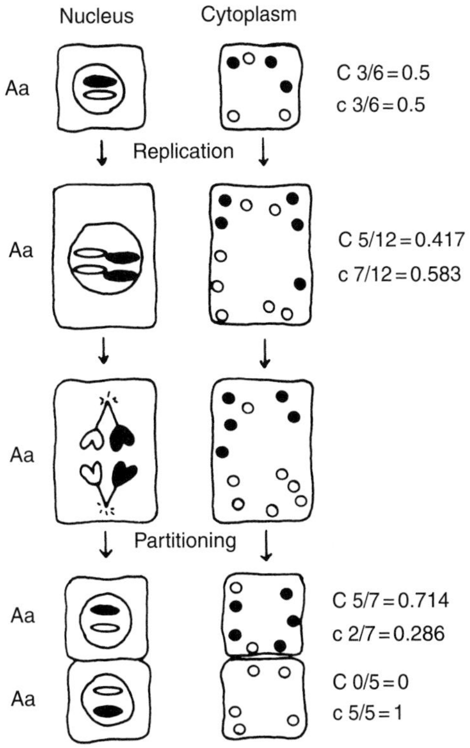

Fig. 2.14. Replication and partitioning of nuclear chromosomes and organelles. During cell division the nuclear chromosomes are partitioned regularly so that each daughter cell gets both homologs; organelles are partitioned randomly and as a result cells can become fixed for one type. (Used with permission from C.W. Birky, © 1983, *Science* 222, 468–475, American Association for the Advancement of Science.)

Interaction Between Forces

As we mentioned at the onset of this chapter, the evolutionary forces of migration, drift and selection rarely act alone. The relative importance of these forces depends on the particular circumstances at hand.

Selection is directional, while drift is not. The relative influence of these two forces is tempered by the strength of the selective coefficient and the population size (Fig. 2.15). Where drift is the predominant force, alleles will be fixed essentially at random; where selection is the predominant force, particular alleles will be favored. All bets are off in the intermediate zones where neither drift nor selection strongly dominates.

Migration can act to supply variability to a population, but it can also act in opposition to drift and selection if the migrants are distinct as to the changes being imposed. In general, very small amounts of migration can block the effects of genetic drift. These relationships can be seen by substituting migration rate for the selective coefficient(s) in Fig. 2.15. Likewise, if the migration rate is greater than or equal to the selection coefficient, then the importance of selection on population differentiation becomes almost insignificant (Ellstrand and Marshall, 1985).

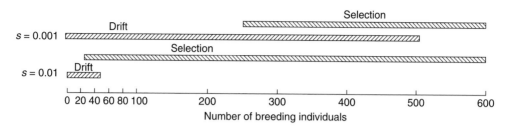

Fig. 2.15. Range of population sizes and selection coefficients where drift or selection will prevail. (Used with permission from V. Grant, © 1963, *The Origin of Adaptations*, Columbia University Press, New York.)

Mutation can act like migration in tempering the effects of selection and drift. However, its rate of occurrence is generally quite low ($<10^{-6}$) and therefore its greatest importance lies as a generator of new variability. Such rare variants can be established much faster in small populations than in large ones even if selective coefficients are quite high.

Sewell Wright (1931) combined migration, drift and selection into a unified process of evolution called the *Shifting Balance Theory*. In it, he viewed most populations as being substructured into small periodically drifting groups and he envisioned fitness as being due to many, interacting genes rather than a few key ones (see Chapter 3).

Wright described evolution as occurring in essentially five stages: (i) a population is "trapped" at a certain fitness level where only a limited number of genotypes are well adapted; (ii) because of drift, a less adapted genotype becomes established by chance in a subpopulation; (iii) subsequent evolution re-assorts the available genes in a new type more fit than the original; (iv) the subpopulation grows and begins to disperse; and (v) nearby neighborhoods receive the new genotype and the whole population reaches an adaptive peak.

It is not known how important the Shifting Balance Model is in nature. A considerable amount of debate has surrounded the frequency with which Wright's phases act in concert and whether simple models of mass selection explain most evolutionary change just as well (Peck *et al.*, 1998; Coyne *et al.*, 2000; Goodenough and Wade, 2000; Weinreich

and Chao, 2005). The Shifting Balance Model is almost impossible to prove in its entirety, but most of the individual aspects have been documented in natural populations (as we have discussed). At the very least, Wright's model is a stimulating attempt to describe the potential complexity of evolution in natural populations.

Summary

Numerous evolutionary forces shape plant populations, including migration, genetic drift and natural selection. Selection causes populations to change in several directed fashions, while genetic drift is synonymous with random change. Numerous factors influence the rate of change in populations. Of primary importance are the amount of variability present and the effective population size. Other important parameters are generation time, the strength of the selection coefficient and the degrees of dominance and epistasis. Selection is generally thought to play the leading role in shaping plant populations, but the critical influence of drift and migration cannot be excluded. Gene flow is generally limited in plant populations and as a result, effective population sizes are often quite small. This substructuring of populations can lead, even in the absence of selection, to patchy distributions of genes and the generation of unique allelic combinations. Both random and non-random forces have probably been important in plant evolution.

3

The Multifactorial Genome

Introduction

Up to this point, we have been describing evolution at primarily the single gene level. It is important to realize, however, that natural selection operates on the whole phenotype of an individual and not only on the product of a single gene. The fitness of an organism is dependent on the interaction of the complete genetic complement. This is alluded to in Wright's "Shifting Balance Theory".

The expression of a trait is regulated by both the genetic background of an individual and the environment that surrounds it (Chapter 1). Some genes have predictable, stable influences on phenotype, while others do not. There are genes with *incomplete penetrance* such that only a proportion of the individuals carrying the gene express the phenotype, and others with *variable expressivity* where the trait is expressed to various degrees in different individuals. The precise reasons for these differential gene effects are rarely known, but they are usually thought to relate to the particular environment of the individual and/or to interactions of the gene to others in the genotype.

The effect of a single locus is often lost in a myriad of genic and cellular interactions. Biochemists have provided us with numerous examples of lengthy, inter-coordinated pathways that are filled with "feedback loops" and internal regulations (Fig. 3.1). In some cases, genes act independently of other genes, so that the phenotype is

the sum of the contributions of the individual loci (additive effects), but frequently the whole is not equal to the sum of its parts, and subtle changes in the product of one locus can have a cascading effect on the overall phenotype.

The evolutionary ramification of this genomic complexity is that alleles are selected at two levels: (i) how well they function in the environment; and (ii) how well they function together. There is selection for harmoniously coordinated alleles across loci. The situation can be compared to an orchestra – the great ones not only have outstanding individual players, but they also perform as a finely tuned, integrated unit.

There is considerable evidence that the genome is cohesive and inheritance is generally multifactorial. Genetic phenomena such as pleiotropy, epistasis and specific combining ability are best explained as intra-genomic interactions, and instances of *coadaptation* have been observed where groups of genes or whole genomes appear to be selected together. We will begin our discussion of genomic complexity by first describing the various types of multigenic interactions that exist, and then evaluating the reported examples of coadaptation.

Intra-genomic Interactions

Pleiotropy occurs when one allele has more than one phenotypic effect. Some clear examples of

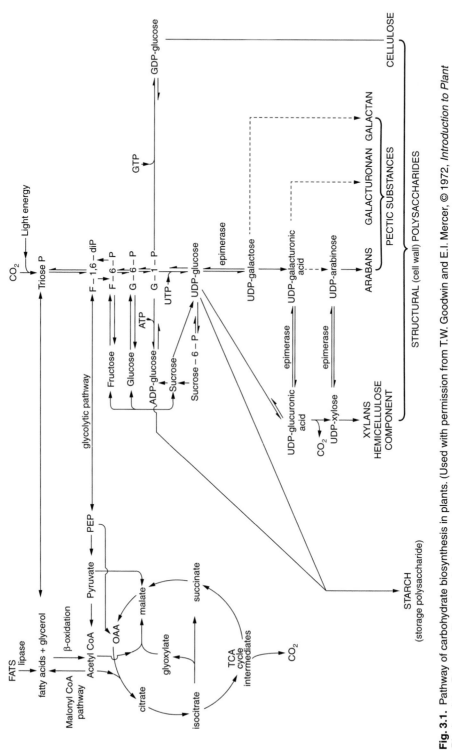

Fig. 3.1. Pathway of carbohydrate biosynthesis in plants. (Used with permission from T.W. Goodwin and E.I. Mercer, © 1972, *Introduction to Plant Biochemistry*, Pergamon Press, London.)

this phenomenon are the gene *S* in *Nicotiana tabacum*, which affects several structures including the shapes of leaves, flowers and capsules (Fig. 3.2), and the color gene *C* in onions, which not only regulates color but also determines resistance to the smudge fungus through the production of specific phenolic acids. The allele *dl* in the tomato (*Lycopersicon esculentum*) reduces the number of hairs on stems, peduncles and stamens. It also produces separated anthers, which diminish levels of self-pollination (Rick, 1947). Particularly strong proof that the genome is tightly integrated has come from transgenic studies where the incorporation of a single gene has been shown to have multiple pleiotropic effects (Wolfenbarger and Grumet, 2003; Little *et al.*, 2009; Miki *et al.*,

2009). The insertion of single genes can affect a wide array of biochemical processes by activating multiple regulatory pathways (for example Gilmour *et al.*, 2000; Fowler and Thomashow, 2002). The fact that these single gene changes influence so many traits argues strongly that the genome must be cohesive.

As we will discuss more fully in Chapter 6, a number of pleiotropic genes were important in the early domestication of beans, maize and wheat. The *fin* gene in dry beans conditions the earliness of flowering, and has significant effects on node number on stems, pod number, and the number of days from flowering to fruiting (Koinange *et al.*, 1996). Several pleiotropic QTL have been identified in *Zea* including: (i) *teosinte glume architecture 1* (*tga1*), which affects internode lengths, inflorescence sex and structure; (ii) *teosinte branched 1* (*tb1*), which also affects internode lengths, numbers and inflorescence sex; and (iii) *suppressor of sessile spiklets1* (*sos1*), which affects branching in the inflorescence and the presence of single versus paired spiklets in the ear (Doebley *et al.*, 1995). The gene *Q* in wheat regulates the tendency of the spike in wheat to shatter, as well as the tightness of the chaff around the grain and whether the spike is elongated or compact (Simons *et al.*, 2006).

In *epistasis*, the allelic constitution at one locus affects the level of expression of alleles at another locus, again illustrating the interactive nature of the genome. For example, the presence of prussic acid in clover requires a dominant allele at both of two loci. Bulb color in onion is also regulated by alleles at two loci – one locus determines whether the bulb will be colored at all and a second locus determines whether it will be red or yellow. The presence of the pungent chemical capsaicin in hot peppers is determined by a dominant allele at one locus, while the degree of heat is regulated by a series of modifiers at several other loci.

Quantitative traits are commonly influenced by epistatic interactions. These can be identified by plotting the trait values associated with each genotype (Fig. 3.3). Suppose you have two loci regulating plant height, A and B, with two alleles each and you plot the values of the genotypes of BB, Bb and bb for each allelic substitution at the A locus. If there is no epistasis and no dominance, the relative values for each genotype will rely solely on the additive combination of alleles

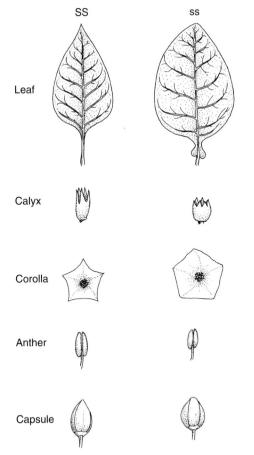

Fig. 3.2. Multiple effects of the gene *S* in *Nicotiana tabacum* (Stebbins, 1959).

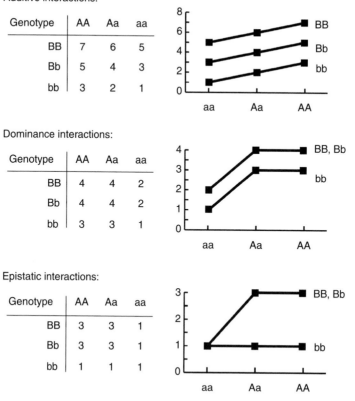

Additive interactions:

Genotype	AA	Aa	aa
BB	7	6	5
Bb	5	4	3
bb	3	2	1

Dominance interactions:

Genotype	AA	Aa	aa
BB	4	4	2
Bb	4	4	2
bb	3	3	1

Epistatic interactions:

Genotype	AA	Aa	aa
BB	3	3	1
Bb	3	3	1
bb	1	1	1

Fig. 3.3. A demonstration of the effects of additive, dominant and epistatic interactions on a quantitative trait.

at each locus, and the slopes of all the lines will equal 1. If there is no epistasis, but there is complete dominance, the values for heterozygotes will equal one of the homozygotes and the trajectories of each line will level off at the same point. If there is epistasis, the alleles will interact in a more complex fashion, and the trajectories of each line will differ.

Statistical analyses have been developed to partition levels of quantitative variation into additive, dominance and epistatic interactions using analysis of variance techniques (Falconer and Mackay, 1996). In one of the simplest analyses, breeders measure complex genomic interactions by calculating *general* and *specific combining ability* (Griffing, 1956). In this type of analysis, the breeder makes a series of crosses among a group of parents and compares their mean performance to that of their progeny. The mean performance of each line in crosses with all the other lines is called general combining ability (GCA) and represents the additive component of variance. The deviation of a particular individual cross from the average general combining ability of the two lines is called the specific combining ability (SCA) and represents intra- and inter-locus interactions (Fig. 3.4). Standard analysis of variance techniques are used to calculate the relative importance of GCA and SCA (Gilbert, 1967). More complex crossing and statistical approaches are required to separate dominance and epistatic interactions, but this example illustrates the general procedure.

Classical statistical studies of quantitative variation have uncovered considerable evidence of epistasis (Falconer and Mackay, 1996), as have more recent studies using molecular markers and QTL analysis (Maimberg and Mauricio, 2005; Zhang *et al.*, 2011). Yamamoto *et al.* (2000) found a significant interaction between two QTL

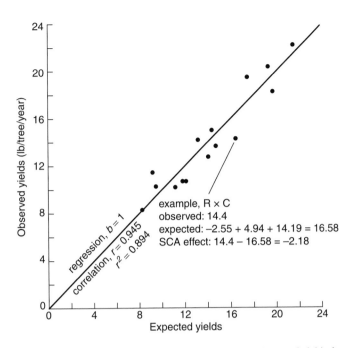

Fig. 3.4. Combining ability for yield in rubber, *Hevea brasiliensis*. The observed yield of progenies is plotted against the expected yields or general combining ability (GCA) of each line. The GCA is calculated as the mean performance of each line in crosses with all other lines. When a regression line is drawn through the various parental values, the deviations above and below the line represent specific combining ability (SCA). (Used with permission from N.W. Simmonds, © 1985, *Principles of Crop Improvement*, Longman, London.)

(*Hd2* and *Hd6*) involved in photoperiod sensitivity in rice. Yu *et al.* (1997) identified 32 QTL associated with four yield traits in rice, and found almost all of them to significantly interact with at least one other QTL. Doebley's group has discovered a significant epistatic interaction in maize between a QTL on chromosome arm 1L (QTL-1L) and another on chromosome arm 3L (QTL-3L) that both influence the number and length of the internodes in both the primary lateral branch and inflorescence (Doebley *et al.*, 1995; Lukens and Doebley, 1999). Alleles at these loci derived from either maize or teosinte had the strongest phenotypic effect in their own species background, further signaling the complexity of genomic interactions in *Zea mays*. QTL-1L was determined to be the locus for *tb1*, and QTL-3L could be *te1*, which we have already described as being highly pleiotropic. Again, selection at these two loci would have had dramatic phenotypic effects during the domestication process.

Coadaptation

If we accept the notion that the genome is highly interactive, then we realize that the whole genotype is the unit of selection and not the individual genes themselves. The simultaneous selection of large blocks of genes has already been referred to as coadaptation. In most cases, it is very difficult to obtain direct evidence of coadaptation since the relationship between most gene products and the phenotype is hazy, but a great deal of circumstantial evidence has been accumulated. Phenomena considered to represent coadaptation include: (i) hybrid breakdowns; (ii) supergenes; and (iii) gametic disequilibria.

Hybrid breakdowns

Hybrids within populations of animal species are usually completely normal, but hybrids between populations are sometimes weak or inviable (Wallace, 1968). Such *hybrid breakdowns* are thought to arise because the gene pools within populations have been selected over time for their harmonious interaction. When individuals are mated from variant populations, their genes may not be well integrated and therefore produce poorly adapted offspring.

Most of the most graphic examples of hybrid breakdowns have been demonstrated in animals. For example, Gordon and Gordon (1957) found that platyfish from different Amazonian river basins had characteristic patterns of small, dorsal spots. When fish from variant origins were mated, progeny were produced with gross, distorted collections of pigment. Backcross individuals proved to have even more unsightly patches.

Few examples of hybrid breakdowns have been described between populations of plant species, perhaps due to the lower developmental complexity of plants (Gottleib, 1984). However, crosses between related plant species often result in the production of weak or sterile progeny. The cross of *Gilia ochroleuca* × *Gilia latifolia* produces hybrids that have mainly abortive pollen grains and their ovules do not develop normally under any conditions (Grant and Grant, 1960). Hybrids of *Gossypium hirsutum* × *Gossypium barbadense* have inrolled leaves, corky stems, bushy growth and are mostly sterile (Stephens, 1946).

In some cases, unfavorable gene combinations do not appear until the F_2 generation and more complete reassortment occurs. Crosses of *Zauschneria cana* × *Zauschneria septentrionalis* (Clausen *et al.*, 1940) and *Layia gaillardioides* × *Layia hieracioides* (Clausen, 1951) produce vigorous, semifertile hybrids but most of the F_2 individuals are weak and dwarfish. We will describe these types of relationships more fully in the chapter on speciation.

Probably the most frequently mentioned case of hybrid breakdown *within* a plant species involves populations of the bean *Phaseolus vulgaris* (Gepts, 1988, 1998). There are large- and small-seeded races from South America and Mexico that, when crossed, produce high percentages of weak, semidwarf progeny. This reduction in hybrid fertility and vigor is associated primarily with two independent loci DL_1 and DL_2, although many other differences exist in the gene pools of the geographic races including

distinct phaseolin seed proteins, electrophoretic alleles, flowering times and floral structures (Shii *et al.*, 1981; Gepts and Bliss, 1985).

Proper development and function depends not only on nuclear interactions but also on nuclear-organelle cooperation. Dramatic differences are found between reciprocal crosses of *Epilobium hirsutum* and *Epilobium luteum* (Michaelis, 1954). *E. hirsutum* × *E. luteum* yields hybrids with stunted growth, narrow yellow-mottled leaves and sterile anthers, while *E. luteum* × *E. hirsutum* produces normal looking plants and fertile anthers. These differences are presumably the result of the egg providing most of the cytoplasm to the fertilized egg. As with most angiosperms, the egg cell of *Epilobium* is rich with organelles, but the sperm cells have little cytoplasm.

Stubbe (1960, 1964) found the plastids of diploid *Oenothera* in the section *Euoenothera* to vary in their functionality in different nuclear backgrounds. He identified five plastome types that differed in their nuclear compatibility (Fig. 3.5). Although the chloroplasts can often survive in the nuclear background of another species, they show varying degrees of bleaching. This dysfunction suggests that the nuclear and plastid genomes coevolved to produce a finely tuned cooperative relationship.

Multigene complexes

There are a number of examples in plants and animals where several genes with related functions are found in close proximity on a chromosome and may represent coadaptation (Ford, 1975). Darlington and Mather (1949) coined the term supergene to describe the case where a series of genes rarely undergo recombination due to tight linkage on a chromosome or association within an inversion. They felt that genes which were originally separate might occasionally migrate together to unify coadapted complexes as selection "capitalized" on cytological aberrations.

Heterostyly in primroses is a clear example of such supergenes (Dowrick, 1956; Crowe, 1964). Successful pollination occurs only between individuals with their stigmata and anthers in the same position (Fig. 3.6) and

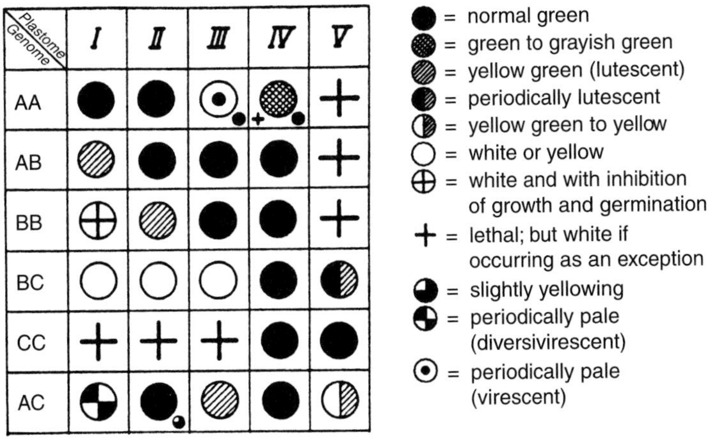

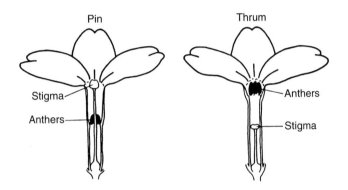

Fig. 3.5. Compatibility between different diploid *Oenothera* nuclear and plastid genomes. Small symbols represent a less frequent occurrence in the A genomes (Stubbe, 1964).

Fig. 3.6. Relative positions of the stigma and anthers in pin and thrum flowers of the primrose. (Used with permission from P.W. Hedrick, © 1983, *Genetics of Populations*, Fig. 5.3, p.173, Jones and Bartlett Publishers, Boston, Massachusetts.)

numerous other characteristics are linked to these traits including size and ornamentation of pollen grains, and the size of stigmatic papillae. Other classic examples of heterostyly are found in the *Polygonaceae, Linoceae, Lythraceae, Oxalidaceae* and *Boraginaceae* (Grant, 1975).

The speltoid mutant in hexaploid wheat is another example of a supergene in higher plants that may even represent a reversion to primitive characteristics. The cultivated bread wheat *T. aestivum* normally has a tough rachis (main fruiting axis) with loose glumes (bracts) that allow the grain to fall out easily. The speltoid mutant mimics the primitive wheat *T. spelta* in which the rachis is brittle and the glumes are tightly attached. *T. aestivum* yields mostly naked

seeds when harvested, while the rachis of the speltoid mutants shatter and the seeds remain attached (Frankel and Munday, 1962). The genes involved in this syndrome of traits are clustered together on one chromosome (IX) and all speltoid mutants have a deficiency for a segment called Q (Sears, 1944; MacKey, 1954).

Chromosomal inversion heterozygosities have also been used to argue for coadaptation. Probably the most complete story has been developed in the fruit fly *Drosophila* (Dobzhansky and Pavlovsky, 1953; Vetukhiv, 1956). Numerous chromosomal arrangements exist in natural populations of this genus and many of these chromosomal types are maintained in relatively stable frequencies. Dobzhansky and Pavlovsky suggested that these polymorphisms represented coadapted blocks of well integrated genes and that the most fit individuals were those heterozygous for chromosomal rearrangements. They tested their hypothesis by comparing the fitness of homozygous and heterozygous genotypes of *Drosophila willistoni* and *Drosophila psuedobscura* from different localities. When experimental populations were initiated with individuals heterozygous for two inversions, high frequencies of inversion heterozygotes were maintained if the populations were begun with individuals from the same locality, but frequencies of inversion heterozygotes diminished in populations initiated with strains of different localities. They suggested that the genes in the inversions from similar populations were well integrated, while those from distinct populations were not. The gross structure of the chromosomal arrangements found in two populations may have been similar, but the alleles they carried were different. This divergence in allelic frequency was later documented by Prakash and Lewontin (1968, 1971) using allozymes.

Such dramatic examples of inversion polymorphism are unusual in plants, but the maintenance of translocation heterozygotes in species like *Oenothera* is thought to reflect coadapted complexes (Chapter 1). Permanent translocation heterozygotes are maintained through a system of balanced lethals where homozygotes are either weak or not produced due to gametic inviabilities. For example, in *Oenothera lamarchiana* the chromosomes form a ring of 14 at meiosis that are oriented in such a way that alternate chromosomes pass to each pole (Fig. 3.7). This results in only two types of gametes being produced, one with chromosomes from only the pollen parent and one with chromosomes from only the mother. A system of balanced lethals then operates in *Oenothera* to allow only heterozygous zygotes to survive. In one case, only one set of chromosomes produces viable pollen and one set viable eggs (gametophytic lethals), while in other instances, both types of gametes are formed, but only heterozygous zygotes survive (zygotic lethals).

It is generally agreed that this complex system of structural hybridity could have arisen as a mechanism to preserve coadapted complexes. To test this possibility, Levy examined electrophoretic variation among chromosomal complexes of several *Oenothera* species that carried

the gametophytic lethal system (Levy and Levin, 1975; Levy *et al.*, 1975; Levy and Winternheimer, 1977). He discovered that the species had very few alleles at individual polymorphic loci, but that individual strains had unique combinations of the whole allelic array. The egg and sperm lines of most strains differed significantly in allelic frequencies at a number of loci and intergenomic linkage disequilibrium accounted for 97.5% of the observed heterozygosity. Thus, particular genic arrays were indeed being maintained in each population – an observation at least consistent with the coadaptation hypothesis.

Gametic disequilibria

In electrophoretic examinations of plant and animal populations, alleles at groups of linked and un-linked genes are often found to be out of Hardy–Weinberg equilibrium. The non-random association of alleles at different loci into gametes is referred to as *gametic phase disequilibrium* or the shortened form, *gametic disequilibrium* (Lewontin and Kojima, 1960; Crow and Kimura, 1970).

As we have already noted, genetic analyses involving multiple loci are quite complex. To measure gametic phase disequilibria, the frequencies of all possible allelic combinations are calculated and then compared to each other. Excesses of one gametic type that do not equilibrate after several generations of mating are thought by many population biologists to represent coadapted complexes. For example, using the simplest case of two loci with the alleles Aa and Bb:

> Gametes carrying AB and ab are said to be coupling gametes
>
> Gametes carrying Ab and aB are said to be repulsion gametes
>
> Gametic phase disequilibrium (D) = Product of coupling genotype frequencies – Product of repulsion genotype

Mathematically:

if

Locus 1	Locus 2
$A = p_1$	$B = p_2$
$a = q_1$	$b = q_2$

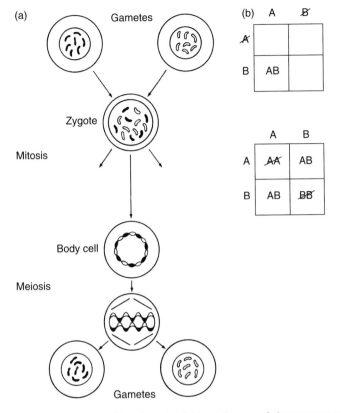

Fig. 3.7. System of balanced lethals in *Oenothera*. (a) Stylized diagram of chromosome segregation in *Oenothera lamarchiana*. The chromosomes form a ring of 14 at meiosis, oriented in such a way that alternate chromosomes pass to each pole. Only two types of gametes are produced, one with all its chromosomes derived from the pollen parent, and the other with a full set from the mother. (b) Two types of balanced lethal systems where only heterozygous zygotes result. Top, gametophytic lethals where only one chromosome set produces viable eggs and pollen. Bottom, zygotic lethals where both types of gametes are formed, but only heterozygous ones survive. The slashes represent mortality or death. (Used with permission from T. Dobzhansky, F.J. Ayala, G.L. Stebbins and J.W. Valentine, © 1977, *Evolution*, W.H. Freeman and Company, San Francisco, California.)

then

$$D = (p_1p_2)(q_1q_2) - (p_1q_2)(q_1p_2)$$

and to weight by gene frequency

$D' = D/pq$
    = relative gametic phase disequilibrium

D' varies from +1 to −1, and

if $D' = 0$, there is no disequilibrium

if D' is negative, there are more repulsion than coupling gametes

if D' is positive, there are more coupling than repulsion gametes

Populations with positive and negative values of D have a surplus of one or the other gametic type and are therefore in gametic phase disequilibrium. This means that alleles at each of the loci are not segregating independently and distinct genic assemblages are being maintained.

The most commonly cited examples of gametic phase disequilibria in plants are those of Allard and his co-workers (Allard *et al.*, 1972; Allard, 1988), although others have been documented (Golenberg, 1989). We described previously the association between allelic frequencies at single loci and moisture levels (Chapter 2). These alleles were further clustered into a few non-random associations across loci. The mesic environments were almost monomorphic for a set of alleles at five loci (denoted 21112), while on the more xeric sites only two sets of alleles predominated (12221, 12211, Table 3.1). Three of these loci were on the same chromosome, while the other two segregated independently (Clegg *et al.*, 1972). When the D' values were averaged across all of the two locus pairs, values ranged from 0.39 to 0.71 in each of the various subdivisions. The direct adaptive benefit of these alleles was not measured, but the most common assemblages were thought to represent coadapted complexes since they were not in the "expected frequencies".

Allard and co-workers (Clegg *et al.*, 1972; Weir *et al.*, 1972) also found associations between alleles at different electrophoretic loci in cultivated populations of barley (*Hordeum vulgare*). They examined seed samples of two highly heterozygous experimental populations (CCII and CCV), which were initially generated by hybridizing 30 parents from the major barley production regions of the world. These populations were propagated in large plots under agricultural conditions for decades without conscious human-directed selection. It was discovered that some allelic combinations had increased over time, while others had declined (Table 3.2). The Allard group hypothesized that natural selection was structuring "the genetic resources of these populations into sets of highly interacting, coadapted gene complexes".

While these gametic disequilibria could indeed be the result of coadaptation, the possibility cannot be excluded that these changes are the result of "hitch-hiking" of allozyme loci with other major adaptive genes (Hedrick and Holden, 1979; Hedrick, 1980). The loci examined by the Allard group may be selectively neutral, but linked to other genes that are selectively important. Since both oats and barley are highly selfed, gametic disequilibria could arise without strong epistatic interactions.

A similar argument can also be made about the electrophoretic variation observed in the chromosomal inversion types of *Drosophila* and *Oenothera*. Over time, neutral allozyme loci within the different chromosomal types

Table 3.2. Most common triallelic four-locus gametic types found in complex hybrid populations of barley (CCII and CCV) over several generations. Only the most common genotypes are shown (Clegg *et al.*, 1972).

Gamete	CCII generation		
	7	18	41
1221	0.071	0.062	0.052
2112	0.109	0.021	0.497
2113	0.115	0.115	0.248
	CCV generation		
	5	17	26
1221	0.009	0.030	0.173
2111	0.129	0.170	0.173
2112	0.038	0.063	0.085

Table 3.1. Changes in five-locus gametic frequencies of wild oats along a moisture gradient. Subdivisions A, B, C and D represent a change from mesic to xeric conditions (Allard *et al.*, 1972).

Gametic type[a]	Subdivision				Location total
	A	B	C	D	
21112	91.5	39.9	16.9	1.9	56.7
12221	1.7	4.1	27.9	31.9	11.3
12211	0.1	1.5	3.2	30.5	4.0
11112	0.7	4.8	2.7	0.6	1.8
21121	0.7	5.8	8.8	5.7	4.0
21221	0.0	4.6	5.6	1.6	2.2
12212	0.4	10.7	1.5	0.3	2.2
22221	0.2	3.9	6.5	2.0	2.5
D'	0.71	0.36	0.52	0.39	0.64

[a]Two alleles (1 and 2) at five loci: esterase loci E4, E9, E10, phosphatase locus P5 and anodal peroxidase locus APX5.

might have gradually diverged due to random forces, since the chromosomal types would be operating as separate gene pools due to the inviability of heterogenetic crossovers (Nei and Li, 1975). Still, the persistence of chromosome heterozygosities in natural populations argues that at least some adaptively important genes are found on the variant chromosomal types.

Complex gene interactions in polyploids

Poor performance in selfed autopolyploids is usually attributed to the loss of higher order allelic interactions in what is known as the overdominance model of inbreeding depression (ID) (Bever and Felber, 1992). Stated in another way, vigor in autopolyploids is positively correlated with the number of different alleles at each locus.

Evidence for the existence of complex inter-allelic interactions in autopolyploids has come from the observation that inbreeding depression in autopolyploid lucerne (Busbice and Wilsie, 1966; Busbice, 1968) and potato (Mendoza and Haynes, 1974; Mendiburu and Peloquin, 1977) is much greater than would be predicted by the coefficient of inbreeding in a two allele model. Busbice and Wilsie (1966) suggested that this rapid loss of vigor is associated with the theoretical rate at which loci with three and four alleles are lost (tri- and tetra-allelic loci). Correlative data have come from comparisons of autotetraploids with different genetic structures. Bingham and his group (Dunbier and Bingham, 1975; Bingham, 1980) produced diploids from natural tetraploids by haploidy, generated diploid hybrids, and then doubled the diploid hybrids using colchicine treatments to obtain defined two-allele duplexes (di-allele loci). These were then crossed to produce double hybrids with presumed tetra-allelic interactions. When the performance of these different structured populations were compared, "progressive heterosis" was observed as the diploid hybrids had higher herbage yield (540 g/plant) than their diploid parents (237 g/plant), and the double hybrids had the highest yield of all (684 g/plant).

Canalization

Individuals of plant and animal species generally maintain a recognizable identity even though they are highly polymorphic and suffer a wide range of environmental extremes. This suggests that their developmental patterns are buffered against a broad amplitude of environmental and genetic conditions. Waddington (1940) and Mather (1943) originally suggested that organisms become resistant to both genetic and environmental perturbations through an extreme type of coadaptation called *canalization*. They evolve to the point where most allelic substitutions and environmental disturbances do not substantially alter developmental patterns. In a sense, canalization is the end result of a complex form of stabilizing selection.

The phenotype is constructed by successive interactions of the genotype with the environment in which development occurs. Due to highly complex inter-allelic networks, there is often a wide range of environments that produce the same phenotype and it frequently takes extreme conditions to drastically disrupt development. This has been referred to as environmental canalization. Likewise, numerous allelic substitutions can occur at several loci without severely affecting development, due to the counterbalancing influences of alleles at other loci. This "hidden" genetic variability has been referred to as *cryptic genetic variation* (Gibson and Dworkin, 2004). The only alleles that persist in a population are those that integrate well under a large range of environments to produce a normal phenotype. This has been referred to as genetic canalization.

Some of the earliest experiments documenting canalization involved *Drosophila melanogaster* (Waddington, 1953, 1957). Waddington took a population of wild type individuals and exposed them to a heat shock. Most were phenotypically normal, but a few showed a crossveinless trait on their wings that did not appear under normal conditions. After a number of generations of heat shock and selection for crossveinless, the trait began to appear in flies without treatment. Subsequent genetic analysis discovered that the trait was being caused by a group of polygenes with small cumulative effects. Waddington suggested that crossveins on wings were canalized and only unusual events such as heat shock and strong selection could lead to the appearance of the abnormal type and the subsequent alteration of the normal phenotype.

As previously described, Huether (1968, 1969) found that most individuals of *Linanthus androsaceus* had five-lobed flowers in nature (>95%). However, he was able to increase the frequency of abnormal types through very strong disruptive selection and by maintaining the plants under various environmental stresses including long days, high temperatures and decapitation. The combination of strong selection and environmental shock produced the highest frequency of anomalous individuals after five generations (Fig. 3.8). Thus, petal number was highly canalized and it took extreme environments and/or strong selection to alter development.

Lauter and Doebley (2002) studied the inheritance of several traits involved in inflorescence and plant architecture that are phenotypically invariant within teosinte, but distinguish it from maize. Using a QTL mapping strategy on the progeny of a cross between teosinte and maize, they identified cryptic genetic variation for the phenotypically invariant traits in teosinte. They argued "that such cryptic genetic variation can contribute to the evolution of novelty when reconfigured to exceed the threshold necessary for phenotypic expression or by acting to modify or stabilize the effects of major mutations".

Obviously, all phenotypic traits are to some extent modifiable or we would not see so much variation in nature. However, the modifiability of a character depends upon the importance of the

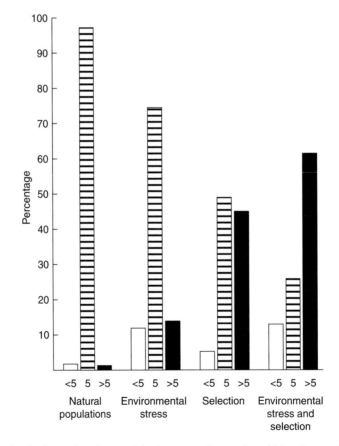

Fig. 3.8. Effect of selection and environmental extremes on the number of *Linanthus androsaceus* corolla lobes after five generations. Abnormal types are rare in natural populations, but they increase in proportion after strong selection or environmental shock (Huether, 1968).

character to the general fitness of the individual. For example, gross physical or developmental abnormalities in reproductive structures could severely limit fitness by reducing attractiveness to pollinators. Such traits would remain constant unless the population is subjected to shifting balance processes or a drastic change in the environment beyond the normal threshold of the trait.

Paradox of Coadaptation

Until a trait or process is canalized, the cost of maintaining coadapted complexes can be extremely high due to recombination (recombinational load; Wallace, 1970). If we consider a complex composed of only five loci with two alleles each, there are 243 possible diploid genotypes generated through recombination (Table 3.3). If only a few of these are well adapted, a lot of individuals will expire before maturity.

Several genetic systems limit free recombination and, therefore, help maintain coadapted complexes. We have already suggested that chromosomal rearrangements and tight linkage can hold coadapted genes together by limiting recombination. Inbreeding and asexual reproduction (apomixis) can also have the same effect, although these systems fix whole genomes rather than "blocks of genes".

Simple linkage reduces recombination since the closer two genes are on a chromosome the less likely there will be a chiasmata between

them at meiosis. Inversions and translocations act as "crossover supressors" because crossovers in heterozygotes result in the production of inviable gametes (see previous section on mutation). The mortality in this system is high, but it is thought to be much lower than if free recombination occurred among coadapted genes.

Selfing limits recombination by promoting homozygosity – homozygous individuals have no variability to reassort. Some species of plants rely on asexual reproduction for proliferation and as a result do not undergo sexual recombination at all. This phenomenon, called apomixis, can occur through agamospermy, where viable seeds arise without fertilization, and through vegetative propagation via structures such as stolons, tubers, rhizomes or suckers. Here, gene assemblages are maintained forever except when spontaneous mutations arise. Numerous genera contain apomicts, including *Taraxicum, Aster, Erigeron, Rudbeckia, Poa, Crepis, Musa, Manihot, Malus, Rubus, Potentilla, Citrus, Allium* and *Tulipa* (Grant, 1981). Some of our most important cultivated crops are asexually reproduced naturally through tubers (white potatoes, yams, sweet potatoes and arrowroot), or artificially via cuttings (banana and cassava).

Apomictic plants cannot undergo free recombination, but a surprising number have collected sufficient variation to possess races with variant ecological requirements (Babcock and Stebbins, 1938; Lyman and Ellstrand, 1984). Most cultivars of banana, manihot and yam arose as distinct apomictic variants

Table 3.3. The number of possible diploid genotypes that can be produced by recombination among various numbers of separate genes, each of which possess various numbers of alleles (Grant, 1963).

Number of alleles of each gene	Number of genes				
	2	3	4	5	n
2	9	27	81	243	
3	36	216	1,296	7,776	
4	100	1,000	10,000	100,000	
5	225	3,375	50,625	759,375	
6	441	9,261	194,481	4,084,101	
7	784	21,952	614,656	17,210,368	
8	1,296	46,656	1,679,616	60,466,176	
9	2,025	91,125	4,100,625	184,528,125	
10	3,025	166,375	9,150,625	503,284,375	
r	$\dfrac{r(r+1)^2}{2}$	$\dfrac{r(r+1)^3}{2}$	$\dfrac{r(r+1)^4}{2}$	$\dfrac{r(r+1)^5}{2}$	$\dfrac{r(r+1)^n}{2}$

(Chapter 4). Solbrig and Simpson (1977) have even described asexual biotypes of dandelion (*Taraxicum officinale*) that are differentially adapted to mowed and unmowed lawns. Such variation arose either through mutation or previous hybridizations between sexual progenitors.

Many plant breeders have exploited clonal reproduction in producing cultivated varieties. Crops like strawberry and blueberry can be sexually crossed to produce variable progeny, but when highly productive, coadapted genotypes arise they can be propagated asexually through runners or cuttings. This eliminates the need to develop homozygous, balanced parental lines to regenerate desirable heterozygous varieties.

Summary

The overall fitness of an individual is generally not dependent on just one gene, but rather on the interactive relationships of all the genes making up the genome. Pleiotropy, epistasis and specific combining ability represent these complex intra-genomic interactions. Genes are essentially selected at two levels: (i) how well their products perform in response to the external environment; and (ii) how well they interact with other gene products. The selection of whole groups of genes together by both internal and external forces is called coadaptation. Evidence for this process has come from several different sources, including: (i) gametic phase disequilibria, where whole groups of alleles are out of Hardy–Weinberg equilibrium; (ii) supergenes, where several genes with related functions are tightly linked; (iii) the maintenance of balanced frequencies of chromosomal inversion heterozygotes: and (iv) hybrid breakdowns, where hybrids within populations are completely normal, but hybrids between populations are weak or inviable. The genetic cost of maintaining coadapted complexes is high until interacting genes become associated through linkage, inbreeding or apomixis.

4

Polyploidy and Gene Duplication

_____

Introduction

As we discussed previously in Chapter 1, gene duplications very commonly arise in plant species after a variety of genetic events that affect different numbers of genes. Unequal crossing over and reciprocal translocations result in one or a small number of genes being amplified. Aneuploidy causes duplication in all the genes of a particular chromosome. Polyploidy results in an amplification of the total genic content. Any type of duplication can have important evolutionary ramifications, but those of polyploidy are often the most dramatic because the whole genome is affected.

One of the most intriguing questions facing plant evolutionists is why there are so many polyploid species. As mentioned earlier, most crop plants have high chromosome numbers and the majority of all angiosperms are polyploid. It has been estimated that 2–15% of all speciation events in flowering plants represent polyploidy (Ramsey and Schemske, 1998; Otto and Whitten, 2000; Wood et al., 2009). All 18 of the world's worst weeds are polyploid (Brown and Marshall, 1981; Clegg and Brown, 1983).

As DNA sequence data accumulate, it is becoming clear that many species that were considered to be diploid based on disomic chromosome behavior may actually be ancient polyploids, whose diploid progenitors have gone extinct. Ancient cycles of genome duplication are evident in the cole crops, cotton, soybean and many important cereals (Wendel, 2000). Even *Arabidopsis*, popular in genetic analysis due to its small genome size, is likely an ancient allopolyploid, based on the degree of gene duplication apparent in the sequence of its whole genome (Blanc *et al.*, 2000). If we use Stebbins' (1950) criterion that $n = 12$ or greater denotes polyploidy, it turns out that a large number of angiosperm genera with very deep phylogenetic roots contain only polyploids (see Chapter 1, Table 1.2).

The majority of all plant species are polyploid, even though the probability of a new polyploid species coexisting with its progenitors or replacing them would be extremely unlikely (Fowler and Levin, 1984). First, their initial numbers would be extremely low and subject to elimination due to chance, and secondly, their fertility would be reduced by the meiotic irregularities caused by genomic duplication or the formation of inviable triploid zygotes after the fertilization of $2n$ eggs by the more abundant haploid pollen. The derived and progenitor species would also be under strong competition, since they would share the same genetic makeup and therefore would have substantial niche overlap. A similar situation would exist for newly emerged gene or chromosomal duplications, except that fertility levels would be generally higher.

Factors Enhancing the Establishment of Polyploids

The probability that a new polyploid will form or any duplication will be established is increased if it is repeatedly synthesized. There would be more individuals present to face chance elimination, and a higher proportion of the polymorphism present in the progenitor species might be captured for subsequent evolution. Since many diploid species produce measurable amounts of unreduced gametes (Bretagnolle and Thompson, 1995), it must be relatively common for polyploid species to have multiple origins. The most common polyploids are those with balanced numbers of genomes ($4x$, $6x$, etc.), although triploids and other aneuploids do occasionally persist and form at least some euploid gametes (Husband and Schemske, 1998; Ramsey and Schemske, 1998).

Many of our crop species produce measurable quantities of unreduced gametes, including potato, cassava, blueberry, cotton, strawberry and cherry. In fact, nearly all polyploids that have been examined with molecular markers have been found to be polyphyletic with multiple origins (Soltis and Soltis, 1993, 1999). In one of the best studied polyploid groups, *Tragopogon mirus*, it has been estimated that there are four to nine lineages, and the estimates for *Tragopogon miscellus* range from 2 to 21 (Soltis and Soltis, 1995; Soltis and Soltis, 2000). Most of the evidence for recurrent formation comes from nuclear genes, but in some cases multiple chloroplast DNA haplotypes have also been documented.

Regardless of a new type's initial numbers, it still must compete with its parental species for space and resources. In some cases, the diploids and polyploids may have sufficiently distinct adaptations to ecologically assort across habitats (Husband and Schemske, 1998). The opening of new disturbed sites may in some instances provide them with an opportunity for establishment. These sites might arise due to natural causes or human intervention. Agricultural disturbances are thought to have contributed to the spread of polyploid wheats in the Old World belt of Mediterranean agriculture (Zohary, 1965). Polyploid forms of *Tragopogon* are not present in Europe, where the genus is native, but they are gradually spreading in the Pacific Northwest of North America, where the group was introduced and unique habitats probably exist (Ownbey, 1950; Soltis and Soltis, 1989a). Polyploids of numerous species are found in previously glaciated areas that their progenitors have not invaded (Ehrendorfer, 1979; Lewis, 1979; Soltis, 1984).

The bottleneck of reduced fertility and low numbers could be partially "solved" by the new polyploid being self-fertile and perennial – this would increase the new variant's chances of producing enough viable offspring to avoid chance elimination and evolving higher levels of fertility. In fact, polyploidy is much more common in perennial species than annual ones and many allopolyploid species are highly self-fertile (MacKey, 1970). Müntzing (1936) and Stebbins (1971) describe numerous examples where polyploidy is more prevalent in perennial than annual species of the same genus. Gustafsson (1948) showed that annuals in general have low percentages of polyploid species. Stebbins (1971) even suggested that "polyploidy in annual flowering plants is almost entirely confined to groups which have a high proportion of self-fertilization in both the polyploids and their diploid ancestors". This statement is much more accurate for allopolyploids than autopolyploids, as we will discuss later.

The simple process of polyploidy can by itself result in a partial breakdown of the self-incompatibility system (Levin, 1983). In the gametophytic system of self-incompatibility, the doubling of genes has been shown to disrupt the recognition system in a wide range of species including members of the *Rosaceae*, *Solanaceae*, *Scrophulariaceae* and *Leguminosae* (Lewis, 1943, 1966; Yamane *et al.*, 2001). The basis of this disruption is unknown, but may be due to competition between pairs of variant S alleles. In those raw polyploids with low self-fertility, Miller and Venable (2000) have proposed that such a breakdown in self-incompatibility could result in the evolution of separate genders as a means of avoiding inbreeding depression.

While high self-fertility would seem to be an advantage in polyploid establishment, evolutionary biologists have not found a consistent relationship between ploidy level and self-compatibility (Marble, 2004; Husband *et al.*, 2008) or inbreeding depression (ID).

Polyploid complexes with lower seed or fruit set in diploids include wheatgrass (Dewey, 1966), maize (Alexander, 1960), clover (Townsend and Remmenga, 1968) and *Epilobium angustifolium* (Husband and Schemske, 1996, 1997). Polyploid complexes with less ID in diploids include orchardgrass (Kalton *et al.*, 1952) and *Amsinckia* sp. (Johnston and Schoen, 1996). ID was found to be comparable in diploid, tetraploid and hexaploid *Vaccinium corymbosum* (Vander Kloet and Lyrene, 1987), while diploid *Vaccinium myrtilloides* had significantly greater ID than tetraploid *V. corymbosum*, but not tetraploid *Vaccinium angustifolium* (Hokanson and Hancock, 2000). These discrepancies between prediction and results, can probably be attributed to the complex nature of inbreeding depression and not a breakdown in theory, as the influence of ploidy on ID is not only dependent on the buffering effect of multiple alleles, but also relative population sizes, levels of dominance, higher order gene interactions, and the amount of time that has been available to purge deleterious alleles and evolve mating systems (Charlesworth and Charlesworth, 1987; Dudash *et al.*, 1997; Cook and Soltis, 1999, 2000).

Evolutionary Advantages of Polyploids

Polyploids have several characteristics that may contribute to their long-term survival and allow them to effectively compete with their parental species. The most commonly implicated advantages are: (i) the effect of nuclear DNA amount on cell size and developmental rate (nucleotypic effects); (ii) the influence of high enzyme levels (dosage effects); and (iii) increased heterozygosity. These factors may play a role in the adaptation of all types of duplication, although the degree of physiological alteration is often dependent on the number of genes involved.

Nucleotypic effects

Nucleus and cell size are positively correlated with DNA content (Ramachandran and Narayan, 1985; Bennett, 1987; Fig. 4.1) and increased cell size frequently translates into larger plants (Grant, 1981). Larger plants sometimes have greater competitive abilities than small ones, but this potential advantage is often balanced by slower developmental rates due to retarded cell division (Bennett, 1972). In most direct comparisons of diploid and polyploid species in the same non-stressed conditions, polyploids are competitively inferior (Levin, 1983).

The slow developmental rate associated with large, polyploid nuclei may occasionally be adaptive in nutrient- and water-poor environments where resources are easily exhausted (Grime and Hunt, 1975; Levin, 1983). When Stebbins (1972, 1980) sowed $2x$ and $4x$ seed of *Ehrharta erecta* at several sites in California, the polyploids became most firmly established on steep, shady hillsides, while the diploids predominated on

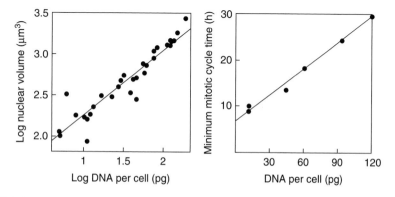

Fig. 4.1. Relationship between DNA content per cell and nuclear volume or mitotic cycle time. (Redrawn with permission from M.D. Bennett, © 1987, *New Phytologist* 106, 177–200.)

more mesic sites. In other studies, tetraploids of *Nicotiana* (Noguti *et al.*, 1940) and *Dianthus* (Rohweder, 1937) were found primarily on dry, calcareous soils where diploids were absent. Such instances of polyploid superiority may be the exception rather than the rule, however, as Stebbins (1971) could find only one study out of nine in western USA where tetraploids were located on more xeric sites than their diploid relatives. Similar inverse relationships have been observed by others (Johnson and Packer, 1965; Price *et al.*, 1981).

Bennett (1976) discovered a positive correlation between DNA amount and latitude among crop species (Fig. 4.2). This cline may relate in some unknown way to environmental constraints or the differential radiosensitivity of large and small chromosomes (Bennett, 1987). The level of ultraviolet light is higher in tropical than temperate latitudes (Sanderson and

Hulbert, 1955) and increases in DNA amount per chromosome make them a more likely target for ionizing radiation.

Dosage effects

The immediate biochemical effect of gene duplications on structural genes can be increased production of a protein or enzyme. Carlson (1972) was able to determine the chromosomal location of enzyme loci by searching for increased activity in trisomics of *Datura stramonium* (Fig. 4.3, Table 4.1). He looked at 12 enzymes and could assign nine of them to specific chromosomes. Levin *et al.* (1979) discovered 1.5- to twofold increases in alcohol dehydrogenase activity in six raw polyploids of *Phlox drummondii*, while Dean and Leech (1982) found regular increases in Rubisco levels across 2*x*, 4*x* and 6*x* wheats (764, 1517 and 2242 pg/cell, respectively). Argoncillo *et al.* (1978) found almost linear increases in protein level as chromosomes were added to nullisomic lines of hexaploid wheat (Fig. 4.4).

Such increases in enzyme activity could have important adaptive consequences if the enzyme plays a critical metabolic role and is rate limiting (Wilson *et al.*, 1975; Gottlieb, 1982). However, the biochemical and physiological consequences of gene duplication are unpredictable. Autopolyploids of *Lycopersicon esculentum* (tomato) were shown to have enhanced activity for four enzymes, decreased activity for one and stable activity for two others (Albrigio *et al.*, 1978). Autotetraploid *Tragopagon miscellus* had alcohol dehydrogenase activities that were intermediate to its diploid progenitors (Roose and Gottlieb, 1980). Rubisco activity and photosynthetic rate were correlated with ploidy level in polyploids of tall fescue (Randall *et al.*, 1977; Joseph *et al.*, 1981), but in castor bean (Timko and Vasconcelos, 1981) and lucerne (Settler *et al.*, 1978) higher ploidies had unchanged or reduced photosynthetic rates even though their Rubisco levels were positively associated with nuclear ploidy. Photosynthetic rates in induced polyploids of *Phlox* varied greatly depending on the diploid progenitor's genotype (Bazzaz *et al.*, 1982).

This unpredictability is probably due to the complexity of the plant genome. As outlined

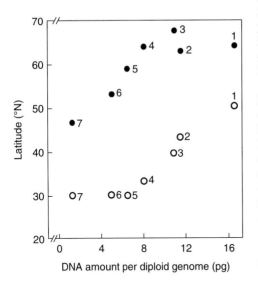

Fig. 4.2. The relationship between DNA amount per diploid genome and the northern limits of cultivation of several cereal grain species. Key to points: (O) for a transect from Hudson Bay to Key West in Florida (approximately 82°W) in winter; (•) for a transect from near Murmansk by the Arctic Ocean to Odessa by the Black Sea (approximately 32°E) in summer. 1, *Secale cereale*; 2, *Triticum aestivum*; 3, *Hordeum vulgare*; 4, *Avena sativa*; 5, *Zea mays*; 6, *Sorghum* spp.; 7, *Oryza sativa*. (Used with permission from M.D. Bennett, © 1976, *Environmental and Experimental Botany* 16, 93–108, Pergamon Press, Elmsford, New York.)

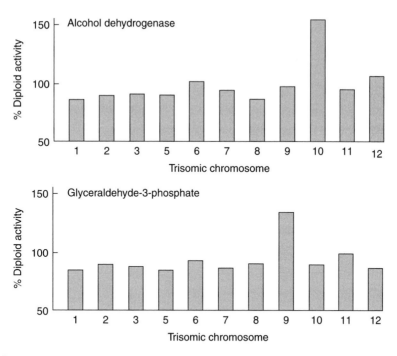

Fig. 4.3. Enzyme activity in trisomics of *Datura stramonium* (data from Carlson, 1972).

Table 4.1. Trisomics of *Datura stramonium* that showed the greatest increase in enzyme activity compared to diploids (Carlson, 1972).

Enzyme	Diploid activity (%)	Trisomic chromosome
Dehydrogenases		
Alcohol	157	10
Glucose-6-phosphate	141	9
Glutamate	162	11
Glyceraldehyde-3-phosphate	135	9
Isocitrate	147	2
Lactate	138	3
Malate	139	5
6-Phosphogluconate	141	9
Hexokinase	140	2

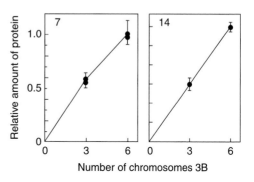

Fig. 4.4. Chromosome-dosage responses for two seed proteins (7 and 14) in allohexaploid wheat. (Used with permission from C.E. Aragoncillo *et al.*, © 1978, *Proceedings of the National Academy of Sciences USA* 75, 1446–1450.)

in the chapter on the multifactorial genome, numerous epistatic interactions regulate the relative importance of each gene duplication. Enzyme levels are presumably influenced by factors separate from the structural genes themselves. Regulatory genes exist that affect timing and expression of enzyme genes, and overall changes in surface–volume ratios of cells have cascading influences as membrane sites become limiting and cellular concentrations change.

A number of recent studies have evaluated the effects of polyploidy on gene expression patterns in allopolyploids, and have discovered that novel patterns of gene expression are commonly observed that are not the mean of the parental species and thus are "non-additive"

(Adams, 2007; Hegarty *et al.*, 2008; Abbott *et al.*, 2010). In microarray comparisons of gene expression patterns in a synthetic allotetraploid of *Arabidopsis suecica* and its parents *A. thaliana* and *A. arenosa*, Wang *et al.* (2006) found about 5% of the genes to be significantly different from the mid-parent values in two independently derived allopolyploids, and many more genes of the *A. thaliana* genome were down-regulated in the allopolyploid than those of *A. arenosa*. When Wang *et al.* (2004) analyzed differential expression patterns of the progenitors during successive selfing generations, they observed changes over time both within and between lines.

In other similar comparisons, Hegarty *et al.* (2005, 2008) found, in a cDNA microarray analysis of allopolyploid *Senecio cambrensis* compared to its parents, that expression levels in about 3% of the genes were significantly outside the parental average. The homoploid hybrid *Senecio × baxteri* was found to have more than twice as many genes with non-additive expression than the allotetraploid *S. cambrensis*, suggesting that allopolyploidy had a buffering effect on gene expression levels (Hegarty *et al.*, 2006, 2008). Pumphrey *et al.* (2009) found approximately 16% of the genes in synthetic *T. aestivum* had non-additive patterns of expression. Buggs *et al.* (2010) determined that 22% of the duplicate genes in *T. miscellus* were differentially expressed.

In the most thorough study on the effects of polyploidy on gene expression, Flagel and Wendel (2010) examined the expression patterns of 42,452 genes in five different allotetraploid *Gossypium* species, an artificial hybrid and their progenitor. They found that most of the genes were not additively expressed in the natural allotetraploids and the artificial hybrid, and there were considerable levels of transgressive up- and down-regulation in the allopolyploids (Fig. 4.5). There was also a homolog expression bias in all species, where D genome genes' expression was favored over A genome ones, with the genomic dominance being the weakest in the artificial hybrid. This work mirrored previous studies by Wendel's group on the expression of genes associated with petal and fiber tissues in *Gossypium hirsutum* (Flagel *et al.*, 2008; Hovav *et al.*, 2008; Rapp *et al.*, 2009). Similar patterns of genomic expression dominance have also been observed in other work on polyploid cotton

(Chaudhary *et al.*, 2009), *Arabidopsis* (Wang *et al.*, 2006), *Tragopogon* and *Triticum* (Pumphrey *et al.*, 2009).

Increased heterozygosity

The presence of duplicated genes can enhance physiological or developmental homeostasis if two genes with distinct properties are maintained (Barber, 1970; Manwell and Baker, 1970). Multiple enzyme forms might minimize variation in substrate affinity or provide catalytic properties adapted to a broader range of environmental conditions. For example, if one gene produced an enzyme that had high activity under one type of condition and another was most active under another set of conditions, the species might survive under a broader set of environments than would be possible with only one gene.

Polyploids will have higher levels of heterozygosity than their diploid progenitors unless they were produced by somatic doubling. Allopolyploids often carry the divergent alleles of their progenitors on non-pairing homologous chromosomes (Chapter 1). Such "fixed heterozygosities" are maintained indefinitely except when heterogenetic (non-homologous) pairings occur. Fixed heterozygosities appear to be the norm in allopolyploid species for a wide range of electrophoretically detectable enzymes (Gottlieb, 1982). In the classic study of Roose and Gottlieb (1976), additive patterns of electrophoretic phenotypes were found for 11 loci in *Tragopogon* diploid–tetraploid pairs (Table 4.2). At all these loci, the polyploid species carried the bands of both progenitors.

Autopolyploids do not carry fixed heterozygosities, but a high percentage of their parents' heterozygosity is transferred to them via unreduced gametes, and polysomic inheritance maintains a greater amount of heterozygosity than disomic inheritance. As discussed in the first chapter, this occurs because more than two alleles constitute a locus and polysomic inheritance generates fewer homozygotes each generation than disomy. For example, selfing of a heterozygous autotetraploid with the genotype AAaa will produce 94% heterozygous progeny, while a selfed diploid genotype of Aa will

Counts of allotetraploid/parental generic expression patterns

Categories	I ♀ Allo ♂	II ♀ Allo ♂	III ♀ Allo ♂	IV ♀ Allo ♂	V ♀ Allo ♂	VI ♀ Allo ♂	VII ♀ Allo ♂	VIII ♀ Allo ♂	IX ♀ Allo ♂	X ♀ Allo ♂	XI ♀ Allo ♂	XII ♀ Allo ♂
G. hirsutum (No change = 18,323)	1581	3240	621	3496	553	523	1855	1563	4444	497	4016	1747
G. barbadense (No change = 14,527)	2067	3750	1097	3453	781	803	3502	2371	3320	733	3902	2153
G. tomentosum (No change = 11,556)	2064	2927	1489	3578	1030	1290	4387	3666	3857	1036	3403	2176
G. mustelinum (No change = 14,793)	1827	3498	978	3601	825	781	3281	2371	3844	800	3863	2024
G. darwinii (No change = 12,561)	2069	3050	1274	4010	924	991	4056	3199	3958	937	3217	2213
F₁ (No change = 23,878)	1581	4888	248	2264	69	168	452	302	1951	60	4629	1951

Fig. 4.5. The number of genes falling into 13 possible states of expression in an F_1 hybrid and five *Gossypium* allotetraploids relative to their maternal (*Gossypium arboreum*) and paternal (*Gossypium ramondii*) progenitors. Each of the 12 variable states is labeled with a Roman numeral and a cartoon depicting the maternal and paternal levels of expression flanking the value for the polyploidy or F_1 hybrid. The 13th category of expression is where there is no difference between the diploid parents and the F_1 hybrid and the allopolyploids. The number of genes falling into this category is listed after each species name. (Used with permission from Flagel, L.E. and Wendel, J.F., © 2010, Evolutionary rate variation, genomic dominance and duplicate gene expression evolution during allopolyploid cotton speciation, *New Phytologist* 186, 184–193.)

Table 4.2. Additive electrophoretic phenotypes of diploid and tetraploid *Tragopogon* (Roose and Gottleib, 1976). Enzymes are identified by their migration from the origin. The parental species are represented by arrows.

Gene	2x porrifolius→	4x mirus	2x →dubius→	4x miscellus	2x →pratensis
EST-2	59	54/59	54	54/57	57
EST-3	53	44/53	44	–	–
EST-4	40	35/40	35, 40	–	–
LAP-1	30	30/32	32	32/35	35
LAP-2	29	25/29	25	25/27	27
APH	52	52/55	55	49/55	49
GDH	16	16/21	21	–	–
G6PD	29	29/31	31	–	–
ADH-1	35	30/35, 35/40	30, 40	30/35	35
ADH-2	–	–	30	18/30	18
ADH-3	–	–	15	5/15	5

produce 50% heterozygous progeny (Table 4.3). Genetic variation has been compared in five native species of diploids and their autopolyploid derivates, and in all cases higher levels of heterozygosity and mean number of alleles per locus were found in the autopolyploids than in the diploids (Table 4.4).

The initial level of heterozygosity transmitted via unreduced gametes is dependent on the process of $2n$-gamete formation. There are a number of events that can result in unreduced gametes (Hermsen, 1984), but the most common are first division restitution (FDR) and second division restitution (SDR). In FDR, homologous chromosomes are not separated during meiosis I, while in SDR sister chromatids remain together in the same gamete due to incomplete meiosis II. Figure 4.6 illustrates the genetic consequences

of FDR and SDR with no genetic crossing over. The reduction cell wall (R) is represented by a horizontal line, while the equatorial cell wall (E) is represented by a vertical line. It can be seen in the figure that normal meiosis produces a tetrad with four $1n$-gametes, FDR leads to a dyad containing two genetically identical $2n$-gametes, and SDR yields a dyad containing two genetically different $2n$-gametes. FDR transmits much more parental heterozygosity to the progeny than SDR because each gamete gets a combination of each parental chromosome rather than just one. Normal crossing over and gene recombination complicate this simplistic picture, but FDR still transmits more heterozygosity than SDR. Assuming one crossover per chromosome, it has been calculated that FDR transmits 80.2% of the parental heterozygosity to the gametes

Table 4.3. Expected segregation groups of balanced heterozygotes in diploids, tetraploids and allopolyploids.

Parental genotype	Possible gametes	Ratio	Gametic frequency	Zygotic phenotypes in F_2	Zygotic frequencies	Heterozygote frequency
Diploid						
Aa	A	1	0.50	AA	0.25	
	a	1	0.50	Aa	0.50	0.50
				aa	0.25	
Allotetraploid						
AA aa	Aa	1	1.00	AAaa	1.00	1.00
Autotetraploid						
AAaa	AA	1	0.17	AAAA	0.03	
	Aa	4	0.66	AAAa	0.22	
	aa	1	0.17	AAaa	0.50	0.94
				Aaaa	0.22	
				aaaa	0.03	

Table 4.4. Genetic variation (mean values) in diploid ($2n$) and autotetraploid ($4n$) populations (Soltis and Soltis, 2000).

Species	P		H		A	
	$2n$	$4n$	$2n$	$4n$	$2n$	$4n$
Tolmiea menziesii	0.240	0.408	0.070	0.237	3.00	3.53
Heuchera grossulariifolia	0.238	0.311	0.058	0.159	1.35	1.55
Heuchera micrantha	0.240	0.383	0.074	0.151	1.14	1.64
Dactylis glomerata	0.700	0.800	0.170	0.430	1.51	2.36
Turnera ulmifolia						
var. *elegans*	0.459	0.653	0.11	0.420	2.20	2.56
var. *intermedia*	0.459	0.201	0.11	0.070	2.20	2.00

P, proportion of loci polymorphic; H, observed heterozygosity; A, mean number of alleles per locus.

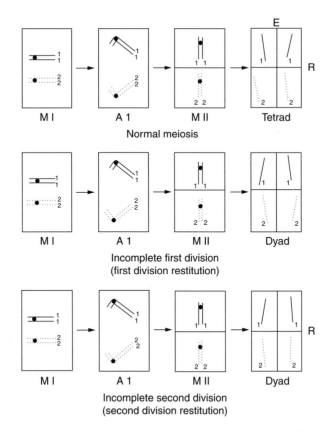

Fig. 4.6. Schematic presentation of first division restitution (FDR) and second division restitution (SDR) assuming one pair of homologous chromosomes and no crossing over. (Used with permission from J.G. Hermsen, © 1984, *Iowa State Journal of Research* 58, 421–436.)

and SDR transmits 39.6% (see Hermson, 1984 for details). Cytogenetic studies of FDR and SDR have been undertaken in a wide range of species, including potato (Mok and Peloquin, 1975; Mendiburu and Peloquin, 1977; Douches and Quiros, 1988), lucerne (Vorsa and Bingham, 1979) and maize (Rhodes and Dempsey, 1966).

Gene duplication can also act to relax stabilizing selection such that new enzyme properties might arise (Lewis, 1951; Stephens, 1951; Ohno, 1970). When there is more than one copy of a gene, an organism can survive an accumulation of mutations in one gene ("forbidden mutations") if the other remains functional. Such mutations could result through random events in the "silencing" of genes, or the evolution of unique adaptive properties or novel functions.

The evidence for beneficial mutations after gene duplication is quite limited. Isozymes generally show little variation in their catalytic properties when compared under laboratory conditions (Roose and Gottlieb, 1980; Weeden, 1983), although a few studies have uncovered differences that might have an adaptive benefit. Chickpea ADH isozymes vary in their heat and acid stability, substrate specificity and inhibitor sensitivity (Gomes *et al.*, 1982). One of the duplicated ADH isozymes of *Gossypium arboreum* is less sensitive to salt than the others (Hoisington and Hancock, 1981). Members of the (CHS) gene family appear to have diverged in function prior to diversification of the *Asteraceae* (Durbin *et al.*, 1995; Huttley *et al.*, 1997). Blackman *et al.* (2010) found that duplicated homologs of the floral inducer *FLOWERING LOCUS T* (*FT*) have differential expression levels and one contains a frameshift mutation that affects flowering time.

Some duplicated genes are differentially expressed in response to developmental or environmental variation. This phenomenon, called subfunctionalization, may signal the evolution of specific adaptations to divergent cellular environments (Scandalios, 1969; Adams, 2007). For example, different alcohol dehydrogenase (ADH) loci are expressed in the seed, pollen and root tissue of a variety of species (Weeden, 1983). One of the ADH loci is commonly induced by anaerobic conditions associated with flooding while the other loci are not (Freeling, 1973; Roose and Gottleib, 1980). A number of duplicated genes also have distinct subcellular locations even though they all originate from nuclear genes (Table 4.5). These enzymes may have specific adaptations to their subcellular environments (Weeden and Gottleib, 1980; Gottleib, 1982). Gaut and Doebley (1997) have identified a pair of duplicated genes in maize, *R* and *B*, that differentially regulate purple pigmentation in variant maize tissues.

Most recently, gene expression studies have been utilized to search for subfunctionalization of duplicate gene pairs in polyploids. Adams *et al.* (2003) examined expression of 40 homologous gene pairs in synthetic and natural *Gossypium hirsutum* and found variable expression levels for a number of them in different organs. Dong and Adams (2011) later compared expression patterns of 60 duplicated genes in leaves, roots and cotyledons of *G. hirsutum* under stress treatments, including temperature, drought, high salt and water submersion, and found numerous examples of the differential expression of duplicated genes in response to these abiotic stresses.

Unique intra-allelic interactions that are of adaptive significance might also appear after gene duplication. In the case of "multimeric" enzymes with more than one subunit, novel hybrid molecules can arise that have unique heterotic properties. Gottleib (1977b) described such a situation in *Clarkia*, where a duplication in PGI produced a hybrid molecule with a maximum velocity that was higher than the parental forms (Table 4.6). Such a change could potentially have a cascading effect on the whole metabolism.

Duplicated genes in plants often remain active after polyploidization for long periods of evolutionary time (Gottleib, 1982; Hart, 1988; Crawford, 1990); however, there are a number of instances where dosage compensations or gene silencing have been documented. Several

Table 4.5. Enzymes of plants whose isozymes are located in different subcellular fractions (Gottleib, 1982; Newton, 1983).

Enzyme	Location
Amylase	Cytosol and plastid
Carbolic anhydrase	Cytosol and plastid
Enolase	Cytosol and plastid
Fructose-1, 6-diphosphate	Cytosol and plastid
Glucose-6-phosphate dehydrogenase	Cytosol and plastid
Glutamate-oxaloacetate transaminase	Cytosol, mitochondria, plastid and microbody
Glutamine synthetase	Cytosol and plastid
Malate dehydrogenase	Cytosol, mitochondria, plastid and microbody
Malic enzyme	Cytosol and plastid
Phosphofructose isomerase	Cytosol and plastid
Phosphoglucomutase	Cytosol and plastid
Phosphoglucoisomerase	Cytosol and plastid
Phosphoglycerate mutase	Cytosol and plastid
6-Phosphogluconate dehydrogenase	Cytosol and plastid
3-Phosphokinase	Cytosol and plastid
Pyruvate dehydrogenase	Mitochondria and plastid
Pyruvate kinase	Cytosol and plastid
Ribulose-5-phosphate epimerase	Cytosol and plastid
Ribulose-5-phosphate isomerase	Cytosol and plastid
Superoxide dismutase	Cytosol, mitochondria and plastid
Transaldolase	Cytosol and plastid
Transketolase	Cytosol and plastid
Triophosphate isomerase	Cytosol and plastid

Table 4.6. Biochemical properties of phosphoglucoisomerase (PGI) enzymes from two species of *Clarkia* (Gottleib, 1977b). PGI is a dimeric enzyme with two subunits. PGI-2B3A is a hybrid molecule composed of one subunit from the ancestral locus PGI -2 and one from the duplicated locus PGI-3, while the other three enzymes are composed of only one type of subunit from a single locus (PGI-1, PGI-2B and PGI-3A). Note that the hybrid molecule PGI-2B3A has a higher V_{max} (maximum velocity) than the parental forms.

Enzyme	K_m (mM)[a]	V_{max} (µmol/min)	E_a (kcal/mol)[b]
Clarkia xantiana			
PGI-1	0.29	33.8	9.74
PGI-2B	0.12	59.3	10.58
PGI-2B3A	0.33	83.2	11.19
PGI-3A	1.12	55.5	12.43
Clarkia rubicunda			
PGI-1	0.37	65.8	
PGI-2	0.17	147.0	

[a]K_m, Michaelis constant (substrate affinity).
[b]E_a, energy of activation.

null or non-active allozymes were found in tetraploid *Chenopodium* (Wilson, 1981; Wilson *et al.*, 1982). An allozyme at a duplicated phosphoglucase isomerase locus, which had a much lower affinity for its substrate, was described by Gottleib and Greve (1981). Hoisington and Hancock (1981) found the recently formed tetraploid *Hibiscus radiatus* to have enzyme and protein levels very similar to the sum of its progenitors, while the more ancient species *Hibiscus acetosella* had several lower levels. In hexaploid wheats, high-molecular-weight glutenin bands are produced from only two of the three genomes constituting this hexaploid (Feldman *et al.*, 1986). Gottleib and Higgins (1984) discovered that *Clarkia* species with and without a duplication of phosphoglucase isomerase had the same levels of cytosolic activity. Roose and Gottleib (1978) also found the number of genes coding for electrophoretically detectable enzymes to remain constant in seven diploid species of *Clarkia* even though their chromosome numbers ranged from $2n = 6$ to $2n = 12$.

Many more examples of silenced genes are emerging as more genomes are comprehensively sequenced and their gene expression patterns examined. When Comai *et al.* (2000) generated an artificial allotetraploid of *A. thaliana* and *Cardaninopsis arenosa* and compared gene expression in the diploids to the polyploid, they found that about 0.4% of the genes were silenced in the polyploid. In an examination of the *PgiC2* gene family of *Clarkia mildrediae*,

18 exons were sequenced and nine of these contained insertions or deletions that resulted in frameshifts and truncation (Gottleib and Ford, 1997). A single mutation in one of six duplicate chalcone synthase (*CHS*) genes was found to block anthocyanin production in the floral limb (Durbin *et al.*, 2000). When Tate *et al.* (2009) examined 13 loci in 84 *T. miscellus* individuals, they found that 11 of the loci showed loss of at least one parental homolog in the neopolyploid. As was mentioned above, there has been preferential silencing of the genes of one parent over the other in allotetraploids of *Gossypium*, *Arabadopsis*, *Tragopogon* and *Triticum*.

Genes can be silenced after polyploidization through the loss of genomic fragments due to recombination (see Chromosomal Repatterning below), spontaneous mutation, methylation and the movement of transposable elements (Flagel and Wendel, 2009; Parisod *et al.*, 2009; Salmon *et al.*, 2010). Polyploidization has been proposed to induce a "burst" in transposon activity (Matzke and Matzke, 1998; Comai, 2000), although the proliferation of transposable elements after polyploidization appears to be relatively restricted (Parisod *et al.*, 2009; Hu *et al.*, 2010).

Genetic bridge

Repeated cycles of hybridization and polyploidization could also add to the amount of variability found in a polyploid group. Any time a new polyploid individual is formed through the unification of unreduced gametes, there is the potential that new combinations of genes will be injected into the polyploid species. The range of a polyploid species might also be increased directly as it comes in contact with diploid species that produce compatible unreduced gametes. Such introgressions have not been directly documented, but we do know that numerous polyploids have multiple origins, including species of *Tragopogon, Tolmiea* and *Triticum*. Much of the extensive variability found in these groups could have been generated by hybridization of polyploids of separate origin and subsequent recombination.

One important evolutionary ramification of genomic duplication is when the polyploids act as a "genetic bridge" between diploid species (Dewey, 1980). In some cases, two diploids that have limited inter-fertility at the diploid level may produce an allopolyploid via unreduced gametes. It can then backcross with the progenitors whenever unreduced gametes from the diploids come in contact with reduced gametes of the allopolyploid. Two newly emerged polyploid types might also hybridize at higher levels than their diploid progenitors (Rieseberg and Warner, 1987). This would afford the allopolyploid with more genetic variability than either of the diploids, since it has access to both gene pools. Numerous examples of polyploid bridges have been described in *Gilia* (Grant, 1971), *Bromus* (Stebbins, 1956) and *Clarkia* (Lewis and Lewis, 1955).

Perhaps the classic example of this phenomenon is found in *Aegilops* (Zohary, 1965). Up to 22 wild species of wheat are located in southwestern Asia and the Mediterranean basin. Diploids contain at least six genomic groups that are sexually isolated, and the polyploids group into three clusters of species, each possessing a common "pivotal genome" (Kihara, 1954; Kihara *et al.*, 1959). Each of the polyploid species clusters also has a common spikelet type. The weedy polyploid *Aegilops* species overlap in their geographical distribution and those with similar pivotal genomes produce partially fertile F_1 hybrids. Zohary and Feldman (1962) and Feldman (1963) examined the linkages between *Aegilops* tetraploids in Israel, Turkey and Greece and found natural hybrids between the seven

polyploids carrying the C genome (Fig. 4.7). They are male-sterile, but are repeatedly exposed to parental pollen such that some backcrossed seed is produced. This seed produces vigorous plants that are relatively fertile, and when selfed they generate an extremely variable progeny. These introgressed genotypes can become genetically fixed due to self-pollination and may act to enlarge the gene pool and ecological amplitude of the species involved.

In many cases, it appears that after the initial polyploidization and diversification, additional episodes of polyploidy occurred to generate what are called *polyploid complexes* (Stebbins, 1971; Grant, 1981). In the example above, the *Aegilops* complex is composed of the various C, M and S genome species and their overlapping allotetraploid derivatives. The genera *Aegilops* and *Triticum* form an expansive polyploid complex involving dozens of polyploid species at three ploidy levels (Chapter 8). Among the cultivated species, the bread wheat *T. aestivum* is an allohexaploid containing three genomes (AABBDD), the durum and emmer wheats *Triticum turgidum* are allotetraploids (AABB) and the einkorn wheat *Triticum monococcum* is a diploid (AA). Similar wide ranging complexes exist for oats and rice (Chapter 8).

One particularly elegant example of a polyploid complex has been described in an Australian species of soybean. *Glycine tomentella* is a perennial whose range is anchored in Australia, but extends into Timor, New Guinea, the Philippines and Taiwan. It is cytogenetically quite complex and is composed of diploids, tetraploids and aneuploids ($2n + 38$, 40, 78 and 80) (Newell and Hymowitz, 1978). The diploids contain several races (D1–D6) that are reproductively isolated from each other and the other species (Brown *et al.*, 2002). Isozyme and Histone H3-D sequences indicate that some diploid races have more than one origin and have undergone lineage recombination. Phylogenetic and network analysis of alleles from diploids and polyploids revealed that polyploid *G. tomentella* is a complex composed of several diploid genomes (T1–T6) and, in many cases, the diploid genomic origins of the various polyploid races could be traced to specific diploid races (Fig. 4.8; Doyle *et al.*, 2002). The origin of the polyploid races is thought to have occurred within the last 30,000 years, during the human occupation of Australia

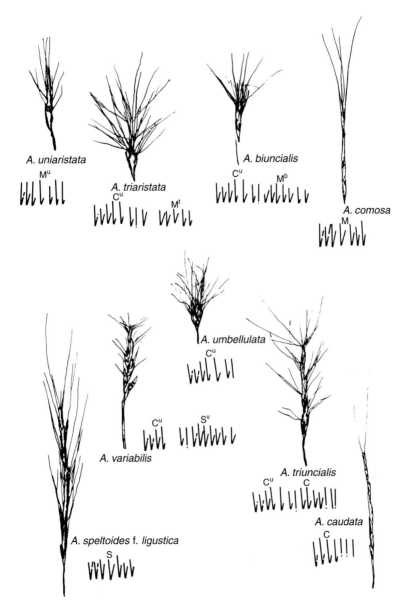

Fig. 4.7. Spikelets and karyotypes of four tetraploid species of *Aegilops* containing the pivotal genome C^u of *A. umbellulata* (center) and of the four modern diploid species that contain counterparts of the other ancestral genomes: M^u, *A. uniaristata* (top left); M, *A. comosa* (top right); S, *A. speltoides* f. *ligustica* (bottom left); C, *A. caudata* (bottom right). The tetraploids are as follows: C^{u}M^t, *A. triaristata*; C^{u}M^b, *A. biuncialis*; C^{u}S^v, *A. variabilis*; C^{u}C, *A. triuncialis*. (Used with permission from G.L. Stebbins, © 1971, *Chromosomal Evolution in Higher Plants*, Addison-Wesley Publishing Company, Reading, Massachusetts.)

and subsequent environmental disruption of the habitat (Doyle *et al.*, 1999).

The origin of agriculture and the subsequent domestication process stimulated the development of many polyploid complexes. The hexaploid wheats are thought to have originated when cultivated tetraploids hybridized with wild populations of diploids. The banana, *Musa*

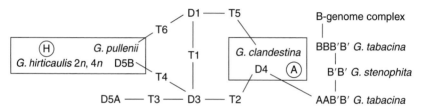

Fig. 4.8. The *Glycine* subgenus *Glycine* polyploid complex. Lines connect tetraploids with diploid progenitors. All taxa lacking names are classified as *Glycine tomentella*. Boxed taxa marked "A" are members of the A-genome group; "H" are members of the H-genome species. (Used with permission from J.J. Doyle *et al.*, © 2002, *Evolution* 56, 1388–1402.)

acuminata (*A* genome), was first cultivated in the Malay region of Southeastern Asia and as its cultivation spread north it hybridized with another species, *Musa balbisiana* (*B* genome), forming an array of edible genomic assemblages (AAB, ABB, AAAB, AABB and ABBB) (Chapter 10). Repeated cycles of hybridization and polyploidy have played an important role in the development of sugarcane, potato and sweet potato (Chapter 10).

Genetic Differentiation in Polyploids

In spite of the many potential advantages associated with polyploidy outlined above, polyploidy has long been considered a conservative rather than a progressive factor in evolution (Stebbins, 1950, 1971; Grant, 1971). Polyploids were considered an "evolutionary dead end" doomed to decline in importance because they had less capacity for genetic differentiation than diploids. It was thought that the presence of multiple alleles reduced the effect of single alleles and genetic differentiation was further restricted by the lack of segregation due to fixed heterozygosities in allopolyploids and the reduced rate of segregation due to tetrasomic inheritance in autopolyploids. Stebbins (1971) provided a number of examples where polyploid complexes appeared to have gone through a pattern of growth and demise. In these examples, the polyploids were initially much rarer than their progenitors; then they predominated, and eventually their importance diminished to the point where they become relictual.

This conservative opinion about the evolutionary potential of polyploids has dramatically changed in recent years. Polyploids may indeed evolve more slowly than diploids due to the buffering effect of multiple alleles, but they may actually have a broader adaptive potential. There is a greater potential dose span between additive alleles in polyploids allowing for a greater range in phenotype (Table 4.7), and the higher levels of genetic variability generally found in polyploids allows for more possible assortments of genes. A number of polyploid species have been shown to have undergone substantial genetic differentiation. The oat populations used as an example of diversifying selection in the second chapter are in fact hexaploid (Hamrick and Allard, 1972). Smith (1946) observed considerable morphological variation in tetraploid *Sedum*, while Adams and Allard (1977) found a large amount of isozyme diversity among hexaploid individuals of *Festuca microstachys* (Fig. 4.9). Among California populations of strawberries, octoploids have a greater ecological range than their diploid progenitor *Fragaria vesca* and they show greater divergence in most morphological traits (Table 4.8). Bringhurst and Voth (1984) were able to increase cultivated strawberry yields by 500% over 25 years by artificial selection even though the crop was an octoploid. Similar dramatic improvements have been made in many other polyploid crops by plant breeders.

Even in species propagated primarily through asexual means, each newly emergent type can potentially contain new gene combinations that are of adaptive or horticultural importance. These hybrid derivatives can then be propagated indefinitely through shoot cuttings, ramets or tubers. Since triploid bananas are sterile, varietal development has been very dependent on the repeated formation of unique polyploids. In taro, numerous cultivated races have been selected from the wild even though

Table 4.7. Potential allelic dose span in diploids, tetraploids and hexaploids. It is assumed that the duplicated loci carry two alleles (A, a).

| Ploidy | Number of A alleles | | | | | | |
	0	1	2	3	4	5	6
Diploid	AA	Aa	aa				
Tetraploid	AAAA	AAAa	AAaa	Aaaa	aaaa		
Hexaploid	AAAAAA	AAAAAa	AAAAaa	AAAaaa	AAaaaa	Aaaaaa	aaaaaa

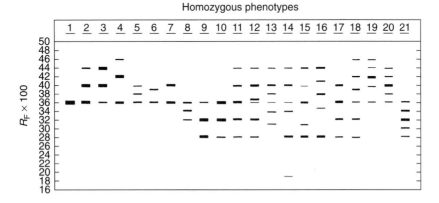

Fig. 4.9. Homozygous PGI phenotypes found in hexaploid *Festuca microstachys*. (Used with permission from W.T. Adams and R.W. Allard, © 1977, *Proceedings of the National Academy of Sciences USA* 74, 1652–1656.)

Table 4.8. Morphological and environmental ranges of diploid and octoploid *Fragaria* in California. A value less than 1.0 indicates that the octoploids had a greater range (Hancock and Bringhurst, 1981).

Character	Range 2x/8x	Character	Range 2x/8x
Environmental		*Morphological*	
January temp.	0.85	Stolon width	0.52
April temp.	0.79	Stolon numbers	1.59
July temp.	0.74	Branch crowns	1.59
Rainfall (mm)	0.95	Petal index	0.82
pH	0.71	Petal area	0.21
Salinity (ppm)	0.52	Peduncle length	1.52
% carbon	0.91	Flower number	0.40
% sand	0.82	Trichome number	0.59
% silt	0.78	Fruit index	1.19
% clay	0.96	First flower	0.86
		Last flower	0.88
		Flowering period	0.65
		Achene weight	0.42
		Fruit weight	0.05
		Leaf area	1.15
		Leaf index	0.55
		Petiole length	1.06
		Sclerophylly	1.93

they rarely flower (Kuruvilla and Singh, 1981). Multiple hybridizations and polyploid formation have played an important role in the development of many other asexually propagated species including cassava, sweet potato, white potato and yam.

Chromosomal Repatterning

A major source of novelty in polyploids may come through genome rearrangement and gene silencing after polyploidization, as was mentioned before. Grant (1966) was the first to experimentally demonstrate this possibility when he produced a hybrid of tetraploid *Gilia malior* × *Gilia modocensis* that had seed fertility of 0.007%. He selfed the lines for 11 generations and was able to isolate a highly self-fertile plant that was reproductively isolated from its parents and had a unique combination of parental traits, presumably through

chromosomal translocation. More recent work has indicated that DNA sequence elimination may be a major, immediate response to allopolyploidization (Adams and Wendel, 2005; Flagel and Wendel, 2009; Tate *et al.*, 2009).

In some of the earliest work demonstrating this phenomenon, Song *et al.* (1995) produced reciprocal hybrids between the diploids *Brassica rapa* and *Brassica nigra*, and *B. rapa* and *Brassica oleracea*. The F_1 individuals were colchicine doubled and progenies were generated to the F_5 generation by selfing. They then conducted an RFLP analysis of F_2 and F_5 individuals of each line using 89 nuclear DNA probes and found substantial genomic alterations in the F_5 generation, including losses of parental fragments and gains of novel fragments (Fig. 4.10). Almost twice as much change was observed in the combinations involving the two most distant relatives, *B. rapa* and *B. nigra* (Table 4.9), and they observed more change in some nuclear/cytoplasmic combinations than others.

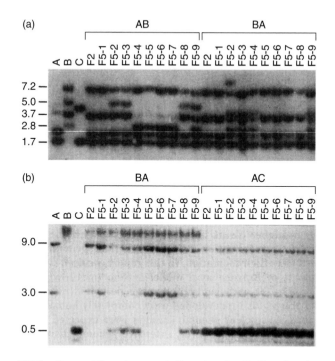

Fig. 4.10. Nuclear RFLP patterns of *Brassica rapa*-A, *Brassica nigra*-B, *Brassica oleracea*-C, F_2 hybrids between them and F_5 populations. (a) *Hind*III-digested DNAs probed with EZ3 that show a loss of fragments and a gain of fragments in some F_5 plants (5.0 kb and 2.8 kb). (b) *Hpa*II-digested DNAs probed with EC3C8 showing a gain of a 0.5 kb fragment in five BA F_5 plants that does not exist in either the A or B parental genome, but is present in the C genome parent and all AC F_5 plants. (Used with permission from K. Song *et al.*, © 1995, *Proceedings of the National Academy of Sciences USA* 92, 7719–7723.)

Table 4.9. Frequencies and types of genomic changes in F_5 progenies of synthetic polyploids of *Brassica* compared with their parents (modified from Song *et al.*, 1995).

	Types of fragment changes				
	Loss/gain of fragments in F_5[b]			Fragments gained in F_2[c]	Fragments found only in F_5[d]
Polyploid line[a]	A	B	C		
AB F5	9/13	25/12		9	19
BA F5	8/12	14/0		5	51
AC F5	7/1		19/4	4	1
CA F5	15/1		16/5	3	4

[a]A, *B. rapa*; B, *B. nigra*, C, *B. oleracea*.
[b]Loss, fragments present in diploid parent and the F_2 but not present in F_5 plants; A, B and C, fragments specific to the various parents.
[c]Fragments found in the F_2 but not in either parent.
[d]Fragments found in F_5 plants but not in the diploids or F_2s.

Feldman *et al.* (1997) examined genomic rearrangements in wheat by examining RFLP patterns in natural diploid and allopolyploid species. They used 16 probes that were from low-copy, non-coded DNA. Nine of these probes were found in all the diploid species, indicating that they were conserved, but when they examined aneuploid and nullisomic lines, they found that each sequence was only retained in one of the allopolyploid genomes. In follow-up work, Liu *et al.* (1998a,b) examined RFLP fragment profiles of both coding and non-coding sequences in synthetic tetraploid, hexaploid and octoploids of *Triticum* and *Aegilops* that had been selfed for three to five generations. They obtained similar results to Feldman *et al.* (1997), observing non-random sequence elimination in all the allopolyploids studied, along with the occasional appearance of unique fragments. They also found that some of the changes were brought about by DNA methylation. By comparing crosses with and without the *PH1* gene that regulates bivalent pairing, they were able to deduce that intergenomic recombination did not play a role in the sequence change, as both types of crosses yielded about the same amount of change.

In two further studies, the Feldman group found that the direction of sequence change in wheat followed a different pattern than that observed by Song's group in *Brassica*, and confirmed that some sequences were silenced by elimination, while others were silenced through methylation. Özkan *et al.* (2001) analyzed diploid parental generations, F_1 progeny and the first three generations (S_1, S_2 and S_3) of synthetic hybrids of several species of *Aegilops* and *Triticum*. When they followed the rate of elimination of eight low-copy DNA sequences, they found that sequence elimination began earlier in the synthetic allopolyploids that most closely resembled naturally occurring ones, and sequence elimination was not associated with cytoplasm. Shaked *et al.* (2001) used AFLP and methylation-sensitive amplification polymorphism fingerprinting (MSAP) to evaluate another set of diploid and tetraploid hybrids within and between genera. They also found considerable sequence elimination after polyploidy that occurred most rapidly in allopolyploids of the same species rather than different species. Further analysis indicated that some of the sequences were eliminated, while others were altered by cytosine methylation.

These studies and a number of more recent ones suggest that genomic "downsizing" is likely a common phenomenon in allopolyploids, where genomic fragments of one homolog or the other are lost. Homologous recombination must play an important role in the loss of these DNA fragments (Udall *et al.*, 2005; Wang *et al.*, 2005; Gaeta *et al.*, 2007; Salmon *et al.*, 2010). Over time, some of the duplicate genes remain functional and perhaps derive new cellular roles, while others move "on the road to diploidization", as coined by Tate *et al.* (2009).

Genome Amplification and Chance

It is important to realize that in all of these discussions, it is assumed that alterations in gene copy number are maintained because they make significant adaptive differences. However, this may not always be the case. Some polyploids may become established simply due to chance, because unreduced gametes are continually produced and chromosome numbers can go up in a species but not down. Simple doublings of genome size may produce individuals with relatively well balanced gene complexes, but going down in size will generally result in genic imbalances as individual chromosomes are lost. In only a few instances have natural polyploids been shown to produce diploid types (DeWet, 1968; Ramsey and Schemske, 1998). Since most plant species produce a small frequency of unreduced gametes, a small percentage of polyploids may be continually produced, and by chance alone, some may become established. If the pressure is always towards higher ploidy and not reductions, the higher numbers will eventually accumulate.

Random processes of gene duplication and deletion can also act within genomes to produce "multigene" families and vestigial sequences. Loomis and Gilpin (1986) have performed computer simulations that generated random duplications and deletions, and found that genome size did not become stable until the amount of "dispensable sequences had increased to the point that most deletions affected vital genes". It may be that genome size fluctuates greatly under "normal" environmental conditions and only becomes important when environmental contingencies demand the amplification or alteration of specific genes.

Orgel and Crick (1980) and Doolittle and Sapienza (1980) propose that much of this amplification involves the proliferation of what they call "selfish" or "parasitic" DNA. Sequences like transposable elements may "insure" their survival in a cellular environment by maximizing their copy number. This scenario is possible only so long as the sequences have little effect on the fitness of the total organism.

Summary

At least 50% of all angiosperm species are polyploid. The adaptive benefit of polyploidy is largely unknown, although the larger nuclear size, higher levels of heterozygosity and greater enzyme content of polyploids may play a role. Polyploids are thought to undergo less ecological differentiation than diploids because they are highly buffered genetically, but they still carry sufficient genetic variability to evolve unique ecotypes. Polyploids may act as a genetic bridge between diploid species when the polyploid accumulates genes from both progenitors and transfers them via partially fertile hybrids. Chromosomal repatterning also may be common after polyploidization. The high number of polyploids found in nature may be due in part to chance, since low levels of unreduced gametes are continually produced by many species and genomic amplifications are generally less disruptive than chromosome reductions.

5

Speciation

Introduction

Up to this point, we have primarily been discussing populations. We are now ready to move up the evolutionary ladder to species. Plants can be thought of as evolving along two dimensions, *anagenesis* and *cladogenesis*. Changes within a specific lineage or species represent anagenesis. Cladogenesis occurs when a lineage splits and the new lines begin to evolve separately. Speciation is the most fundamental cladogenetic process.

Traditionally, most evolutionists felt that lineages evolved at essentially the same rate before and after speciation. This view, called "phyletic gradualism", assumed that evolution was a slow, continuous process. Patterns of change in the fossil record have led some scientists to suggest that species may often undergo rapid changes as they come into existence, but then remain largely unchanged (Mayr, 1963; Eldridge and Gould, 1972; Gould and Eldridge, 1977). Thus, speciation occurs in spurts rather than through gradual change. This view, called "punctuated equilibrium", is one way to explain some of the abrupt discontinuities found in the fossil records. The general consensus is that punctuated patterns can fit into the framework of Darwinian evolution as long as we accept variations in rate due to environmental and genetic perturbations. Darwin himself was

primarily a gradualist, but he also "accepted the influence of both local, episodic speciation and migration" (Rhodes, 1983).

What is a Species?

Before we can begin to discuss speciation, we must first attempt to define a species. Commonly, populations that can be distinguished by prominent morphological differences are considered species. This "Taxonomic Species Concept" is often successful in identifying separately evolving groups, but in some instances can lead to artificial groupings. Quite distinct morphotypes can retain the ability to freely interbreed, making them in reality one genetic entity; and large differences in morphology are sometimes influenced by only a few genes that do not necessarily reflect the divergence of the whole genome (Chapter 1). These ambiguities have led to frequent revisions of important crop assemblages such as oats, rice, wheat and sorghum as different taxonomists have reviewed the available data on natural populations.

Another way to distinguish species is by directly testing their capacity to successfully reproduce. Mayr (1942) defined the "Biological Species" as "groups of actually or potentially interbreeding natural populations, which are reproductively isolated from each other." If taxa are reproductively isolated, they are evolving separately and therefore must have an identity of their own. This concept has gained widespread approval and is currently the most popular.

While the biological species concept allows for the unambiguous delineation of species, it is still not without occasional problems. Strongly divergent groups of plants often maintain some degree of inter-fertility even though they differ at numerous loci and are effectively evolving on their own. As we will discuss below, sunflower and violets provide particularly striking examples. Plants also show great ranges in fertility from obligate outcrossing to complete selfing to apomixis (uniparental). In a highly inbred or apomictic group, every individual would be a species according to the biological species concept. As we already mentioned, many of the grain and legume species are highly inbred, and most of the starchy staples like banana, cassava, potato, sugarcane, sweet potato, taro

and yam are only propagated through asexual means.

Harlan and de Wet (1971) developed the "Gene Pool System" to deal with varying levels of inter-fertility between related taxa (Fig. 5.1). They recognized three types of genic assemblages:

1. Primary gene pool (GP-1): hybridization easy; hybrids generally fertile.
2. Secondary Gene Pool (GP-2): hybridization possible, but difficult; hybrids weak with low fertility.
3. Tertiary Gene Pool (GP-3): hybrids lethal or completely sterile.

The primary gene pool is directly equivalent to the biological species. The recognition of GP-2 and GP-3 allows other levels of inter-fertility to be incorporated into the overall concept of species. These are related taxa which share a considerable amount of genetic homology with GP-1, but are divergent enough to have greatly reduced

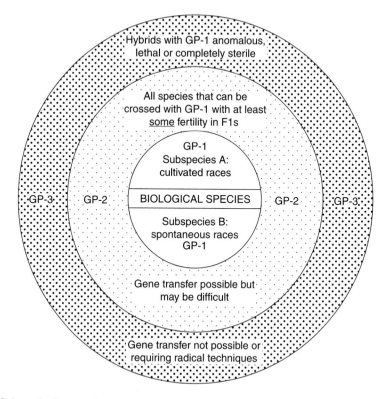

Fig. 5.1. Schematic diagram of primary gene pool (GP-1), secondary gene pool (GP-2) and tertiary gene pool (GP-3). (Used with permission from J. Harlan, © 1992a, *Crops and Man*, American Society of Agronomy, Madison, Wisconsin.)

inter-fertility. Several agronomically important groups have been described using this system including legumes (Smartt, 1984), wheat (Fig. 5.2) and most of the other cereals (Table 5.1).

Simpson (1961) developed the idea of an "Evolutionary Species" to minimize the problems associated with uniparental species. He suggested that a species must meet four criteria: (i) is a lineage; (ii) evolved separately from other lineages; (iii) has its own particular niche or habitat; and (iv) has its own evolutionary tendencies. This definition fits uniparental species better than the biological species concept, but we are still left with deciding on what constitutes a lineage. Templeton (1981) expanded this theme by using molecular data to construct phylogenies in his "Cohesion Species Concept". He defined a species as an "evolutionary lineage, with the lineage boundaries arising from the forces that create reproductive communities (i.e. cohesive mechanisms)".

Numerous other concepts have been developed to include ecological with reproductive criteria in defining species (Levin, 2000; Schemske, 2000). Nevertheless, it is clear that no model solves all of the potential problems in trying to define separate evolutionary units. Each has its own strengths and weaknesses. Levin (2000) suggests that "the choice of a species concept has to do, in part, with the perspective that gives one satisfaction". Probably the most definitive definition is the concept of biological species since it can be directly tested: two individuals can either successfully reproduce or they cannot. Of course, environmental and genotypic variation can cloud even this approach, but it does minimize the number of subjective judgments. Regardless, in a survey of phenetic and crossing relationships in 400 genera of plants and animals, Rieseberg *et al.* (2006) found that 70% of plant taxonomic species are also reproductively independent lineages.

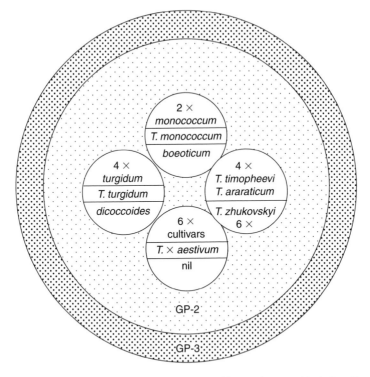

Fig. 5.2. The gene pools of wheat. The secondary gene pool is very large and includes all species of *Aegilops*, *Secale* and *Haynaldia*, plus at least *Agropyron elongatum*, *Agropyron intermedium* and *Agropyron trichophorum*. The tertiary gene pool includes several species of *Agropyron*, several of *Elymus* and *Hordeum vulgare*. (Used with permission from J. Harlan, © 1992a, *Crops and Man*, American Society of Agronomy, Madison, Wisconsin.)

Table 5.1. Primary and secondary gene pools of the major cereals (Harlan, 1992a).

Crop GP-2	Ploidy level	Primary gene pool, GP-1			Secondary gene pool
			Spontaneous subspecies		
		Cultivated subspecies	Wild races	Weed races	
Wheats					
Einkorn	2x	Triticum monococcum	Triticum boeoticum	T. boeoticum	Triticum, Secale, Aegilops
Emmer	4x	Triticum dicoccum	Triticum dicoccoides	None	Triticum, Secale, Aegilops
	4x	Triticum timopheevi	Triticum araraticum	T. timopheevi	Triticum, Secale, Aegilops
Timopheevi					
Bread	6x	Triticum × aestivum	None	None	Triticum, Secale, Aegilops
Rye	2x	Secale cereale	S. cereale	S. cereale	Triticum, Secale, Aegilops
Barley	2x	Hordeum vulgare	Hordeum spontaneum	H. spontaneum	None
Oats					
Sand	2x	Avena strigosa	Avena hirtula; Avena wiestii	A. strigosa	Avena spp.
Ethiopian	4x	Avena abyssinica			
		Avena vaviloviana	Avena barbata	A. barbata	Avena spp.
Cereal	6x	Avena sativa	Avena sterilis	A. sterilis; Avena fatua	Avena spp.
Rices					
Asian	2x	Oryza sativa	Oryza rufipogon	O. rufipogon	Oryza spp.
African	2x	Oryza glaberrima	Oryza barthii	Oryza stapfii	Oryza spp.
Sorghum	2x	Sorghum bicolor	S. bicolor	S. bicolor	Sorghum halepense
Pearl millet	2x	Pennisetum americanum	Pennisetum violaceum	P. americanum	Pennisetum purpureum
Maize	2x	Zea mays	Zea mexicana	Z. mexicana	Tripsacum spp., Zea perennis

Reproductive Isolating Barriers

There are many different ways that plants can be reproductively isolated (Table 5.2). These *reproductive isolating barriers* (RIBs) are generally broken into two classes: (i) premating or prezygotic mechanisms that prevent the formation of hybrid zygotes; and (ii) post-mating or zygotic isolating mechanisms that reduce the viability or fertility of hybrid zygotes. Prezygotic barriers typically contribute more to total reproductive isolation in plants than do postzygotic barriers (Reiseberg and Willis, 2007).

In eco-geographic isolation, the habitats of two species are sufficiently different that they rarely have the opportunity to interbreed. The low-bush blueberry, *Vaccinium angustifolium*, is generally found on dry hillsides and gravelly barrens, while the highbush blueberry *Vaccinium corymbosum* is located in boggy wetlands. The common bean, *Phaseolus vulgaris*, lives in warm, temperate areas in Mexico–Guatemala, in contrast to the runner bean, *Phaseolus coccineus*, which inhabits the cool, humid uplands of Guatemala. *Tradescantia canaliculata* grows on rocky slopes with full sun, while *Tradescantia subaspera* subsp. *typica* prefers rich soil and deep shade (Anderson and Hubricht, 1938). In all these cases, the species remain distinct from each other as long as their habitats are not in close proximity. When they

Table 5.2. Types of reproductive isolating barriers (RIBs) found in sexually reproducing organisms.

Pre-mating – formation of hybrid zygotes is
    prevented
        Eco-geographic – habitats are distinct and
            rarely come in contact
        Temporal – flowering times have little overlap
        Floral – flowers attract different types of
            pollinators
        Gametic incompatibility – foreign pollen grains
            fail to fertilize ovules
Post-mating – viability or fertility of zygotes is
    reduced
        Hybrid inviability – hybrids are weak and have
            poor survival
        Hybrid sterility – hybrids do not produce
            functional gametes
        Hybrid breakdown – F_2 and backcross hybrids
            have reduced viability or fertility

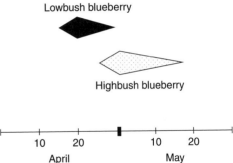

Fig. 5.3. Bloom dates of lowbush and highbush blueberries at Otis Lake, Michigan in 1986. The bar widths represent the proportion of individuals in full bloom on each date.

come in contact with each other, they can hybridize and form viable hybrids that survive when an intermediate habitat is present.

Species that are isolated temporally have distinct flowering times that do not have sufficient overlap to allow for hybrid formation. The blueberry and *Tradescantia* species mentioned above are not only separated by habitat, but also by bloom date (Fig. 5.3). This further reduces the production of hybrids. The weedy lettuces, *Lactuca canadensis* and *Lactuca graminifolia* live together in the southeastern USA but rarely hybridize because they bloom in different seasons (Whitaker, 1944). *L. graminifolia* flowers in the early spring, while *L. canadensis* blooms in summer.

Floral isolation occurs when species have flowers that attract different types of pollinators. *Penstemon* spp. in California have flowers of different shapes, colors and sizes that are pollinated by four completely different animals including hummingbirds, wasps and two different-sized carpenter bees. Species of columbines in western North America have distinct-looking flowers with nectar at the base of spurs that can only be reached by specific pollinators (Fig. 5.4). *Aquilegia formosa* has short, stout spurs that provide nectar to hummingbirds whose bill is about as long as the spurs. *Aquilegia longissima* and *Aquilegia chrysantha* have much longer, thinner spurs that can be successfully harvested by hawk moths, which have a much longer proboscis. *Aquilegia pubescens* has an intermediate sized

spur that hummingbird bills can barely reach, so it hybridizes freely with the normally hawk moth-pollinated *A. formosa* (Grant, 1952).

Schemske and Bradshaw (1999) directly tested the genetic basis of pollinator discrimination in bee-pollinated *Mimulus lewisii* and hummingbird-pollinated *Mimulus cardinalis*. They developed a QTL-based map of the key traits associated with pollinator preference and then tracked the activity of bees and hummingbirds in an F_2 population set in the field. They found that bees preferred large flowers with low pigment content, and hummingbirds favored nectar-rich flowers with high anthocyanin content. Most remarkably, they were able to uncover a single allele that increased petal carotinoids and reduced bee visitations by 80%, and another allele that increased nectar production and doubled hummingbird visits.

One of the strongest RIBS involves gametic incompatibility where foreign pollen grains cannot germinate on another species' stigmata, or they cannot successfully grow down the style to the ovaries. Species show a broad range of interaction from no hint of germination to successful pollen growth but failed fertilization. Fruit trees in the subgenus *Amygdalus* of *Prunus* freely cross, but gene transfer outside this section is generally obstructed by pollen and/or ovule sterility barriers (Chapter 11). Crosses of *Gilia splendens* × *Gilia australis* (Latimer, 1958) and *Iris tenax* × *Iris tenuis* (Smith and Clarkson, 1956) often fail because the foreign pollen tubes grow more slowly than native pollen tubes; native pollen of *Iris* species reach the ovule in 30 h, while foreign pollen takes 50 h.

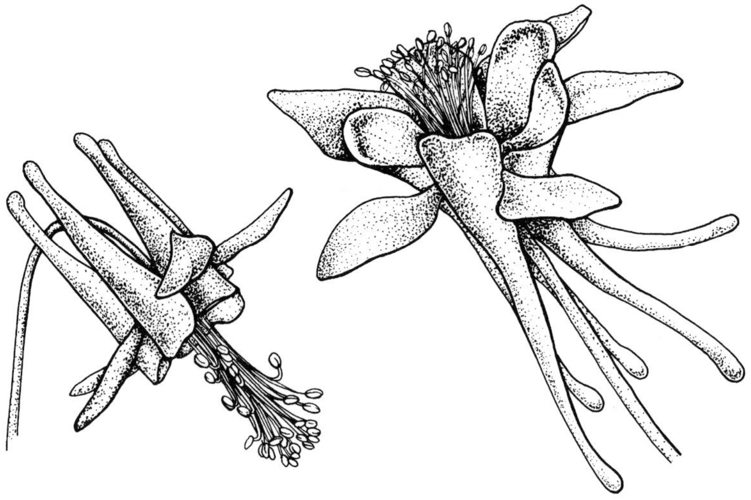

Aquilegia formosa		Aquilegia pubescens	
Species	Spur length (cm)	Primary pollinator	Bill/proboscis length (cm)
A. formosa	1–2	Hummingbirds	1–3
A. pubescens	3–4	Celerio lineata	3–4.5
A. chrysantha	4–7	C. lineata	3–4.5
		Phlegethontius sexta	8.5–10.0
A. longissima	9–13	P. sexta	8.5–10.0
		Phlegethontius quinquemaculatus	10.0–12.0

Fig. 5.4. Flower structure and pollinators of *Aquilegia* species in California. (Modified with permission from V. Grant, © 1981, *Plant Speciation*, Columbia University Press, New York.)

I. tenuis pollen tubes sometimes even burst in the style of *I. tenax*. Crosses between white- and purple-flowered species of *Capsicum* display a range of crossing barriers from a lack of pollen germination, to the restriction of pollen tube growth, to an inability to penetrate the egg cell (Fig. 5.5).

Hogenboom (1973, 1975) suggested that such prefertilization barriers between divergent taxa often arise due to poor intergenomic coadaptation. One species may lack the genetic information necessary to properly coordinate the critical functions of the other. This phenomenon, called "incongruity", arises due to disrupted gene regulation, the absence of a gene or poor genomic-cytoplasmic interactions. A number of studies have provided circumstantial evidence of this phenomenon by showing that the vigor of interspecific crosses can be improved through subsequent breeding, presumably through selec-

tion for the most coordinated genes (Haghighi and Ascher, 1988).

In some cases, there is genotypic variation within a species such that some genotypes will cross with another species, while others cannot. Diploid species of *Vaccinium* are generally isolated from each other, but combinations can be found that produce some viable seed (Table 5.3; Ballington and Galletta, 1978). Often, the success of an interspecific cross is dependent on the genotype used as the maternal parent. For example, interspecific backcrosses in *Phaseolus* are much more successful when the F_1 is used as the maternal parent (Hucl and Scoles, 1985). Such variations have allowed interspecific hybridizations between otherwise strongly isolated species.

Sometimes fertility blocks arise after fertilization in the form of hybrid inviability. Here, zygotes either do not develop completely, or they

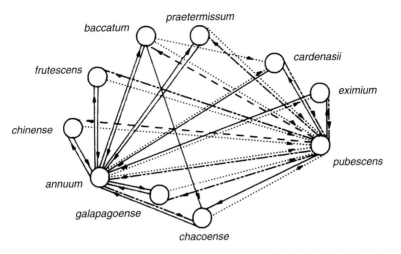

Fig. 5.5. Pollen tube growth in crosses between *Capsicum*. Arrows point in direction of female parent. The connections between species denote: no penetration into the egg cell (—); barrier in ovary (– -- –); barrier in style (----); and barrier in stigma (·····). (Used with permission from S. Zijlstra, C. Purimahua and P. Lindhout, © 1991, *HortScience* 26, 585–587.)

Table 5.3. Cross-compatibility and cross-fertility (poorest and best) among diploid *Vaccinium* species (Ballington and Galletta, 1978).

Species hybridization		Return/100 pollinations			
		Fruit	Seed	Germinating seeds	Vigorous seedlings
atrococcum–	Poorest	7	7	0	0
caeseriense	Best	76	798	551	523
atrococcum–	Poorest	0	0	0	0
darrowi	Best	91	807	544	483
atrococcum–	Poorest	6	8	0	0
tenellum	Best	8	92	44	22
caesariense–	Poorest	14	27	0	0
darrowi	Best	55	277	145	109
caesariense–	Poorest	4	6	0	0
tenellum	Best	58	705	336	263
darrowi–	Poorest	14	72	4	4
tenellum	Best	76	274	228	223

are weak and have reduced viability. Crosses between the cultivated rices *Oryza sativa* and *Oryza glaberrima* produce mostly sterile hybrids. In *G. australis* × *G. splendens* (Grant and Grant, 1954) and *Papaver dubium* × *Papaver rhoeas* (McNaughton and Harper, 1960), many weak plants are produced, which grow very slowly and remain dwarfed. Most of the GP-2 species described by Harlan and de Wet fall into this class along with some of those in GP-3 (Figs 5.1 and 5.2; Table 5.1). In most cases, the embryos abort before they can be successfully grown, but in some cases, tissue culture techniques can be used to "rescue" them. These techniques have been successfully employed by breeders in such diverse groups such as *Triticum*, *Gossypium*, *Solanum* and *Prunus* (Briggs and Knowles, 1967).

Hybrid sterilities occur when hybrids do not produce functional gametes. The basis of these sterility barriers can be genic in nature or the result of meiotic irregularities due to chromosomal imbalances. Crosses between the wild chickpea species *Cicer echinospermum* and *Cicer reticulatum* yield very few hybrids and those that are produced are sterile due to poor chromosome pairing. The weak F_1 plants of *Gilia* and *Papaver* mentioned above are generally sterile even though their chromosome pairing is regular. This sterility is probably due to genic discordances affecting normal flower development (Grant and Grant, 1954). Most hybrids of *Gilia ochroleuca* × *Gilia latifolia* are vigorous, but they are largely sterile due to poor pairing of chromosomes between the two genomes (Grant and Grant, 1960).

Inter-ploidy crosses generally result in hybrid sterility as unbalanced sets of chromosomes are distributed to the different gametes. The pentaploid hybrids arising from the natural hybridization of the strawberries *Fragaria vesca* ($2x = 14$) × *Fragaria chiloensis* ($2x = 56$) produce a whole range of gametes containing 19–70 chromosomes with varying levels of fertility (Bringhurst and Senanayake, 1966). Most banana cultivars are triploid and produce no viable gametes. This trait is critical to edibility, as normally pollinated diploids have flinty, teeth-breaking seeds (Chapter 10).

The final class of isolating mechanisms is hybrid breakdown, where the F_2 or backcross hybrids have reduced viability or fertility. Low vigor is found in many of the F_2 populations of interspecific bean crosses (Hucl and Scoles, 1985). As we discussed in Chapter 3, *Zauschneria cana* × *Zauschneria septentrionalis* produces a vigorous, semifertile hybrid, but when Clausen *et al.* (1940) examined 2133 F_2s, all were weak and dwarfish. Crosses between *Layia gaillardioides* × *Layia hieracioides* (Clausen, 1951) and *Gilia malior* × *Gilia modocensis* (Grant, 1966) also produce high percentages of subvital, sterile F_2 individuals. A preponderance of weak *Gilia* plants is still carried in the F_6 generation. Hogenboom's concept of incongruity probably applies to these cases, since non-harmonious gene combinations are presumably segregating and producing poorly adapted individuals. Disruptive selection would have to be performed to purge the populations of these associations.

Any kind of RIB reduces the amount of gene flow between species, but combinations of mechanisms result in the tightest isolation. Most "good" species are separated by multiple combinations of RIBS. For example, the leafy-stemmed *Gilia* of central California, *G. millefoliata* and *G. capitata*, are isolated in five ways according to Grant (1963): "1) Ecological isolation: *G. capitata* occurs on sand-dunes and *G. millefoliata* on flats. 2) Floral: *G. capitata* is large-flowered and bee pollinated, while *G. millefoliata* is small flowered and self-pollinating.... 3) Seasonal isolation: *G. millefoliata* blooms earlier than *G. capitata*. 4) Incompatibility: Hybrids are very difficult to produce by artificial crosses in the experimental garden. 5) Hybrid sterility: The F_1 plants when they can be obtained are chromosomally sterile to a high degree, producing only about 1 percent of good pollen grains and no F_2 seeds."

Another clear example of multiple RIBs is found in crosses between *P. vulgaris* and *P. coccineus*. They rarely hybridize in the first place because they have different habitats (as mentioned earlier), and when they do cross, there is poor seed set, high seedling mortality and the surviving hybrids have a "crippled" morphology represented by dwarfism and abnormal development. In addition, the F_1 plants produce only a low proportion of viable gametes (Hucl and Scoles, 1985). In cases like these, it is highly unlikely that any successful hybridization will occur at all, even if the species are in proximity.

Nature of Isolating Genes

Rieseberg and Blackman (2010) were able to find 41 speciation genes in the literature that have been cloned and sequenced. Among these, seven were involved in pre-pollination reproductive isolation, one to post-pollination reproductive isolation, eight to hybrid inviability and 25 to hybrid sterility. The most common genes isolated were those associated with the anthocyanin pathway and its regulators (pollinator isolation), *S* RNase self-incompatibility genes, disease resistance genes that caused hybrid necrosis, and genes regulating cytoplasmic male sterility. They found that regulatory changes were most common among the mutations associated with prezygotic mechanisms, while a mix of both regulatory and structural gene mutations were involved in postzygotic mechanisms. Loss-of-function mutations and copy number variation were found to frequently play a role in reproductive isolation.

Modes of Speciation

New species are thought to arise through a number of different pathways (Table 5.4, Fig. 5.6). One way the different modes are distinguished is by the degree of separation between the speciating populations. The populations are geographically well isolated in *allopatric* speciation, there is no separation between populations in *sympatric* speciation, and the populations touch along one axis in *parapatric* speciation.

Other important criteria used to describe speciation pathways are: (i) how large the speciating groups are; and (ii) how much genetic differentiation precedes the formation of strong RIBs. Allopatric speciation is broken into two groups: *geographic* and *peripatric*. In geographic speciation, large populations are thought to gradually diverge and form RIBs, while in peripatric speciation, isolating barriers are thought to appear quickly in small populations without much differentiation. Sympatric speciation occurs when a single individual or small group arises that is reproductively isolated from the surrounding population. Parapatric speciation is closely related to sympatric speciation, except that the gradually diverging crowd of genotypes is on one side of the parent population rather than surrounded by it.

Numerous other speciation taxonomies have been proposed, but the four depicted in Table 5.4 are the ones most commonly discussed

Table 5.4. Factors distinguishing major modes of speciation in sexual plants.

Mode of speciation	Separation	Population size	Differentiation before RIBs
1. Allopatric			
a. Geographic (Type 1)	Wide	Large	Much
b. Peripatric (Type 2)	Wide	Small	Little
2. Parapatric	Touching	Large	Much
3. Sympatric	None	One	Little

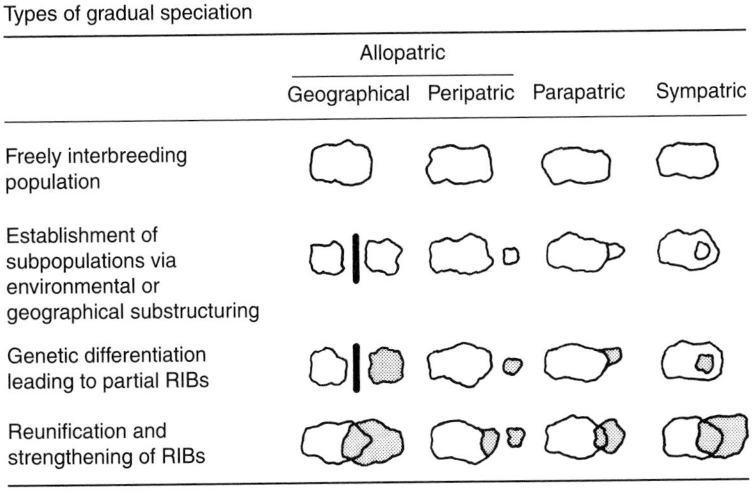

Types of gradual speciation

	Allopatric		Parapatric	Sympatric
	Geographical	Peripatric		
Freely interbreeding population				
Establishment of subpopulations via environmental or geographical substructuring				
Genetic differentiation leading to partial RIBs				
Reunification and strengthening of RIBs				

Fig. 5.6. Diagrams illustrating the different modes of gradual speciation. The drawings found below each of the individual headings (geographical, peripatric, parapatric and sympatric) represent the various stages of species development. See the text and Table 5.4 for more details.

in the evolutionary literature. We will describe synonyms and related concepts in the text where appropriate.

Geographic speciation

The most widely recognized type of speciation is geographic speciation. In its first stage, there is a single population found in a large homogenous environment. The environment then becomes partly diversified due to physical or biotic factors and populations become isolated. These populations begin to diverge genetically and eventually acquire sufficient variation to become reproductively isolated from each other. Further changes in the environment allow some of the newly evolved groups to come back in contact, but they do not produce successful hybrids because of past differentiation. Natural selection against the formation of weak or sterile hybrids promotes the reinforcement of RIBS through additional differentiation.

Most of the evidence for this type of speciation is circumstantial in nature due to the slow speed of the process. Few scientists' careers are long enough to follow the whole scenario. However, populations in the various stages of speciation have been identified by individual investigators. Probably the most complete story in plants has been accumulated by Grant and Grant (1960) in their oft cited *Gilia* studies. In this group, they were able to find races and species in all the hypothesized stages of geographical and ecological differentiation (Fig. 5.7), and they were able to show a subsequent build-up of isolating mechanisms as the species became more divergent (Table 5.5).

Peripatric speciation

This type of speciation is very similar to the geographic mode, except the speciating population is much smaller. Other terms used for this type of speciation are quantum (Grant, 1981),

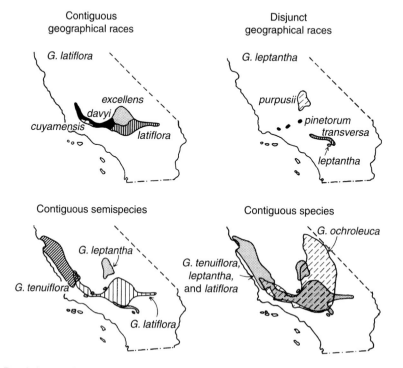

Fig. 5.7. Populations and species of *Gilia* representing different stages of allopatric speciation. (Redrawn with permission from V. Grant, © 1963, *The Origin of Adaptations*, Columbia University Press, New York.)

Table 5.5. Relative ease of crossing diploid cobwebby *Gilia* at different levels of divergence (Grant and Grant, 1960).

Type of cross	Entities crossed	Number of flowers pollinated	Average no. plump seeds per flower	Number of hybrid individuals per ten flowers pollinated
Inter-individual	Different individuals belonging to the same population	116	17.8	22
Inter-racial	Different geographical races of the same species	562	15.2	12
Inter-specific	Different diploid species of cobwebby *Gilia*	2016	3.7	3
Inter-sectional	Diploid species of cobwebby *Gilia* with diploid species of leafy-stemmed or woodland *Gilia*	528	0.004	0.032

speciation by catastrophic selection (Lewis, 1962) and founder-induced speciation (Carson, 1971; Carson and Templeton, 1984). Because of reduced population size, genetic drift becomes more important and the rate of speciation is accelerated. Peripatric speciation can result in reproductively isolated species that are otherwise quite similar to their progenitors or they may become morphologically quite distinct depending on how many genes are affected. Most of the evidence for peripatric speciation is also circumstantial, although the process can occur during the lifetime of an individual investigator.

One of the most completely documented cases of peripatric speciation associated with few genetic changes concerns two *Clarkia* species in southern California (Lewis, 1962). *Clarkia biloba* is a relatively widely distributed species, while *Clarkia lingulata* is rare and is found on only two sites at the extreme edge of the *C. biloba* range. The two species are very similar electrophoretically and morphologically except for flower petal shape (Fig. 5.8); however, they are reproductively isolated by a translocation, several paracentric inversions and a chromosomal fusion. Lewis suggested that *C. biloba* arose when an isolated *C. lingulata* population crashed during a drought to only a few individuals that inadvertently contained the chromosomal rearrangements due to chance. White (1978) used the term "stasipatric speciation" to describe the formation of new species due to the fixation of chromosomal rearrangements.

The dramatic reduction of a population due to an environmental catastrophe or the establishment of a "founder population" of a few individuals can also lead to a morphologically distinct species when the remaining sample of the gene pool is unbalanced and "undergoes a genetic revolution" (Mayr, 1954) or "genetic transilience" (Templeton, 1981). Genetic drift and changes in selection pressure can result in a shift of many genes into new coadapted complexes. A wide range of distinct plant and animal species in the Hawaiian islands are thought to have arisen in this manner (Carson and Templeton, 1984). The most thoroughly documented cases involve representatives of the *Drosophila*, but the silversword alliance of the *Compositae* includes giant herbs, small trees and ecologically diverse shrubs (Carr and Kyhos, 1981). Small founder populations do not always undergo dramatic alterations, however, as most of our crop species are based on relatively few genotypes and still retain a strong resemblance and inter-fertility with their progenitors (Chapter 7).

Parapatric speciation

In this mode of speciation, RIBs evolve without geographical separation. The diversifying population is adjacent to the progenitor population (neighboringly sympatric; Grant, 1985). The process occurs when a subgroup diverges in response to environmental challenges and isolating barriers begin to form as a by-product of ecological differentiation.

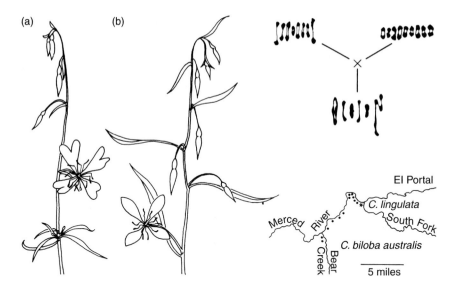

Fig. 5.8. Morphology, cytology and geographical range of *Clarkia biloba* (a) and its peripatric derivative species *Clarkia lingulata* (b). (Used with permission from H. Lewis, © 1962, *Evolution* 16, 257–271.)

Whether populations can diverge sufficiently without geographical separation has long been a matter of debate. Mayr (1942, 1963) contended that portions of populations are unlikely to differentiate enough to become genetically isolated in the face of strong gene flow. This would be particularly true among sympatric subgroups where the diverging race is completely contained within the parent population and is bombarded on all sides by pollen and seed. There are, however, numerous examples where parapatric populations have undergone substantial differentiation in the face of one-dimensional gene flow. We have already described the mine tailing experiments of Bradshaw where the differentiated populations varied not only in heavy metal tolerance but also in flowering date (Chapter 2). Zheng and Ge (2010) have documented significant ecological divergence in the closely related rice species *Oryza rufipogan* and *Oryza nivara*, even in the presence of gene flow. McNeilly and Antonovics (1968) have cataloged numerous similar scenarios where ecologically divergent populations have different bloom dates (Table 5.6). Artificial selection experiments have also shown that relatively strong RIBs can arise over a few generations when strong selection is placed on unrelated characteristics. Paterniani (1969) planted white flint and yellow sweet maize in

the same field together. Each generation he selected the purest ears for subsequent planting, and after six generations less than 5% of the seeds resulted from an outcross (Fig. 5.9). These experiments demonstrate that parapatric speciation is at least theoretically possible; it is up to the field biologists to document the complete process in nature.

Sympatric (instantaneous) speciation

Occasionally, new species arise through spontaneous mutation without any ecological or geographical separation. A new isolated type appears in only a few generations without substantial genetic differentiation. Most of these types face a high likelihood of going rapidly extinct due to their low numbers, but even against these odds, many species are known to have gotten their start in this manner. Probably the most frequently cited example of sympatric speciation occurred in the apple maggot, *Rhagoletis pomonella* (Bush, 1975). The original host plant of *R. pomonella* in the USA was hawthorn (*Crataegus*), but the fruit fly began to infest introduced populations of apples in the mid-1800s. Apparently, there was a change in a single gene trait affecting

Table 5.6. Differences in flowering time of ecotypes compared to the "normal" type (McNeilly and Antonovics, 1968).

Flowering species	Ecotype	Time	Reference
Gilia capitata	Sand dune	Later	Grant (1952)
Madia elegans[a]	Subsp. *vernalis*	Spring	Clausen (1951)
	Subsp. *aestivalis*	Summer	
	Subsp. *densifolia*	Autumn	
Layia platyglossa[a]	Maritime	Later	Clausen and Heisey (1958)
Hemizonia citrina[a]		April	Babcock and Hall (1924)
Hemizonia lutescens[a]		Aug.–Sept.	
Hemizonia luzulaefolia[a]		April	
Hemizonia rudis[a]		Aug.–Sept.	
Lactuca graminifolia[a]		Early spring	Whitaker (1944)
Lactuca canadensis[a]		Summer	
Ixeris denticulata	Subsp. *typica*	Spring	Stebbins (1950)
	Subsp. *sonchifolia*	Autumn	
	Subsp. *elegans*	Summer	
Pinus attenuata[a]		Later	Stebbins (1950)
Pinus radiata		Earlier	
Lamium amplexicaule[a]	Vernal race	Earlier	Bernstrom (1952)
Viola tricolor	Sand dune	Later	Clausen (1926)
Silene cucubalis		Earlier	Marsden-Jones and Turril (1928)
Silene maritima		Later	
Geranium robertianum	Shingle beach	Later	Böcher (1947)
Mimulus guttatus	Coastal	Late	Vickery (1953)
	Mountain	Latest	
	Valley and foothills	Early	
Geum urbane[a]		Later	Clausen and Heisey (1958)
Geum rivale[a]		Earlier	
Succisa pratensis	Northern race	Earlier	Turesson (1925)
Ranunculus acer	Alpine	Earlier	
Solidago virgaurea	Alpine and coastal	Earlier	
Rumex acetosa	Alpine	Earlier	
Leontodon autumnale	Coastal	Earlier	
Clarkia xantiana[a]	Self-compatible race	Earlier	Moore and Lewis (1965)
Salvia mellifera		Early spring	Grant and Grant (1964)
Salvia apiana		Late spring	

[a]Some evidence given by author that "ecotypes" are closely adjacent.

host recognition that isolated the two populations with minimal genetic change. The genus *Rhagoletis* has a large number of very similar species that infest fruits of different plant families:

R. *pomonella* – Rosaceae

R. *mendax* – Ericaceae

R. *carnivora* – Cornus

R. *zephyria* – Caprifoliaceae

Examples of instantaneous speciation abound in plant species. We have already discussed polyploidy at length, where a chromosomal duplication instantly isolates a progeny plant from its parents – at least half of all plant species are polyploid. Even the appearance and fixation of simple chromosomal rearrangements can result in a new species. *Stephonomeria malheurensis* is a self-compatible species that was probably derived from the self-incompatible *Stephonomeria exigua* subsp. *coronaria* (Gottleib, 1974, 1977a).

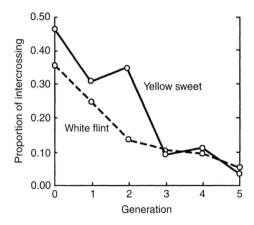

Fig. 5.9. Development of reproductive isolation in mixed plantings of maize when only the purest ears were selected each generation. (Used with permission from E. Paterniani, © 1969, *Evolution* 23, 534–547.)

They are morphologically and electrophoretically quite similar and share a single habitat, but *S. malheurensis* carries a chromosomal translocation that reproductively isolates it from *S. exigua*. The appearance of self-compatibility in the new type probably allowed the translocation to become homozygous through selfing rather than being lost in an outcrossed flood of parental chromosomes.

Genetic Differentiation During Speciation

During the course of this chapter, it has been repeatedly suggested that RIBs arise when "sufficient" genetic variability has occurred. We are now left with the question of just how much genetic variation is necessary to begin the speciation process. We have shown that genetic differentiation is important, but have not yet defined what is "sufficient". This question has been posed as whether reproductive isolation occurs at the genome-wide or genic level (Wu, 2001).

Many investigators have been concerned with quantifying the number of genes involved in gradual speciation through differential adaptation. Much of the available information has come from populations that have diverged to the point of ecotypes, but have not speciated. Clausen and Heisey (1958) demonstrated that

the races of *Potentilla glandulosa* found at different elevations in the Sierra Nevada mountains of California were genetically distinct by growing them in a common garden. They then made crosses to determine the minimum number of alleles separating the subalpine and timberline ecotypes and found dozens of allelic differences for 19 morphological and physiological traits (Table 5.7). Similar high numbers of differences were also discovered between coastal and inland *Viola tricolor* (Clausen 1922, 1926, 1951) and winter and summer annual races of *Lamium purpureum* (Muntzing, 1932; Bernstrom, 1952).

Examinations have also been conducted on closely related species that can still be successfully hybridized. Baur (1924) crossed two species of snapdragon, *Antirrhinum majus* and *Antirrhinum molle*, and found considerable phenotypic variability in the F_2 population. Most individuals had combinations of parental characteristics, but a few had unique characteristics that must have come from novel genic assortments. Baur estimated that the two species were separated by more than 100 allelic differences.

Gel electrophoresis has also been used to access levels of variation during the speciation process. Comparisons have been made between populations at various levels of evolutionary divergence in a wide range of plants and animals (Ayala, 1975). Local populations of plants have average genetic identity values of I = 0.95, while conspecific populations average I = 0.67 (Gottleib, 1981). If we take the natural logarithm of the conspecific value, the genetic distance value of 0.40 is obtained, which represents about 40 changes per 100 loci in an average pair of divergent plant species.

These studies show that extensive amounts of genetic differentiation can be associated with speciation, but it is important to realize that not all of this variability is directly related to the speciation process. For taxonomic species, only those genes that control the characters used to distinguish the species are associated with speciation, and in the case of biological species, RIBs can arise at any time during the process of divergence regardless of the level of differentiation. There is increasing evidence that relatively strong RIBs can be regulated by relatively few genes. When Bradshaw *et al.* (1998) searched for QTL loci associated with 12 differences in

Table 5.7. Minimum number of alleles governing the inheritance of characters in hybrids of subalpine and foothill races of *Potentilla glandulosa*. (From Clausen and Hiesey, 1958 as summarized by D. Briggs and S.M. Walters, © 1984, *Plant Variation and Evolution*, Cambridge University Press, New York.)

Character and action in genes	Estimated number of pairs of alleles at unlinked gene loci
1. Orientation of petals: 2 erecting, 1 reflexing	3
2. Petal notch: 1 producing notch, 2 inhibiting	3
3. Petal color: 2 whitening, 2 producing yellow, 1 bleaching	5
4. Petal width: 4 widening, 1 complementary,[a] 1 narrowing	6
5. Petal length: 4 multiples[b]	*c.*4
6. Sepal length: 3 or 4 multiples for lengthening, 1 for shortening, 1 complementary	*c.*5
7. Achene weight: 5 multiples for increasing, 1 for decreasing	*c.*6
8. Achene color: 4 multiples of equal effect	4
9. Branching, angle of	*c.*2
10. Inflorescence, density of	*c.*1
11. Crown height	*c.*3
12. Anthocyanin: 4 multiples (1 expressed only at timberline), 1 complementary	5
13. Glandular pubescence: 5 multiples, in series of decreasing strength	5
14. Leaf length: transgressive[c] segregation; many patterns of expression in contrasting environments; possibly different sets of multiples activated	*c.*10–20
15. Leaflet number in bracts	*c.*1
16. Stem length: transgressive segregation, 5–6 multiples plus inhibitory and complementary genes; many patterns of expression in contrasting environments	*c.*10–20
17. Winter dormancy: 3 multiples of equal effect	3
18. Frost susceptibility: slight transgression toward resistance	*c.*4
19. Earliness of flowering: strongly transgressive; many patterns of altitudinal expression; possibly different sets of genes activated	Many

[a]Complementary effects: factor A or B no effect but combination A+B produces phenotypic difference.
[b]Multiples: factors with comparable effects.
[c]Transgressive effects: in F_2 segregants values for "extreme" individuals exceed parental values.

floral morphology assumed to reproductively isolate *Mimulus lewisii* and *Mimulus cardinalis*, they found only 1–6 QTL for each trait, and most traits (9/12) were regulated by a single QTL that influenced over 25% of the total phenotypic variability. Major QTL have also been shown to control differences in inflorescence architecture in *Zea* (Doebley and Stec, 1993), flowering time in *Brassica* (Camargo and Osborn, 1996) and the growth and flowering of *Arabidopsis* (Mitchell-Olds, 1996). Similar results have been obtained in studies elucidating the genetics of the domestication process (Chapter 7). This body of work implies that a relatively small number of genes with cumulative effects can have rather dramatic effects on the evolution of reproductive isolation and speciation.

Hybridization and Introgression

Thoughts about the relative importance of interspecific hybridization in plant evolution have switched back and forth over time. In the middle of the last century, the role of hybridization was considered to be substantial, based on a number of morphological studies of native and crop species (Anderson, 1949). In the 1970s, enthusiasm waned (Heiser, 1973), but with the advent of molecular markers in the 1980s support grew dramatically (Rieseberg, 1995; Rieseberg *et al.*, 2000). Most plant systematists now believe that hybrid speciation is at least common, with estimates of the percentages of hybrid species in different floras ranging from 25 to 80% (Whitman *et al.*, 1991; Abbott, 1992; Masterson, 1994).

Ellstrand *et al.* (1996) found 16–34% of the families in five biosystematic floras to have at least one pair of species that hybridize locally, and 6–16% of the genera.

Hybridizations between native and introduced species have often led to the development of new taxa and have even been implicated in the evolution of a number of new invasive species. Abbott (1992) estimated that 45% of the British flora was alien, and 7% of those introduced species were involved in the production of hybrids now prominent in the native flora. One of the most widespread examples is *Senecio vulgaris* var. *hibernicus*, a hybrid of native *S. vulgarus* var. *vulgaris* and introduced *Senecio squalidus*, which escaped from the Oxford Botanical Garden in 1794 (Abbott, 1992). Highly invasive thistles from Europe have widely hybridized in Australia (O'Hanlon *et al.*, 1999). Ellstrand and Schierenbeck (2000) found 28 examples "where invasiveness was preceded by hybridization" and at least half of these hybrid lineages were the product of native × non-native hybridizations.

As we have discussed previously in this chapter, many plant species retain the ability to hybridize with their relatives, even when they are quite distinct and have relatively strong RIBs. Numerous hybrid populations or "hybrid swarms" have been identified where closely related species come into contact. These hybrid zones are often narrow and stable when hybrid fitness is low or the species are adapted to very distinct habitats (Levin and Schmidt, 1985; Campbell and Waser, 2001); however, interspecies hybridization can act as the nucleus for evolution if hybrids are at least partially viable and suitable habitats exist to support them (Buerkle *et al.*, 2000).

Hybridization can stimulate evolutionary change in two ways: (i) the adaptive potential of one or both parents might be increased through backcrossing, or "introgressive hybridization" (Fig. 5.10; Anderson, 1949), and this trickle of genes might expand the adaptive potential of the recipient population; and (ii) the hybrid population itself may evolve unique adaptations through genetic differentiation and genomic reorganization or "recombinational speciation" (Stebbins, 1957; Grant, 1958). This process is generally thought to occur over a number of generations, although Rieseberg and Ellstrand (1993) have identified many examples where F_1 hybrids had unique phenotypes that could be a nucleus for speciation. Most hybrids are less fit than their progenitors, but several cases have been found where hybrids had higher fitness and unique adaptations (Emms and Arnold, 1997; Burke and Arnold, 2001; Johnston *et al.*, 2001).

Most hybrid species are polyploid, but speciation via homoploid hybridization has also been documented in several instances (Gross and Rieseberg, 2005; Soltis and Soltis, 2009;

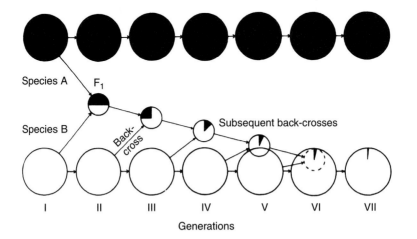

Fig. 5.10. Diagram illustrating introgression between two species. Backcrossing of the F_1 hybrid to species B ultimately results in the absorption of some genes from species A into at least some individuals of species B. (Used with permission from D. Briggs and S.M. Walters, © 1984, *Plant Variation and Evolution*, Cambridge University Press, New York.)

Abbott *et al.*, 2010). In one of the earliest studies of hybridization, Riley (1938) described natural hybridization between *Iris fulva* and *Iris hexagona* in Louisiana, USA using a "hybrid index". He compiled a list of differences between the two species and one was arbitrarily selected at the low end (Fig. 5.11). Plants in the natural environment were then scored for each character and the grand total was calculated for each individual. Many populations contained individuals with intermediate scores, suggesting that they were indeed hybrids. Randolph (1966) later identified a population of *Iris* called 'Abbeville Reds' that had a bright red-purple color distinct from any other *Iris* species, different shaped capsules and a unique habitat of deep shade and high water. He gave it a new species name, *Iris nelsonii*, and speculated that it was derived from the hybridization of *I. fulva* and other local species.

Arnold (1993) used a broad array of nuclear and cytoplasmic markers to confirm that the Louisiana irises were indeed hybridizing. He found individuals with many different combinations of the cpDNA and nuclear genes of three species, *I. fulva*, *I. hexagona* and *Iris brevicaulis* (Fig. 5.12). He also found that *I. nelsonii* carried markers from all three species, although every individual had the *I. fulva* cytotype and only a few individuals carried species-specific RAPDs from the other two species. This suggested that the taxa was the product of repeated backcrossing of the original F_1 hybrid to *I. fulva*, a likely scenario, as the appearance of hybrids in nature is rare and the species have several prezygotic and postzygotic RIBs (Carney *et al.*, 1994; Carney and Arnold, 1997; Emms and Arnold, 2000).

Using morphological and cytological data, Heiser (1947, 1949) documented widespread hybridization and introgression among *Helianthus* taxa in the southeastern USA. He proposed two instances of introgression, *H. annuus*

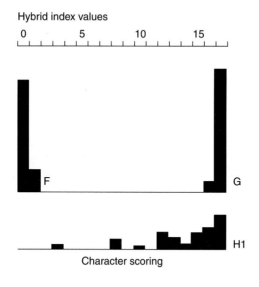

Character scoring

	Tube color	Sepal blade color	Sepal length	Petal shape	Exertion of stamens	Stylar appendage	Crest
Like *I. fulva*	0	0	0	0	0	0	0
Intermediates	1	1, 2 or 3	1, 2	1	1	1	1
Like *I. hexagona*	2	4	3	2	2	2	2

Fig. 5.11. Hybridization between *Iris fulva* and *Iris hexagona* in Louisiana, USA. Two relatively pure populations (F and G) and one hybrid population are shown (Riley, 1938).

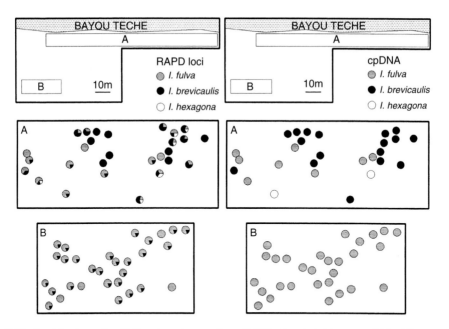

Fig. 5.12. Combinations of cpDNA and nuclear markers (RAPDs) in a hybrid swarm of *Iris* at Bayou Teche, Louisiana, USA. The relative proportion of *I. fulva*, *I. hexagona* and *I. brevicaulis* markers in each plant is represented by the pie charts. (Used with permission from Arnold, © 1993, *American Journal of Botany* 80, 577–583.)

subsp. *texanus* arising due to hybridization with *H. debilis* subsp. *cucumerifolius* (Heiser, 1951), and a weedy race of *H. bolanderi* being the product of introgression with *H. annuus* (Heiser, 1947). He also suggested that three species, *H. paradoxus, H. deserticola* and *H. neglectus*, were the stabilized products of hybridization between *H. annuus* and *H. petiolaris* (Heiser, 1958). Later molecular work by Rieseberg and associates (Rieseberg *et al.*, 1988, 1990; Rieseberg, 1991) verified Heiser's prediction of a hybrid origin for *H. paradoxus* and *H. deserticola* and the proposed introgression between *H. annuus* and *H. debilis*. They also identified another species, *H. anomalus*, as a hybrid derivative and found evidence of the introgression of cpDNA from *H. annuus* into southern California populations of *H. petiolaris*. However, the hybrid origin of *H. negectus* was not supported, nor was any evidence found of gene transfer between *H. annuus* and *H. bolanderi*.

Rieseberg *et al.* (1995) went on to show how the genomes of *H. annuus* and *H. petiolaris* were rearranged to produce *H. paradoxus*. They developed a molecular linkage map of the two progenitor species and then compared it to the proposed derivative. They discovered six linkage groups that were conserved across all three species (collinear), and another 11 chromosomal regions whose gene order differed across taxa. Of these 11, *H. paradoxus* shared four of them with one or the other parent, while the other seven were distinct from either parent, suggesting that substantial reorganization had occurred within these linkages (Fig. 5.13).

To determine if particular gene assemblages were maintained by selection during the formation of *H. paradoxus*, Rieseberg *et al.* (1996) produced three hybrid lineages through artificial crosses and compared them to the ancient hybrid species using a map-based approach. The genomic composition of the ancient and synthesized genomes were highly concordant, indicating that the particular gene assemblages maintained in the hybrids were probably under selection. Some parts of the genome where less subject to introgression than others, suggesting that there was coadaptation between blocks of parental species genes. To further confirm the role of coadapted complexes in hybrid species formation, Rieseberg *et al.* (1999) examined 88 markers across 17 chromosomes in three natural hybrid zones of *H. petiolaris* and

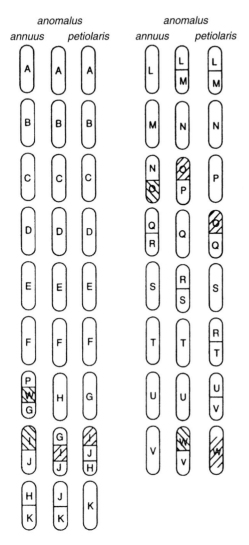

Fig. 5.13. Major linkage relationships between *Helianthus annuus, Helianthus petiolaris* and its hybrid derivative *Helianthus anomalus*. Note that six linkage groups are collinear between the two species and 11 are structurally different. Lines of shading indicate inversions. (Used with permission from Rieseberg *et al.*, © 1995, *Nature* 375, 313–316.)

H. annuus. They found patterns of introgression to be quite similar across all three hybrid zones, with 26 chromosomal locations having significantly lower levels of introgression than neutral expectations. They were even able to link pollen sterility with 16 of these segments.

In subsequent work on hybrid speciation in *Helianthus*, Rieseberg's group (2007)

has demonstrated that considerable levels of chromosome repatterning have occurred in all the hybrid sunflower species (Burke *et al.*, 2004; Lai *et al.*, 2005), although the rearranged regions do not appear to be more divergent than collinear ones (Yatabe *et al.*, 2007; Strasburg *et al.*, 2009). They have shown that *H. anomalus*, *H. paradoxus* and *H. deserticola* have unique ecological adaptations from their parents (Lexer *et al.*, 2003b, 2004; Gross *et al.*, 2004; Donovan *et al.*, 2010) and the group has identified QTL related to those adaptations (Lexer *et al.*, 2003a). They have also shown that the various hybrid species were likely formed more than once (Schwarzbach and Rieseberg, 2002; Gross *et al.*, 2003), and while most of the populations with different origins appear to be evolving in concert, there are exceptions (Gross and Rieseberg, 2007). Even though the homoploid hybrids and their parental species are isolated by chromosomal incompatibilities, gene flow between them can still be significant and contribute to their subsequent evolution (Rieseberg *et al.*, 2007; Scascitelli *et al.*, 2010).

Other recent studies have documented significant changes in gene expression patterns associated with hybrid speciation. We have already described the work of Hegarty *et al.* (2005, 2006, 2008, 2009) where they found 475 genes showing significant up- or down-regulation in *Senecio squalidis* compared to its progenitors *Senecio aethnensis* and *Senecio chrysanthemifolis*. They also discovered that gene expression patterns were much more dramatically altered in the diploid homoploid hybrid *S. squalidis* than the allotetraploid *Senecio cambrensis*. In *Helianthus*, when Lai *et al.* (2006) compared the expression patterns of 2897 genes in *H. deserticola* compared to its parental species, they found that 12.8% of the genes had significantly different levels of expression between the three species. Overall, 154 genes were differentially expressed in *H. deserticola* compared to its parent *H. annuus* and 174 compared to its other parent *H. petiolaris*.

Hybridization and Extinction

In some instances, hybridization between native species can result in the extinction of one of the species (Levin *et al.*, 1995; Buerkle *et al.*, 2000). There can be "genetic assimilation", where the hybrids are fertile, and they replace pure conspecifics of either or both hybridizing taxa. This is thought to be happening on Santa Catalina Island, where *Cerocarpus betaloides* appears to be assimilating *Cerocarpus traskiae*, which is represented by a single population (Rieseberg and Gerber, 1995). *Clarkia speciosa polyantha* is replacing *Clarkia nitens* in regions of overlap in the Sierra Nevada foothills of central California (Bloom, 1976). Substantial introgression has occurred between *Gossypium darwinii* on the Galapagos Islands and the crop *Gossypium hirsutum* (Wendel and Percy, 1990). The cultivated radish, *Raphanus sativus*, and the jointed charlock, *Raphanus raphanistrum*, have completely merged in California (Hegde *et al.*, 2006). There can also be "demographic swamping", where population growth in a numerically inferior taxon is retarded by the formation of hybrid seed, and population growth falls below the replacement level. Hybridization with cultivated rice is thought to have led to the near extinction of the endemic Taiwanese taxon *Oryza rufipogon* subsp. *formosana* (Kiang *et al.*, 1979). In fact, most populations of native Asian subspecies of *O. rufipogon* may be endangered through hybridization with the crop (Chang, 1995; Ellstrand *et al.*, 1999).

A number of factors influence the risk of extinction after hybridization, including the strength of the RIBs separating the taxa, habitat differentiation, hybrid viability and fertility, population sizes and growth rates and environmental stochasticity (Wolf *et al.*, 2001). Interspecific hybridization is a particular threat to rare species, where extinction can occur whether the hybrid seeds are viable or abort. Small populations can be so swamped by alien pollen that they produce too few non-hybrid individuals to replace themselves.

Crop–Weed Hybridizations

As human beings carried domesticants far from their origins, the crop species occasionally came into contact with previously isolated relatives. When RIBs were not complete, the domesticated and native species could then hybridize and produce novel progeny. The adaptations of the native congeners could be altered through subsequent introgression or backcrosses to the crop, and unique types might be selected by humans. The natural disturbances caused by farming would have greatly facilitated the hybridization process, by providing novel habitats for the hybrids to gain a foothold (Anderson, 1949).

Hybridization still occurs between many domesticated and wild species (Anderson, 1961; Harlan, 1965; Zhukovskii, 1970). In many cases, the RIBs are so minimal that the wild and domesticated taxa are considered subspecies. Evidence for crop introgression into wild populations of congeners has been provided for 31 species, including lucerne, barley, beets, cabbages, canola, cotton, carrots, cassava, chili, cocona, common bean, cowpea, finger millet, foxtail millet, hemp, hops, lettuce, maize, oats, pearl millet, pigeon pea, potato, quinoa, radish, raspberry, rice, rye, sorghum, soybean, squash, sunflower, tomato, watermelon and wheat (Ellstrand *et al.*, 1999; Jarvis and Hodgkin, 1999).

Most F_1 hybrids of cultivated and wild congeners have reduced fitness in nature and the genes of domesticated plants rarely travel far from narrow hybrid zones. However, some wild plants have acquired crop genes that make them more effective agronomic weeds. In some instances, genes have introgressed into wild plants that allow them to "mimic" the habit of domesticated ones, and thus escape removal in agronomic sites by farmers. The weed may look identical to the crop until seed dispersal, or the weed seeds may be impossible to separate from the agronomic source. Crop mimicry has been particularly prevalent among the grains (Harlan *et al.*, 1973; Harlan, 1992a). A classic example of crop mimicry can be found in sugarbeet fields in Europe, where weed introgressants bolt and scatter seeds before crop harvest (Viard *et al.*, 2002). There are weedy bolters, which probably resulted from the contamination of seed producers by pollen from wild individuals, and bolters that emerge from the seed bank containing wild/crop introgressants. The bolters carry the dominant B allele, which cancels any cold requirement. A number of additional examples are described in Chapter 7.

Wild/crop hybridizations may also have occasionally resulted in alterations of the crop itself, when farmers noticed new, useful genetic combinations. Farmer selection of crop/weed introgressants may have played a particularly important role in the early development of crops, as agriculture spread out of the centers of origin. Local adaptations would have been greatly enhanced by hybridization with native populations. It is difficult to document the historic introgression of native genes into crops, but Jarvis and Hodgkin (1999) have identified nine examples where today's farmers are selecting crop/wild introgressants. Such farmer-based selection is probably widespread even today, as a large percentage of the world's agriculture is still conducted by subsistence farmers that grow traditional crop varieties in the native range of their antecedents.

Risk of Transgene Escape into the Environment

The possibility of transgene flow from engineered crops to wild relatives has been one of the primary concerns associated with the unrestricted release of genetically modified crops (Colwell et al., 1985; Dale, 1992; Conner et al., 2003). It has been pointed out that such gene flow could result in the evolution of increased invasiveness in wild relatives (Rissler and Mellon, 1996; Snow and Palma, 1997; Hails 2000). While the early consensus was that hybridization between crops and their wild relatives occurred infrequently, more recent research has shown that crop/wild hybridizations are relatively common, as was described above (Ellstrand et al., 1999; Ellstrand, 2003; Watrud et al., 2004).

Concerns about the deployment of genetically engineered (GE) crops have now shifted to whether transgenes will persist in native environments and have negative consequences (Ellstrand, 2001; Hancock, 2003; Snow et al., 2003). It has been generally assumed that crop/weed hybrids would be poorly adapted in nature and as a result transgenes would not spread and persist; however, a few experiments have now shown that F_1 hybrids do occasionally have fitness equal or superior to the wild antecedents, and that crop alleles often persist for long periods of time in natural populations (Klinger and Ellstrand, 1994; Arriola and Ellstrand, 1996; Linder et al., 1998; Spencer and Snow, 2001).

Ultimately, the impact of transgene escape into wild populations will be strongly associated with the phenotypic effect of the gene itself and the invasiveness of its wild progenitors. Decisions on the risk of transgenic crops to native relatives should be based on three questions about risk factors: (i) Is a compatible relative present in the areas of deployment? (ii) Is the native relative highly invasive? (iii) Will the engineered trait significantly affect the invasiveness of the native relative?

It has been suggested that the escape of transgenes into native populations could have a negative impact on levels of genetic diversity in natural populations (Rissler and Mellon, 1996). However, the addition of the transgene would actually increase genetic diversity slightly and any subsequent loss in genetic diversity would occur at only those loci tightly linked to a selectively beneficial transgene. The alleles that are adjacent to such a transgene would be "dragged" along, possibly replacing any native diversity at these loci. The relative impact on native diversity would be small, as the loci tightly linked to the transgene would comprise only a small fraction of the species genome.

There has also been concern expressed that gene flow from transgenic crops could negatively impact on the landraces or indigenous crops grown by farmers (Gepts and Papa, 2003; Gepts, 2004). The degree of gene flow between GE maize and indigenous races in Mexico has been an area of active debate (Quist and Capela, 2001; Ortiz-Garcia et al., 2005; Soleri et al., 2006). The major concerns are that the native races will be displaced by GE crops or that levels of genetic diversity in the landraces will be reduced.

The challenge that landraces face from transgenic crops is the same as that presented by conventionally bred crops. Alterations in the genetic content of landraces are to a large extent under the control of the farmer, who can choose whether to plant seeds from hybrid types. Hybridization with a transgenic crop will initially raise the level of genetic diversity in a farmer's field, and is unlikely to diminish that genetic variability unless the farmer stops growing the traditional varieties altogether.

A number of studies have documented that indigenous farmers have long maintained the purity of traditional varieties grown in the same field with other traditional and conventional types (Perales *et al.*, 2003; Bellon and Berthaud, 2004; Gepts, 2004). Landraces are generally evolving entities subject to the vagaries of nature and the desires of the farmer. The farmer sometimes improves his/her landraces through introgression, by planting some hybrid seed and selecting from each generation those individuals truest to type that also carry beneficial traits. Farmers commonly share seeds and it is not unusual for them to plant mixed seed populations or different varieties side by side to generate hybrid populations. Farmers have been likely conducting such informal breeding since domestication began (Jarvis and Hodgkin, 1999; Hancock, 2003).

Summary

Many different classes of reproductive isolating barriers exist in plant species including both pre-mating (eco-geographic, temporal, floral and gametic incompatibility) and post-mating mechanisms (hybrid inviability, sterility and breakdown). Numerous different modes of speciation have been proposed based on the size of the speciating populations, how far they are separated and how much they differentiate before RIB formation. The most widely recognized type of speciation is called geographic, where the speciating populations are well separated geographically and undergo substantial differentiation before they become reproductively isolated. There are two types of geographic speciation: (i) allopatric, where the speciating population is large; and (ii) peripatric, where the speciating population is small. Other forms of speciation are called parapatric and sympatric. In parapatric speciation, adjacent populations gradually differentiate enough to become reproductively isolated, while in sympatric speciation reproductively isolated individuals arise within the borders of a population. Sympatric speciation can occur instantaneously when mutations arise that immediately isolate an individual from its progenitors. High levels of gene flow prohibit gradual sympatric speciation. In some instances, interspecific hybridization has increased the adaptive potential of both parents through backcrossing or introgression, and in other cases the hybrid population itself evolved into a new species. This phenomenon is most commonly associated with polyploidy, but has also occurred among homoploids. Crop–weed hybridizations have contributed to the adaptations of crops as they were dispersed by humans and led to the evolution of weeds that mimic the crop in such a way that their removal becomes difficult. A recent concern is that transgenes' genes will escape from engineered crops into their wild progenitors and alter their adaptive characteristics. Before the deployment of genetically modified crops, the invasive biology of a crop species and the nature of the transgene need to be carefully evaluated.

6

Origins of Agriculture

―――――――――――

Introduction

Up to this point, we have concerned ourselves primarily with evolutionary mechanisms and have placed little emphasis on the emergence of humans and their crops. We are now ready to expand our discussion to the development of land plants and people. Today, our landscape is dotted with farms, from small garden plots to huge corporate giants. Virtually everyone on earth relies on farms or farmers for their daily sustenance, and in only a few remote corners of Africa, Australia and North America do humans still rely on the ancient hunter-gatherer strategy. Even these societies are greatly endangered, tainted by their use of industrial technologies such as guns and snowmobiles, and backed into an ever diminishing corner due to deforestation. Remarkably, humans did not start farming until about 13,000 years ago, and domesticated plants and animals did not become the major source of food until the last few thousand years. In this chapter, we will outline the series of changes that led to contemporary plants and humans, and speculate on the circumstances associated with their appearance.

Rise of Our Food Crops

The angiosperms provide most of our food crops. They first appeared in the early Mesozoic or late Paleozoic Era about 200–250 million years ago (Mya), but fossil evidence of them is extremely limited until they began to dominate during the Cretaceous (136–190 Mya). The angiosperms were the first plants to have double fertilization and the enclosure of seeds into fruit. Double fertilization provided zygotes with copious resources to help them get established and fruits attracted animals for dispersal. It was the appearance of the angiosperms that set the stage for the development of our mammal ancestors.

One of the most intriguing mysteries left by an incomplete fossil record is the sudden widespread appearance of the flowering plants or angiosperms in the Cretaceous period. There is evidence of earlier origins (Sun *et al.*, 1998), but most of the record of angiosperm diversification is among Cretaceous fossils (Sporne, 1971). Dramatic variations arose in flower shape, symmetry, arrangement, part number, location and aggregation (Stebbins, 1974; Dilcher, 2000). We are forced to conclude that the evolution of flowering plants was extraordinarily rapid or that, for some unknown reason, most pre-Cretaceous fossils of angiosperms have disappeared.

Dicotyledons were the first angiosperms to appear, followed by monocotyledons. Both groups are thought by most experts to be monophyletic in nature, with the monocots being derived from primitive dicotyledons 135–75 Mya. Both dicots and monocots have undergone considerable genetic differentiation – there are currently 200,000 living species of dicots and 50,000 monocots (Simmonds, 1985).

A wide range of hypotheses implicating selection has been presented for the rapid emergence and diversification of the angiosperms (Beck, 1976). The most popular hypothesis is that the concomitant rise of pollinating insects led to powerful divergent selection as foragers and hosts developed complex relationships (Takhtajan, 1969; Proctor and Yeo, 1973; Faegri and van der Pijl, 1979; Armstrong *et al.*, 1982). Whitehouse (1950) and De Nettancourt (1977) proposed that the appearance of closed carpels and self-incompatibility restricted self-pollination without reducing cross-pollination and as a result increased genetic diversity and the potential for genetic divergence. Mulcahy (1979) took this hypothesis a step further and suggested that the pollen tube growth associated with reaching a closed carpel "enhanced the ability of natural selection to act on the gametophytic phase of the life cycle". Poorly balanced genomes would be selected in the style according to their "metabolic vigor". Other hypotheses involving selection, stress the overall competitive ability of plants containing closed carpels as a defense against predation, the value of an endosperm in seedling establishment, the improved water conducting ability of vessel elements, the defensive nature of higher plant alkaloids and the wide dispersion of fruit eaten by animals (Mulcahy, 1979).

Drift may also have played an important role in the refinement of angiosperms via the shifting balance process. As with the emergence of land plants, the population crashes and genetic reorganizations associated with invading new environments must have catalyzed the emergence and establishment of these totally unique types.

It was the angiosperms that ultimately provided us with most of our crops and their emergence pre-dated the appearance of our species, *Homo sapiens*. In fact, most of our food families or their close relatives were in existence long before we began farming. The only completely new crop type to appear after the advent of agriculture was maize, *Zea mays*, which has an ear and tassel arrangement unique from its progenitors (Chapter 8). In most cases, human beings did not influence the overall structure of crop species, only the size of their edible organs and their ease of harvest. This topic will be more fully discussed in Chapter 7.

Human beings now consume a diverse array of plant structures (Table 6.1), and at least 64 families of angiosperms and 180 genera are utilized as crops (Simmonds, 1985). This is a broad systematic group, but represents only a small fraction of the total number of angiosperm families (300) and genera (3000). The dicots provide the highest number of crop plants (Table 6.2); however, the bulk of the world is fed by a few monocotyledonous grains (maize, rice and wheat).

Table 6.1. Diversity of plant structures eaten by *Homo sapiens*.

Plant part		Example
Root		Beet, radish and carrot
Above ground stem		Sugarcane
Underground stems	Tuber	Potato, yam and cassava
	Corm	Taro
	Bulb	Onion
Leaf		Cabbage, lettuce and tea
Inflorescence		Cauliflower and broccoli
Fruit	Multiple	Pineapple, fig and breadfruit
	Aggregate	Raspberry and strawberry
	Pome	Apple and pear
	Drupe	Peach, olive, coconut and mango
	Hesperidium	Orange and lemon
	Pepo	Cucumber, watermelon and cantaloupe
	Nut	Walnut and hazelnut
	Grain	Wheat, rice, maize and barley
	Achene	Sunflower and safflower
	Legume	Bean, pea, groundnut and soybean

Table 6.2. Selected food families.

Family	Crop
Dicotyledoneae	
Anacardiaceae	*Mangifera* – mango
Camelliaceae	*Camellia* – tea
Caricaceae	*Carica* – papaya
Chenopodiaceae	*Beta* – sugarbeet
Compositae	*Carthamus* – safflower
	Helianthus – sunflower, Jerusalem artichoke
	Lactuca – lettuce
Convolvulaceae	*Ipomoea* – sweet potato
Cruciferae	*Brassica* – kale, cabbage, turnip, rape
Cucurbitaceae	*Cucumis* – cucumber, melons
	Cucurbita – squash, gourds
Euphorbiaceae	*Manihot* – cassava
Lauraceae	*Persea* – avocado
Leguminosae	*Arachis* – groundnut
	Cajanus – pigeon pea
	Cicer – chickpea
	Glycine – soybean
	Lens – lentil
	Phaseolus – beans
	Pisum – peas
	Vicia – field bean
	Vigna – cowpeas
Moraceae	*Artocarpus* – breadfruit
	Ficus – fig
Oleaceae	*Olea* – olive
	Phoenix – date palm
Pedaliaceae	*Sesamum* – sesame
Rosaceae	*Fragaria* – strawberry
	Prunus – cherry, peach, almond
	Malus – apples
	Rubus – raspberries, blackberries
Rubiaceae	*Coffee* – coffee
Rutaceae	*Citrus* – orange, lime
Solanaceae	*Capsicum* – peppers
	Lycopersicon – tomato
	Solanum – aubergine, potato
Vitaceae	*Vitus, Muscadinia* – grapes
Monocotyledoneae	
Araceae	*Colocacia* – taro
Bromeliaceae	*Ananas* – pineapple
Dioscoreaceae	*Dioscorea* – yams
Gramineae	*Avena* – oats
	Eleusine – finger millet
	Pennisetum – bulrush millet
	Hordeum – barley
	Oryza – rice
	Saccharum – sugarcane
	Secale – rye
	Sorghum – sorghum
	Triticum – wheat
	Zea – maize

Continued

Table 6.2. Continued.

Family	Crop
Liliaceae	*Allium* – onion
	Musa – bananas
Palmae	*Cocos* – coconut
	Elaeis – oil palm

Evolution of Our Genus *Homo*

The earliest fossils that are commonly attributed to our genus have been found in East Africa, eastern Ethiopia and South Africa, and represent a diverse group (Leakey and Lewin, 1992; Haviland, 2002). Louis Leakey named one of these creatures *Homo habilis*, or "handy person". Other early collections of bones have been referred to as *Homo ergaster* and *Homo rudolfensis*. Archaic *Homo* were the earliest hominoid toolmakers, producing what are called "Oldowan" tools by battering rocks together to remove a few flakes (Washburn and Moore, 1980) (Fig. 6.1). These primitive tools were used to perform a wide array of daily tasks such as cutting, chopping, scraping and perhaps defense.

The early *Homo* were hunters of small and medium-sized animals, including rodents, snakes, antelopes and pigs. They probably supplemented their diets by scavenging larger animals and gathering plant material, but there is still no evidence that they practiced agriculture. They lived together in base camps of temporary shelters, evidenced by semicircular concentrations of tools and marrow-bearing bones, and circles of stones, which may have been used to support branches. These early people were unlikely to have well developed speech, but they had clearly begun down the human path, as they made tools, shared food and returned to a base camp.

The first widely dispersed taxon of *Homo* to appear was *H. erectus*, which arose as early as 1.5 Mya (Pfeiffer, 1978; Brace *et al.*, 1979). Their brain was larger than the earlier *Homo* and they had larger teeth and jaws. In many respects, they looked like a short, rugged version of modern people. The tools of *H. erectus* consisted of a much more efficient hand axe, which was modified into cleavers, scrapers and other tools that could serve a multitude of jobs from skinning and

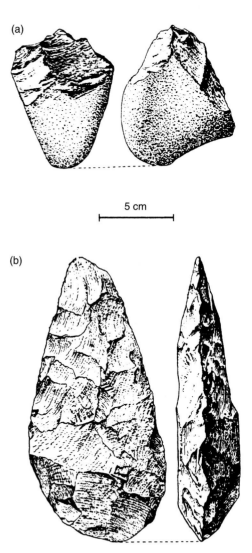

(a)

5 cm

(b)

Fig. 6.1. Examples of early tools: (a) Oldowan chopper of archaic *Homo*; (b) Acheulian hand axe of *Homo erectus*. (Used with permission from V. Grant, © 1977, *Organismic Evolution*, W.H. Freeman and Company, San Francisco, California.)

butchering wild animals to digging up wild roots (Fig. 6.1). These tools were superficially similar to the Oldowan choppers, but were much more pointed and sharp, and have been given another name, "Acheulian", after a city in France where they were originally discovered. They were produced from flint by chipping off flakes with a hammer stone (Marshack, 1976).

H. erectus was clever enough to begin using fire for warmth and cooking, and had developed systematic means of herding and slaughtering local large animals, such as elephants, rhinoceros, bears, horses, camels and deer. It seems likely that H. erectus knew well the habits of every animal they hunted and the seasonal cycles of the plants they gathered, but there is no evidence that they chose to domesticate them. It is not known whether H. erectus could speak, but their level of cooperative activity, particularly in hunting large game, suggests they must have had good communication skills.

H. erectus was the first hominoid to migrate out of Africa, and by 1 million years before present (BP) had dispersed throughout Europe, India and Southeast Asia (Fig. 6.2). Their increased body size and shifted emphasis towards meat eating may have contributed to this dramatic expansion in range. Regional game scarcities and seasonal animal migrations surely must have stimulated movement, and their success as hunters and gatherers could have resulted in large enough population densities to encourage dispersal. This movement into novel environments may have led to considerable population differentiation as our ancestors adapted to their new challenges, and may even have accelerated the development of being human.

The population structure of H. erectus, and all the later hominid species, would have encouraged rapid evolution and diversification. There were numerous, small groups of hominids scattered all over the world that were isolated from each other, except for occasional contact. Mutations would have periodically arisen in some of these hominid populations, and their small population size would have been conducive to the rapid establishment of novel adaptations.

The earliest evidence of our species, Homo sapiens, appears about 200,000 years ago in Africa (McDougal et al., 2005; Tattersall, 2009). What are considered modern forms of our species, or "Cro-Magnon", emerged in the Middle East about 50,000 years ago and by 30,000 BP fossils of these people are found all across Europe and Asia (Campbell, 1982). There has been much discussion over how Cro-Magnon could have appeared so quickly across so many places (Balter, 1998; Haviland, 2002). One theory suggests that our direct ancestors evolved simultaneously at several locations. This seems unlikely, as one would not expect such widely dispersed populations to follow such similar patterns of

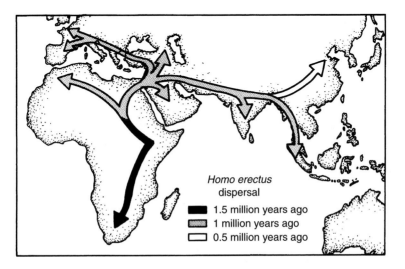

Homo erectus
dispersal

■ 1.5 million years ago
▨ 1 million years ago
□ 0.5 million years ago

Fig. 6.2. Spread of *Homo erectus* from Africa. (Used with permission from Sally Black, © 1988, Discover Publications, New York [in Shipman, 1988].)

evolution. Another theory is that Cro-Magnon originally emerged in the Middle East and, after a period of local adaptation, dispersed rapidly across the world. This seems more likely than the multiple origin idea, as it identifies a more plausible single origin and there is ample evidence that early humans could spread vast distances. As we will discuss later, the entire New World was inhabited within 2000 years of the first arrival of *H. sapiens* in Alaska.

For thousands of years, *H. sapiens* cohabited the earth with a related species, *Homo neanderthalensis*, which likely appeared about 100,000 years ago in Europe (Brace *et al.*, 1979; Lasker and Tyzzer, 1982). The so called "Neanderthal man" was first found in the Neander valley of Germany, but has subsequently been located at many points across the range of *H. erectus* (Constable, 1973). Their first discovery in the mid-1800s caused quite a stir, as it was the first evidence of a prehistoric being that resembled us, but was definitely different. Neanderthal had much larger jaws and eye sockets, and particularly strong brow ridges. It still had a low forehead like its ape ancestors, but its cranium had

reached proportions that were slightly larger than those of humans today.

There has been considerable debate about whether Cro-Magnon evolved from Neanderthal or diverged from a common ancestor (Leakey and Lewin, 1992; Cavalli-Sforza and Cavalli-Sforza, 1995; Tattersall, 1998), but the consensus appears to be that they were of separate lineages (Fig. 6.3). A remaining question about these two groups is why Neanderthal went extinct while Cro-Magnon succeeded (Gibbons, 2001a). Some have suggested that the two groups merged through interbreeding, but there is no fossil evidence of intermediate forms in the late periods of Neanderthal existence, and the cultural and physical differences between the two subspecies probably made sexual contact infrequent. Recent molecular data derived from DNA from fossils supports this contention (Serre *et al.*, 2004). It has also been suggested that Cro-Magnon displaced Neanderthal by conquest and slaughter, but, again, there is no evidence of large-scale slaughter or imprisonment of Neanderthal groups. Probably the simplest explanation for Neanderthals' disappearance

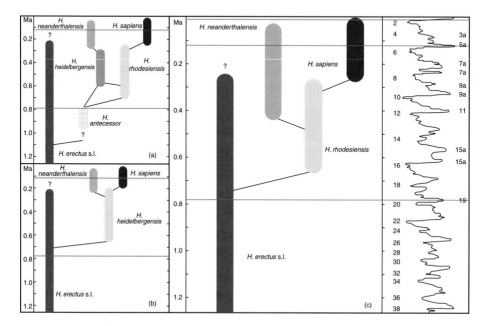

Fig. 6.3. Three of the evolutionary scenarios for the origin of *Homo sapiens* and *Homo neanderthalensis*. (Redrawn from J.J. Hublin, © 2009, The origin of Neanderthals. *The Proceedings of the National Academy of Sciences USA* 106, 16022–16027.)

may be that the two groups began to compete for the same resources, and Neanderthals were the losers. Neanderthal would not have been the first hominid line to go extinct.

The early Cro-Magnon strongly resembled modern human beings and had greatly developed tool making and art, but still no agriculture. They lived in large groups of cooperating families and the sharing of culture was very important to them (Prideaux, 1973). As more and more was learned, it was stored in the "human data bank" and each new generation was able to build on the previously collected knowledge, rather than relearning the old. Cro-Magnon were proficient big game hunters who lived in small communities. They probably had semipermanent camps where most people remained and satellite camps where small foray groups would operate. They learned to make a diverse array of multi-piece tools of stone, wood and bone (Clark, 1967). Antlers were gathered and used for a multiplicity of purposes. Among their food gathering implements were nets and snares, fishhooks made of bone, harpoons, spear-throwers, bows and arrows. They were aesthetic and spiritual, as they buried their dead with ceremony and made jewelry, musical instruments, statues and produced magnificent cave paintings. They also had sewn clothing, which ultimately allowed them to move into regions even colder than Neanderthal could tolerate.

So what led to this apparent leap forward in culture? It was not simple brain size, as Neanderthal actually had slightly larger brains than Cro-Magnon. The voice box may have developed to the point that modern language became possible, and this allowed the further development of culture. It is also possible that there were changes in brain organization that improved cognitive abilities, but left no fossil record. Since Neanderthal had already developed the rudiments of culture, it would not have taken much biological change to greatly accelerate the evolution of human society. In fact, at Châtelperronian, France there is evidence that Neanderthal was using a number of tools very similar to Cro-Magnon and may have been making ornaments of teeth and ivory beads. The debate rages as to whether Neanderthals made these advancements on their own, or were copying Cro-Magnon. Regardless, it is clear that Neanderthals' culture was at the threshold of modern humans.

Spread of *Homo sapiens*

Once they were established in Europe and the Middle East, *H. sapiens sapiens* rapidly spread over the rest of the world (Edwards and Cavalli-Sforza, 1964) (Fig. 6.4). Japan was settled by 20,000 BP and Australia more than 30,000 years ago. Humans arrived in North America by around 20,000 BP and by 12,000 BP had migrated to the tip of South America. When

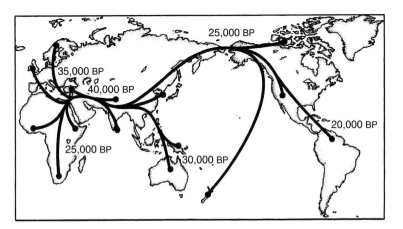

Fig. 6.4. Possible migration routes of *Homo sapiens* estimated from blood groups of existing races of people. (Used with permission from L.L. Cavalli-Sforza and W.F. Bodmer, © 1971, *The Genetics of Human Populations*, W.H. Freeman and Company, San Francisco, California.)

agriculture appeared 12,000 years ago, modern peoples were scattered over most of the world. Major mesolithic cultures had appeared in Europe, Australia, Siberia, Greece, northern Africa, the Near East, India, China, Southeast Asia, South America, Mesoamerica, and eastern and central North America.

The diffusion across the contiguous land mass stretching from Africa to Europe and Asia was probably relatively straight forward as people spread out in response to crowding and in search of food. Similarly, movement on to what are now the Indonesian islands of Borneo, Sumatra, Java and Bali was unimpeded, as the seas between them were dry when the early migrations occurred during the Ice Age (Diamond, 1998). The further migrations across open water to the other Indonesian islands would have taken the invention of watercraft, but was probably encouraged by the distant sight of land. The very long range migrations across open ocean to places like Australia and beyond, took much more faith, courage, and perhaps an extreme desire to escape crowding or warfare. The long oceanic migrations into the Pacific were begun by seafaring people from the Bismark Archipelago north of New Guinea about 3000 years ago, and continued until 1000 BP with the settling of New Zealand and the central Pacific islands. These seafarers had their origins from Southeast Asia (Gibbons, 2001b; Kayser et al., 2001).

North America was probably first reached by people from Siberia who crossed the Bering Strait on the land bridge exposed during the last great Ice Age (Marshall, 2001). The first unquestioned remains of humans in Alaska date to about 14,000 BP, but a much earlier arrival of humans is likely. The movement may have been enticed by a much milder climate than we know today. Once people arrived in the New World, they spread out rapidly, leaving fossil evidence all over North America and reaching the tip of South America by 12,000 BP, a migration rate of almost 10 miles a year.

It is also possible that some people arrived in North and South America by routes other than the Bering Strait (Dillehay, 2000; Marshall, 2001). Ancient human skulls have been found in Brazil, Central America and the Pacific Northwest that resemble those of South Asians, Polynesians and even African bushmen. If this is true, some early Americans must have arrived by

canoe from Asia and/or Europe, hugging the frozen coastlines of North America to the beyond.

As people spread across the world, they learned to use the resources available in each area and became specialized hunters and gatherers. In some parts of Europe, people were primarily big game hunters who followed a diverse array of herd species, including deer, pig, cattle, horses, elk and foxes. In other parts of Europe, people were more stationary, relying on fishing for the bulk of their calories. Along the coasts of both the New and Old World, people collected shellfish. Across broad expanses of western parts of California and northern Mexico, people became specialized gathers of plant food. Hunter gatherers in the Zagros mountains of Iraq relied on a combination of wild goats, sheep and wild cereal grasses for their sustenance. The stage was now set for the emergence of a diverse agriculture, one that was specifically tuned to the natural resources available in each region.

Agricultural Origins

Agriculture arose independently at several locations across the world, beginning about 13,000 years ago. Vavilov (1926, 1949–1950) originally identified eight centers of domestication based primarily on patterns of crop diversity. These were modified by Harlan (1967), who used a combination of archeological evidence and the native ranges of crop progenitors. He identified three relatively small geographical regions, which he called centers, and another three rather diffuse regions, which he called non-centers. He felt the centers were probably independent of each other, but the non-centers may have had some contact with adjacent ones. The three centers he envisioned were the Near East (parts of Jordan, Syria, Turkey, Iraq and Iran), Mesoamerica (Mexico and Central America) and north China. He considered Africa, Southeast Asia and South America to be non-centers. In the late 1980s, Bruce Smith added eastern North America as a center of crop origin.

Archeological data emerging over the last 10–20 years has really expanded this list of centers and non-centers (Fig. 6.5) (Doebley et al., 2006; Pickersgill, 2007; Purugganan and Fuller, 2009). Recognized primary centers of

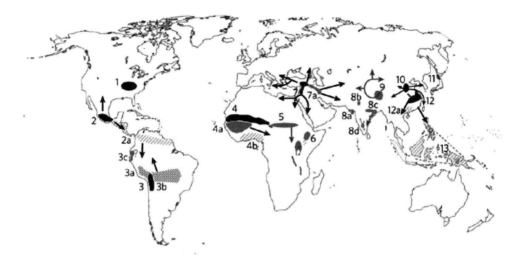

Fig. 6.5. Centers of plant domestication. Solid-shaded areas and hatched areas indicate regions of important seed-crop domestication and vegecultural crops, respectively. Accepted primary domestication centers are shown in black, and potentially important secondary domestication centers are shown in gray. Arrows indicate major trajectories of spread of agriculture and crops out of some centers. Areas are numbered as follows: 1, eastern North America; 2, Mesoamerica; 2a, northern lowland neotropics; 3, central mid-altitude Andes; 3a, north and central Andes; 3b, mid-altitude and high altitude lowland southern Amazonia; 3c, Ecuador and northwest Peru; 4, West African sub-Sahara; 4a, West African savannah and woodlands; 4b, West African rainforests; 5, east Sudanic Africa; 6, East African uplands and lowland vegeculture; 7, Near East; 7a, eastern fertile crescent; 8a, Gujarat, India; 8b, Upper Indus; 8c, Ganges; 8d, Southern India; 9, eastern Himalayas and Yunnan uplands; 10, northern China; 11, southern Hokkaido, Japan; 12, Yangtze, China; 12a, southern China; 13, New Guinea and Wallacea. (Used with permission from M.D. Purugganan and D.Q. Fuller, © 2009, The nature of selection during plant domestication. *Nature* 457, 843–848.)

origin now include: (1) eastern North America; (2) Mesoamerica; (3) central mid-latitude Andes; (4) West Africa sub-Sahara; (5) east Sudanic Africa; (6) East African uplands; (7) Near East; (8) India; (9) eastern Himalayas and Yunnan uplands; (10) northern China; (11) southern Hokkaido, Japan; (12) Yangtze, China; and (13) New Guinea and Wallacea. Secondary centers of origin, perhaps stimulated by the primary centers include: (2a) northern lowland neotropics; (3a) north and central Andes; (3b) lower southern Amazonia; (3c) Ecuador and northwest Peru; (4a) West African savannah and woodlands; (4b) West African rainforests; (7a) eastern Fertile Crescent; (8a) Gujarat, India; (8b) Upper Indus; (8c) Ganges; (8d) Southern India; and (12a) southern China.

Most evidence indicates that human beings first began farming about 13,000 BP in the hills above the Tigris River on the western edge of what is now Iran. So many ancient agricultural

sites have been located in this region that Harlan and Zohary (1966) have suggested that the Near East be thought of as a "nuclear area" where events at one location influenced others. Among the first crops domesticated in this region were emmer and einkorn wheat (*Triticum* spp.), barley (*Hordeum vulgare*), pea (*Pisum sativum*), lentil (*Lens culinaris*), chickpea (*Cicer arietinum*) and flax (*Linum usitatissimum*) (Table 6.3). People in the West African sub-Sahara were responsible for pearl millet (*Pennisetum glaucum*) about 4500 BP. East African uplands were the origin of finger millet (*Eleusine coracana*) and the east African lowlands yielded the yam (*Dioscorea cayenensis*) about 4000 BP. Cowpea (*Vigna unguiculata*) and African rice (*Oryza glaberrima*) were domesticated 2000–4000 BP in the West African savannah and woodlands. From the West African rainforests came the yam (*Dioscorea rotundata*). People in New Guinea and Wallacea were responsible for domesticating the yam (*Dioscorea esculenta*),

Table 6.3. When and where our major crops were probably first domesticated.

Place	Years before present			
	13,000 to 8,000	8,000 to 4,000	4,000 to recent	Recent
Early centers				
Africa (except Egypt)	Yam	Cowpea Finger millet Muskmelons Pearl millet Sorghum Watermelon	Coffee Oil palm	
China	Broomcorn millet Foxtail millet Rice	Hemp Peach Pear Soybean Tea	Onion	
Mesoamerica	*Lagenaria* gourds *Cucurbita* squash	Avocado Chili peppers Common beans Cotton (*Gossypium hirsutum*) Grain amaranth Maize	Agave Cocoa Tobacco Tomato	
Near East	Barley Chickpea Garden pea Lentil Wheats	Date palm Faba bean Fig Flax Olive Grape		
North America		*Cucurbita* gourds	Goosefoot Sumpweed Sunflower	Blueberry
Southeast Asia	Yam	Coconut Sugarcane Taro	Banana Citrus	Rubber
South America	Chili peppers Common bean Sweet potato	Cassava Cotton (*Gossypium barbadense*) *Cucurbita* squash Guava *Lagenaria* gourds Groundnut Potato Sweet potato Quinoa	Pineapple Quinine Strawberry	
Later centers				
Central Asia		Apple	Black mustard Pistachio	
Europe		Rye	Oats Rape	Strawberry
Indus Valley		Cotton (*Gossypium arboreum*) Cotton (*Gossypium herbaceum*)	Aubergine Black pepper Pigeon pea	

Continued

Table 6.3. Continued.

Place	Years before present			
	13,000 to 8,000	8,000 to 4,000	4,000 to recent	Recent
Mediterranean (including Egypt)		Cucumber Sesame Almond Lettuce Onion Radish	Cabbage Celery Sugarbeet Turnip Safflower	

banana (*Musa acuminata*) and taro (*Colocasia esculenta*) about 7000 BP. Broomcorn (*Panicum miliaceum*) and foxtail millet (*Setaria italica*) were domesticated in northern China about 8000 BP and rice (*Oryza sativa*) in central China about 8000 BP.

Farming probably began independently in the New World, 1000–2000 years later than in the Old World. There was a relatively compact Mesoamerican center extending from Mexico City to Honduras, while South American crops emerged in a broad area covering most of coastal and central South America. Coastal Peru is often described as a focal point of early South American agriculture, but the data are somewhat biased since most archeological work has been performed on dry coastal sites, where plant material is readily preserved. The center in eastern North America spanned the area between the Appalachian Mountains and the western borders of Missouri and Arkansas (Smith, 1989).

A diverse group of crops was originally developed in the Americas (Table 6.3). The squash (*Cucurbita pepo*) and maize (*Zea mays*) were domesticated in Mesoamerica 10,000 to 9,000 years ago. The northern lowland neotropics gave us the squash (*Cucurbita moschata*), sweet potato (*Ipomoea batatus*) and the common bean (*Phaseolus vulgaris*) about 9000 to 8000 BP. The north and central Andes (mid- and high elevation) were the domestication sites about 8000 BP of the potato (*Solanum tuberosum*) and Cañihua (*Chenopodium pallidacaule*). From the central Andes mid-elevations came quinoa (*Chenopodium quinoa*) and Inca wheat (*Amaranthus caudatus*) about 5000 BP. Lowland southern Amazonia yielded the origins of cassava (*Manihot esculenta*)

and groundnut (*Arachis hypogaea*) about 8000 BP. Ecuador and northwest Peru produced the lima bean (*Phaseolus lunatus*) about 10,000 BP. Eastern North America was the domestication site of sunflower (*Helianthus annuus*), sumpweed (*Iva annua*) and goosefoot (*Chenpodium berlandieri*) about 4000–4500 years ago.

Early Crop Dispersals

Near East grain culture spread rapidly to Europe, West Africa and the Nile Valley (Ammerman and Cavalli-Sforza, 1984; Zohary, 1986). By the 8th millennium BP, Near East crops appeared in Greece and Egypt, along the Caspian Sea and in Pakistan. Central Europe was heavily farmed less than 1000 years later, and by 5000 BP farming communities spanned from coastal Spain to England to Scandinavia. Wheat and barley reached China about 4000 BP (Ho, 1969). The Chinese literature of 2000–3000 BP mentions the "five grains": millet, glutinous millet, soybean, wheat and rice (Whittwer *et al.*, 1987).

Most of the Near East founder crops (emmer wheat, einkorn wheat, barley, lentil, pea and flax) traveled across Europe as a group (Fig. 6.6), picking up other crops along the way. Oats and flax began as weeds moving with the Near East assemblage, but were eventually exploited and became secondary crops (Harlan, 1992a). Many of the vegetables that appeared in Europe and the Mediterranean regions were probably developed through this route (Zohary, 1986).

The spread of agriculture across the Middle East and Europe could have been caused by *cultural diffusion*, where the new techniques were

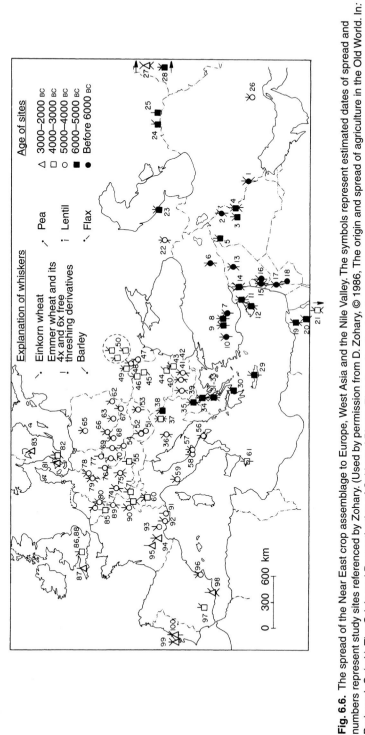

Fig. 6.6. The spread of the Near East crop assemblage to Europe, West Asia and the Nile Valley. The symbols represent estimated dates of spread and numbers represent study sites referenced by Zohary. (Used by permission from D. Zohary, © 1986, The origin and spread of agriculture in the Old World. In: Barigozzi, C. (ed.) *The Origin and Domestication of Cultivated Plants*, Elsevier, Amsterdam.)

transmitted by simple learning, or by *migration*, where the transfer was associated with population expansion and intermating (Ammerman and Cavalli-Sforza, 1984). Sokal *et al.* (1991) tested these two hypotheses by examining 26 polymorphic blood proteins of extant people from 3373 locations across the Near East and Europe. They found a clinal trend in the allelic frequencies at six loci that was significantly correlated with the local dates of agricultural settlement. This lends support to the migration hypothesis, by which the original stock of Near Eastern agricultural peoples was slowly diluted as their descendants moved west and mated with local peoples along the way.

Indigenous African agriculture was considerably more diffuse than the Near East agricultural complex, and the African crops lacked cohesion. However, the early agriculture of Africa was associated primarily with the savannah and it probably spread out from the Sakel and Guinea zones southward toward East Africa (Vavilov, 1949–1950; Harlan, 1992b). Sorghum, pearl millet (*Pennisetum americanum*), finger millet (*Eleusine corocana*) and cowpea reached India about 4000 years ago from Africa, along with cotton, sesame and pigeon pea. While these crops became very important across all of Asia, native Asian crops were of little significance in Africa except for Asian rice (*Oryza sativa*), which was utilized where African rice (*Oryza glaberrima*) was already established. Most of the dispersals out of Southeast Asia were seaward, towards the Malay Archipelago (2000 BP) and then to the far off South Pacific islands. Agriculture reached remote Hawaii and New Zealand from Southeast Asia about 1000 years ago (Emory and Sinoto, 1964).

Agriculture gradually spread across China over a period of several thousand years (Wittwer *et al.*, 1987). Rice was probably introduced into Southeast Asia from South China 4000–5000 years ago (Crawford, 1992; Smith, 1998). Soybean remained close to home until recent history. The millets, *Setaria indica* and *Panicum miliaceum*, are found in Neolithic villages in Europe, and may have been introduced from China. However, the possibility of independent domestications cannot be excluded, as the Chinese centers were so remote and no other crops show similar distributions.

Several new crops reached the Near East from Asia and Africa approximately 2000–3000 BP. Sorghum, sesame and the Old World cottons came from Africa, while common rice entered from Asia. These initiated summer crop agriculture as an integral part of food production (Zohary, 1986). Fruit trees also arrived from the east about 2000 years ago, including apricot, peach and citron (Bailey and Hough, 1975; Hesse, 1975).

From the Mesoamerican center, a maize–bean–squash assemblage gradually move northward, picking up sunflower and numerous other native species on the way to eastern North America where it was well established by 4500 BP (Chomkos and Crawford, 1978; Berry, 1985). Upon its arrival, it displaced the indigenous crops of sumpweed (*Iva annua*) and chenopod. There is an ongoing debate whether these Mesoamerican crops diffused across the West Gulf Coast Plain or through the American Southeast on the march west (Story, 1985). Movement south from Mesoamerica is difficult to trace for most crops, but at least maize had arrived in Central America (Piperno *et al.*, 2000) and the Amazonian basin by 4000–5000 BP (Pickersgill, 1969; Bush *et al.*, 1989). The South American domesticants potato, groundnut and lima bean had reached north to Mexico by 2000–3000 BP, traveling through either the Caribbean islands from Venezuela or via Central America. American crops were not known in the Old World until the ocean explorations of Columbus (Harlan, 1992a).

Transcontinental Crop Distributions

Until the last 500 years virtually all crop dispersal was within continents and not between. There was spread into Oceania from Southeastern Asia, but little movement occurred between the two hemispheres. The complete homogenization of world crops began with Columbus in 1492 in his discovery of the New World and his subsequent attempts to settle Hispaniola. Columbus introduced banana, cabbage, carrot, chickpea, citrus, cocoa, cucumber, grape, melon, olive, onion, radish, rice, sugarcane and wheat to the New World and took back sweet potato and maize.

The Spanish and Portuguese sailors finished the bulk of world crop homogenization, through their routes of exploration and trading in the 1500s and 1600s. Ultimately, their settlers tried to transfer the entire European agricultural system to the New World, with varying levels of success depending on the climate of the settled region (Butzer, 1995; Simmonds, 1995c). Into Mesoamerica, Central America and the Andean region, the Spanish introduced barley, chickpea, cucumber, fig and wheat, which were originally domesticated in the Near East, citrus, pear and peach from China, melons from Africa, and cabbage, lettuce, grapes and onions from the Mediterranean. They brought back to Europe cotton, potato, chili pepper and maize from Mesoamerica, and tomato, beans, maize and groundnut from South America. The Spanish also carried maize, sweet potato and groundnut to the Philippines, and from there they ultimately found their way to China. The Portuguese introduced a wide array of crops into Brazil including chickpea, faba bean, fig and wheat from the Near East, sugarcane and banana from Southeastern Asia, peach and citrus from China, sorghum from Africa and grapes from the Mediterranean.

They moved cassava, common bean, cotton, lima bean, groundnut, maize and sweet potato from the New World to Africa. In the late 1600s, the English and French explorers/colonists also introduced the full array of European crops to North America and brought back sunflower and the wild strawberry, *Fragaria virginiana*.

Today, the major crops grown in many parts of the world are far from their origins and, in fact, many regions are now dependent on alien crop species (Table 6.4). Sorghum, millet and yam were originally dominant in Africa, but now the most widely grown crops are maize from Mesoamerica, cassava and sweet potato from South America, and banana from Southeast Asia. Banana was an ancient introduction from Southeast Asia, but the other three were post-Columbian. Europe and North America are now almost totally dependent on crops from elsewhere, including wheat and barley from the Near East, maize from Mesoamerica, potatoes from South America and soybean from China. Rice and soybean have remained important in China, but maize, sweet potato and potato are now almost as important. At almost all locations in the world,

Table 6.4. Dependence of various regions of the world on outside crops. Dependence is based on percentage of total production (source: Kloppenburg, 1988).

Region	Dependence	Original crops	Major imports	Origin of imports	Period of introduction
Africa	87%	Millet	Banana	Southeast Asia	Ancient
		Sorghum	Cassava	South America	Post-Columbian
		Yam	Maize	Mesoamerica	Post-Columbian
			Sweet potato	South America	Post-Columbian
China	60%	Millets	Maize	Mesoamerica	Post-Columbian
		Rice	Peanut	South America	Post-Columbian
		Soybean	White potato	South America	Post-Columbian
			Sweet potato	South America	Post-Columbian
Europe	90%	Oats	Barley	Near East	Ancient
			Maize	Mesoamerica	Post-Columbian
			White potato	South America	Post-Columbian
			Wheat	Near East	Ancient
S. America	56%	Cassava	Maize	Mesoamerica	Ancient
		Sweet potato	Wheat	Near East	Post-Columbian
		White potato			
		Yam			
N. America	80%	Beans	Barley	Near East	Post-Columbian
		Maize	Wheat	Near East	Post-Columbian
		Squash	White potato	South America	Post-Columbian
			Soybean	China	Post-Columbian

people are now dependent on crops originally domesticated at distant locations.

Summary

Land plants arose from the sea as improved methods of utilizing and conserving water evolved. Some of the earliest adaptations that appeared were cuticles, roots and resistant spores. Later adaptations were conducting systems, stomata, leaves and land-based reproductive systems. The first angiosperms with double fertilization and fruit appeared hundreds of millions of years ago. The history of human evolution was intimately associated with that of plants. The appearance of our primate ancestors was dependent on the emergence of angiosperms, and most of our crop genera had evolved long before we began agriculture. All our ancestors had to do was learn how to effectively exploit the available plant foods. Human evolution went through several stages involving face and teeth structure, mode of locomotion and cranial capacity. The earliest primates lived in trees and had flexible hands and feet, but could not swing. Gradually, quadrupedal monkeys and apes appeared with the ability to brachiate and chew hard foods. *Australopithecus* evolved from apes at the edge of the forest, with the ability to occasionally stand upright and use tools. Finally our early hominid ancestors arose, with a continuous upright posture and the ability to make tools. Modern humans, *Homo sapiens sapiens*, emerged about 50,000 years ago and coexisted for thousands of years with an even earlier form of humans, *Homo sapiens neanderthalensis*. Agriculture first began in the Near East about 10,000–12,000 years ago, but at least five other areas soon followed, including China, Southeast Asia, Africa, South America and Mesoamerica. A separate origin is also postulated for the eastern portion of North America. Crops were gradually moved across continents in antiquity, but it was not until the last 500 years that inter-continental movement began.

7

The Dynamics of Plant Domestication

Introduction

The scattered appearance of agriculture all across the globe suggests that farming was an important step in the evolving culture of human beings. A question that has intrigued anthropologists and ethnobotanists alike is why it took so long for farming to emerge (Pringle, 1998). It seems likely that people had the wherewithal to farm long before they actually began doing it. Our ancestors surely gained considerable knowledge about plants and animals through the very acts of hunting and gathering. They had observed seasonal patterns of plant development and animal migrations, and noticed seeds germinating and growing on their dump heaps. Our antecedents burned fields to drive game, and must have noticed the subsequent plant regenerations. They had developed intimate knowledge of how countless plant species could be used for food and medicine, and knew how to detoxify otherwise poisonous food sources.

Probably the oldest formal idea about why humans began cultivation is Childe's "Oasis Theory" (1952). He suggested that after glaciation, North Africa and southwest Asia became drier and humans began to aggregate in areas where there was water. People first learned how to domesticate the animals that congregated around them and then, as human populations grew, they learned how to raise crops to avoid starvation.

While this theory is an appealing explanation for agriculture at xeric sites, it is now known that mesic areas in Southeastern Asia and tropical South America also spawned agriculture. In addition, the climate may not have been as harsh as Childe imagined. Evidence suggests that even the dry Zagros Mountains of the Near East may have been shifting from a cool steppe to a warmer and perhaps moister savannah when agriculture was beginning in that area (Wright, 1968).

Sauer (1952) suggested that farming first arose among fishermen in Southeastern Asia. They had a dependable food source and were sedentary, and therefore had the time and strength to experiment with new food production systems. Again, this theory works well in areas where fish and crustaceans were readily available, but it does not explain the origin of agriculture in dry places without seafood such as Mesoamerica and central Africa. In addition, there is evidence that not all fishing people took up agriculture readily, as the Natufion fishing people of the Near East were one of the last to take up agriculture in the Fertile Crescent, even though they actively gathered wild plant material (Harlan, 1992a).

Many anthropologists have related population growth with the rise of agriculture. It is thought that as populations grew, food requirements rose to the point where alternative sources were needed to supply adequate resources (Cohen, 1977). Cities are thought to have been possible because of agriculture, and the subsequent growth of populations led to specialization in jobs and the need for a farming class. While this theory holds great appeal, it is difficult to eliminate the possibility that population growth was often stimulated by the advent of agriculture rather than the reverse. There is also evidence that quite large cities could arise with only the most rudimentary developed agriculture (Balter, 1998).

Flannery (1968) and Binford (1968) combined the population pressure and sedentary hypotheses into what is called the "marginal zone hypothesis". They envisioned communities of fishermen who were initially sedentary, but as populations grew they moved out to more marginal regions. They were knowledgeable botanists who were used to gathering food in a restricted area, and developed agriculture as a means of feeding themselves. They became farmers not only because of population pressure, but because they were sedentary and in competition with the original gathering people. Again, this theory is plausible for many locations, but not all agricultural people have fishing ancestors.

There are suggestions that agriculture arose as a by-product of religious ceremony (Hahn, 1909; Anderson, 1954; Heiser, 1990). Plants providing ritualistic drugs were gathered and perhaps grown. Seeds may have been scattered on burial mounds. Animals could have been domesticated for sacrifice. While religion would have been a strong impetus for Neolithic peoples to apply what they knew about the life cycles of plants and animals, we still are left with our original question of why it took so long for people to begin the farming process. As was outlined earlier, there is considerable evidence that people were spiritual long before they began domesticating plants and animals.

The simple answer to why it took us so long to begin farming is probably that hunting and gathering was a very comfortable way of life, and humans had to have a very good reason to give it up. Juliet Clutton-Brock (1999) states that "with the abundance of food and excellent raw materials of wood, bone, flint and antler it is difficult to see what the Mesolithic people of Europe lacked". Food gathering did not need to be an intense daily activity. Richard Lee (1968), in an examination of !Kung Bushmen of the African

Kalahari desert, found that they spend an average of 2.3 days a week in food gathering. They had sufficient food to consume 2140 calories a day, which is above the USDA Recommended Daily Allowance for small vigorous people (1975/day). Jack Harlan (1992a) has shown that enough wild wheat can be collected in a few hours in the Near East to provide adequate nourishment for over a week. Even casual gardeners know that farming is hard work, and early crop production must have been just as subject to the vagaries of nature as gathering, particularly before the advent of irrigation.

Paleolithic people were complex, intelligent creatures who could readily adapt to the situations at hand. Sauer (1952) suggested, "We need not think of ancestral man as living in vagrant bands, endlessly and unhappily drifting about. Rather, they were as sedentary as they could be and set up housekeeping at one spot for as long as they might." They liked hunting and gathering, and were pushed towards farming only by a variety of regionally specific forces, including population growth, climatic change, overhunting, religion, or a simple desire for more of something in short supply, be it food, spice, oil, ceremonial color or fiber. Food production is only one of the possible reasons for bringing plants under cultivation. Harlen (1992a) has referred to this possibility as the "No-model model", where the reasons for farming are as diverse as the people and environments found at the focal points of agriculture. Farming began gradually in a number of diffuse areas as something desirable became scarce, and only began in earnest when natural plant and animal populations were not sufficient to feed the minions of people. Already sedentary, fishing peoples may have made this transition more easily than nomadic ones, but there is no reason to assume that all human populations could not make the transition when given good reasons. We waited so long to farm because we could.

Evolution of Farming

The shift from the hunter-gatherer strategy to farming probably occurred in stages (Fig. 7.1). For millions of years, our ancestors subsisted on the bounty provided by our natural environment.

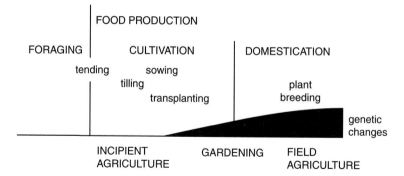

Fig. 7.1. Development of food production methods over time. (Used with permission from R.I. Ford, © 1985, *Prehistoric Food Production in North America*, Museum of Anthropology, University of Michigan, Ann Arbor, Michigan.)

Our earliest upright ancestors may not have had a particularly orderly approach to finding food, but by *H. erectus*, hominoids were surely *collectors* who planned the use of resources whose locations were known and monitored. By the time Cro-Magnon appeared, *H. sapiens* must have had considerable savvy about how plants and animals developed, and were returning to the same areas year after year to harvest and hunt dependable sources. This probably led to significant changes in the plant populations, a topic we will return to later.

Once Cro-Magnon were returning regularly to the same spot, it may not have taken them long to become *cultivators*, who enhanced the productivity of native fields by weeding, pruning and burning. They probably began tilling with a digging stick or hoe to reduce competition and encourage germination. They may also have discovered at an early stage that crops did better the following year if the soil was turned after harvest. This may have been particularly apparent in sunflower, which produces allelochemicals that inhibit germination, but readily detoxify if exposed to air (Wilson and Rice, 1968).

Eventually, Cro-Magnon became *producers*, who transplanted small numbers of plants and held a few animals captive. These early gardens were probably very small and in close proximity to residences, and remained small until humans decided to make a major commitment to agriculture. Edgar Anderson (1954) has suggested that the original idea for planting may have come from waste dumps, where seeds were observed to germinate and grow. Larger farms may have first appeared when a specific farmer

class emerged. During the early stages of the domestication process, *H. sapiens* may not have been consciously selecting superior plant types, but it would not have taken us long to become domesticators who saved seeds and clonal material of superior types for replanting.

Early Stages of Plant Domestication

Evidence for plant domestication comes from a variety of sources including: (i) carbonized remains formed by high temperature "baking" under low oxygen; (ii) impressions pressed into pottery and bricks; (iii) parched plant remains produced under extreme dry conditions; (iv) plant material sunk in peat bogs or mud under anaerobic conditions; (v) impregnations of metal oxides; (vi) mineralization where cell cavities are replaced by minerals; and (vi) fecal remains (Smith, 1998). Domestication is thought to be signified by substantial increases in the size of seeds (Fig. 7.2), dramatic reductions in seed or fruit coat thickness (Fig. 7.3) and the apparent loss of dispersal organs. This evidence is often supplemented by human artifacts that give clues about diet, such as sickles and grinding wheels.

Soil from ancient settlements is sometimes passed through screens to obtain small objects, but this technique often allows valuable material such as seeds to be lost (Smith, 1998). Flotation techniques are now widely employed where the archeological soils are poured into water, so that organic materials will float to the top and can be

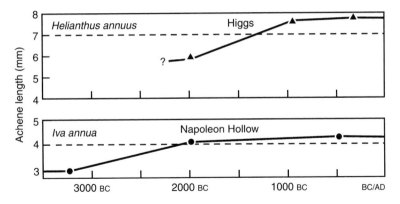

Fig. 7.2. Evidence for domestication of indigenous crops in eastern North America based on increases in achene size. The dashed line represents the baseline used for domestication. (Used with permission from B.D. Smith, © 1989, Origins of agriculture in eastern North America, *Science* 246, 1566–1571.)

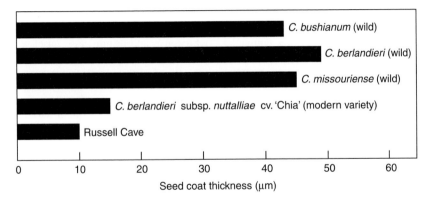

Fig. 7.3. Seed-coat thickness in wild and domesticated species of *Chenopodium*. The seed-coats of domesticated *C. berlandieri* subsp. *nuttalliae* from the Russell Cave were much thinner than the wild forms (data from Smith, 1998).

recovered. Cloth filters are utilized to catch the materials as the water is poured off.

In some cases, phytoliths and starch grains can be used to distinguish domesticated from wild forms of plant species, and even plant organs (Piperno, 2006). Phytoliths are stone cells composed of silica that are found in the cells of many kinds of plants. They remain well-preserved through millennia, long after plants die and decay. Starch grains are commonly recovered from grinding stones and stone tools, while phytoliths are present not only in the grinding stones that yield starch grains but also in sediments associated with artifacts and column samples spanning a site's occupation (Piperno *et al.*, 2009).

Dating of remains is done by analyzing the ratio of the isotopes ^{12}C and ^{14}C. After incorporation into organic material, ^{12}C remains stable while the radioactive isotope ^{14}C gradually disintegrates, with a half-life of 5568 years. Most estimates are based on the assumption that ancient atmospheres were similar to those of today, but corrections have been made by evaluating carbon isotope ratios in the annual rings of trees. Ratios are available for the last 8000 years from bristlecone pines in California and for the last 9000 years from oaks in Ireland.

At most of the early agricultural sites, the transition from hunter-gather to farmer was a gradual one that took thousands of years. A very early record of this slow transition is found in the excavations of Richard MacNeish and his group (MacNeish *et al.*, 1967) in the Tehuacán Valley of Mexico (Fig. 7.4). He excavated 12 sites and uncovered 12,000 years of agricultural history in the area. Initially, the people lived on wild plant food and small animals like jack rabbits, deer, peccary and lizards. They collected plant foods on a scheduled round of annual activities. About 9000 years ago, game became more scarce and the people began to shift more of their energy into the collection of wild plants, including squashes, chili peppers and avocados. They scattered out in small foraging groups during the dry season, and came together during seasons of plenty. They may have begun the sporadic cultivation of wild plants during this period, but the effort was minimal.

Over the next 5000 years, the people of the Tehuacán Valley gradually increased their use of domesticated plants, such that by 7000 BP, about 10% of their diet came from cultivated plants. They were outside the original areas of domestication, but by this time were growing a large group of presumably introduced crops including maize, amaranth, beans, squashes and chilies. The maize ears were only about the size of a pencil eraser, but the plant now existed in its modern form. The dog appeared about 5000 years ago. As time went on, the people continued to devote more and more effort into farming, and by 3000 BP, the majority of their food came from domesticated sources, with maize being grown, along with avocados, amaranth, squash and cotton. Turkeys were domesticated about 2000 years ago.

Similar evidence of transitions from hunters to farmers can be found at numerous locations across the Near East. One such site is Jericho in the Jordan Valley, where a continuous record of 9000 years of habitation was left as people built new mud huts on top of others as they deteriorated over time. In the earliest period, the settlement consisted of Natufians, who were primarily hunters of gazelles and foxes and tended a few cereals, but had no domestic animals (Hopf, 1969). About 9000 years ago, they

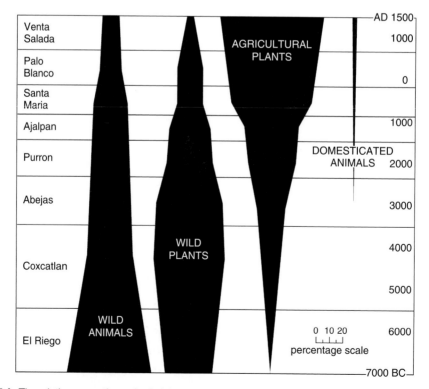

Fig. 7.4. The relative proportions of subsistence components in the diets of Tehuacán Valley inhabitants over time (MacNeish *et al.*, 1967).

began to raise cereals in earnest, and there is the first evidence that sheep and goats were being domesticated. A similar long-term record of successive settlement is recorded at Catalhöyük, Turkey, where people were initially foragers who raised a few cereals on the side, but by 10,000 BP had domesticated cattle and were large scale farmers (Balter, 1998).

Origins of Crops

Almost all of our major crops had been domesticated by 5000 BP, with the earliest grains and legumes appearing in the Near East over 10,000 years ago. The first crops were as diverse as the people and places where agriculture began. Different native plant assemblages were located in each of the early farming areas, and as a result, variant plant and animal species were domesticated in each of them. In the Middle East, there were huge natural stands of wheat and barley, and as a result the early farmers in this region exploited these as their staple crops. In Southeast Asia, wheat and barley were absent, but large-grained rice was plentiful, and as a result, rice became one of the crops of choice. Wheat, barley and rice were not present in Mesoamerica and Africa, so people exploited the locally abundant monocots: sorghum in Africa and maize in Mesoamerica. No large-grained species of any kind existed in South America, and as a result the early farmers there domesticated the tuberous species potato, sweet potato and cassava, and the pseudograins chenopodium and amaranth.

Starchy staples were among the first domesticants at all the centers of crop origins, but they were always complemented with a high protein vegetable and fiber crop. Vegetables in the legume family were domesticated in all the major regions including, cowpea (Africa), soybean (China), groundnut (South America), lentil and chickpea (Near East). Amaranth and chenopod were also very important sources of vegetable protein in the New World. Fiber was provided by different cotton species in Africa and South America, flax in the Near East and hemp in China.

To the core group of crops in each region, additional leafy vegetables, spices, oil crops and fruits were gradually added. Among the last groups of plants to be domesticated were the fruits. While the grape and fig are very ancient and may have been cultivated for 10,000 years, most of our other woody fruit crops were among the last additions to farming. This may be due, in part, to tree fruits taking so long to mature, since after planting the farmer must wait 5–10 years for a harvest. In addition, the fruit crops are outcrossed species, whose seedlings would frequently be inferior to the mother plant due to cross-pollination. Complex pruning and grafting techniques had to be developed to fully exploit their potential.

Characteristics of Early Domesticants

Among the earliest crops, three were grains in the *Gramineae* (barley, rice and wheat), two were fleshy tubers in *Convolvulaceae* (sweet potato) and *Dioscoreaceae* (yams), four were seeds in the *Leguminoseae* (chickpea, garden pea, lentil and common bean) and two were fruits in the *Solanaceae* (chili peppers) and *Cucurbitaceae* (squashes and gourds). Some species were domesticated more than once (dry beans and chili peppers), while others had relatively restricted origins (maize). Some were domesticated for more than one plant part (beets, mustard and squash).

This early group of domesticants represents a broad array of families, but most of them are herbaceous annuals, capable of selfing. It is not hard to speculate on why so many of these early crop species are annual and self-pollinating. Annuals allocate a high proportion of their energy to reproduction, and as a result produce unusually large seeds, fruits or tubers that are readily harvested. As Sauer suggests in his book *Historical Geography of Crop Plants* (1993), "the annual habit offered farmers a syndrome of advantages – quick results after planting, heavy seed production in a sudden burst for a single harvest, escape from unfavorable cold or dry seasons by storage of seed indoors allowing spread to diverse climatic regions, and periods of fallow or rotation preventing buildup of pests and parasites". Self-pollinating species breed true-to-type and any seed collected will be similar to the parent. Also, they do not need a pollinizer for successful reproduction, increasing the dependability of their reproduction and assuring seed production even if only a few genotypes were selected by early farmers. They also would be at least partially isolated from their wild progenitors, which would aid in the maintenance of their integrity. In most cases, the annual, selfing habit of individual crops was derived directly from native progenitors; however, the cultivated, annual cottons evolved from wild perennial shrubs and the self-pollinating soybean was converted from a wild outcrosser to a cultivated inbreeder.

The only major crop species that are not annual selfers are: (i) maize, pearl millet and quinoa, which are annual but outcrossed; (ii) cassava and sugarcane, which are both outcrossed and perennial; and (iii) a number of woody fruit crops that are also both outcrossed and perennial. Maize, pearl millet and quinoa were domesticated in areas where there were few large-grained grasses available, and so primitive farmers had no other option but to work with outcrossed grains. Pearl millet is not large seeded, but does produce a large inflorescence. Cassava and sugarcane, while perennial-outcrossers, are relatively easy to propagate by cuttings and as a result were easily domesticated. Woody perennials were very difficult to propagate, but their fleshy fruits offered great rewards to the people who figured out how to proliferate them.

Both Harlan (1992a) and Diamond (1998) have made the observation that most of the large-seeded grain and tuber crops were domesticated in the savannah or Mediterranean climates of the Middle East, China, Mesoamerica and South America. These climates have long annual periods of drought, which would have favored plants with large annual seeds or tubers. Seeds can survive long periods of drought and germinate when rains come, and likewise tubers can remain quiescent until moisture levels encourage sprouting. The few grains that were domesticated in eastern North America were not drought adapted, but large seeds are advantageous in areas with periods of extreme cold. Here seeds remain dormant during the winter, and then germinate when temperatures warm up in the spring.

Changes During the Domestication Process

Once humans began sowing seed, they began to alter plant species through both conscious and unconscious selection. The process of domestication is commonly described as occurring in three stages: (i) hunting and gathering; (ii) cultivation of wild plants (pre-domestication cultivation); and (iii) fixation of domestication syndromes (emergence of full agriculture) (Allaby, 2010). Simmonds has made the point in his book *Principles of Crop Improvement* (1985) that "probably the total genetic change achieved by farmers over the last 9,000 years was far greater than that of scientific breeders in the last 100 years".

At the beginning of the domestication process, a number of changes began to appear in the genetic and physiological makeup of many crop species (Table 7.1). Some of these changes were due to conscious selection, such as increases in palatability and color, but many of the others were the unconscious by-product of planting and harvesting. Harlan *et al.* (1973) recognized a whole syndrome of traits associated with inadvertent selection due to the broadcasting and harvesting of grain crops (Table 7.2). Domestications of other broadcast seed crops such as legumes produced similar patterns (Zohary, 1989). Harvesting resulted in the selection of the non-shattering trait, more determinate growth, more uniform ripening and increased seed production. All of these characteristics would have increased the likelihood that the seed of a genotype would be collected and subsequently planted. Other characteristics

Table 7.1. Traits commonly associated with the domestication process.

Increased reproductive effort
Larger seeds and fruit
More even and rapid germination
More uniform ripening period
Non-dehiscent fruits and seeds
Self-pollination
Trend to annuality
Increased palatability
Color changes
Loss of defensive structures
Increased local adaptations

Table 7.2. Adaptation syndromes resulting from automatic selection due to planting and harvesting seed of cereals. (Adapted from J.R. Harlan, J.M.J. DeWet and E.G. Price, © 1973, Comparative evolution of cereals, *Evolution* 27, 311–325.)

Selection pressure	Response	Adaptation
Harvesting	Increase in percent seed recovered	Non-shattering
		More determinant growth
	Increase in seed production	Increase in seed set
		Reduced or sterile flowers become fertile
		Increase in inflorescence size
		Increase in inflorescence number
Seedling competition	Increased seedling vigor	Increase in seed size
		Reduction in protein content of seeds and an increase in carbohydrate
	More rapid germination	Loss or reduction in inhibitors of germination
		Reduction in glumes and other appendages

associated with enhanced harvests would be: selection towards erect types with synchronous tillering, and an increase in the size of inflorescences, number of fertile florets per inflorescence and the number of inflorescences.

Seedling competition caused by planting in close proximity, probably increased seedling vigor and rate of germination, as individuals with these characteristics would have been most likely to win the race to reproductive age. Under natural conditions, plants would have evolved a delay in germination from seed maturation until favorable conditions the next growing season. Also, prolonged seed dormancy would guarantee that at least a few seeds would be available to germinate after prolonged stretches of suboptimal weather. Once seed were harvested and stored away from the natural environment, these adaptations were no longer necessary. Greater seed size likely contributed to seedling vigor, and a loss of germination inhibitors would have allowed faster germination. Thin seed coats also evolved under domestication, as a reduced seed coat is more permeable to water and results in more rapid germination.

Of course, not all crop domestications followed the same evolutionary path as the broadcast grains and legumes. In those crops harvested for some other plant part than seeds, increases in seed size would have been much more modest than in the legumes and grains, as humans selected genotypes partitioning more energy into larger leaves or tubers. In the seeded crops like lettuce and radish, non-dehiscence

and high fertility would have remained important, but among the vegetatively propagated crops, the ability to sexually reproduce would have been far less critical. In fact, many vegetatively propagated lines of taro, banana and potato cannot sexually reproduce at all. These genotypes presumably devote more energy into their harvested parts than sexually reproducing ones and, in the case of banana, there may even have been human selection against flinty, tooth-breaking seeds.

Sometimes the primary reason for domestication changed as humans began to consciously improve a crop (Anderson, 1954). Squashes and pumpkins started out with small fruits and bitter flesh. They may have first been used as rattles in ceremonies and dances, as dishes and as storage vessels. Only much later were they used as food, first for their seeds and then for their flesh. Other crops that came to have multiple uses include flax (oil and fiber), hemp (oil, fiber and stimulation) and chenopodium (seeds and leafy vegetables). One of the most dramatic examples of a species used in multiple ways is *Brassica oleracea*, whose flowers came to be used as cauliflower and broccoli, its leaves became kale, cabbage and Brussels sprouts and its fleshy corms became kohlrabi (Chapter 11).

As humans began to save grain seeds and plant the same field every year, local adaptations would have gradually increased over time as the highest proportion of seeds were gathered from the most vigorous individuals. An experimental documentation of such change was briefly described in Chapter 2, where Clegg *et al.* (1972) studied evolutionary change in a synthetic population of 28 worldwide barley varieties. The mixture of seeds was initially sown in a large plot in 1929 and was allowed to reproduce by natural crossing without any artificial selection, and a random sample of seeds was collected and sown annually. Over the ensuing decades, they documented dramatic changes in gene frequency resulting in higher and more stable grain yields with more compact, heavier spikes with larger numbers of seeds (Allard, 1988).

Genetic Regulation of Domestication Syndromes

Many of the traits associated with plant domestication are regulated by only a few genes, making rapid evolutionary change possible (Table 7.3). For example, seed non-shattering is controlled by only one or two genes in a broad range of grains and legumes. A single recessive mutant transforms two-rowed barley into six-rowed ones. Determinate versus indeterminate growth in maize and bean is regulated by one or two genes, as is branching versus not branching in sunflower and sesame.

Even in cases where traits are thought to be regulated by a large number of genes or quantitative trait loci (QTL), it is not unusual to find that a few major genes influence a large amount of the genetic variability (Knight, 1948; Hilu, 1983; Gottlieb, 1984). Koinange *et al.* (1996) found major genes associated with mode of seed dispersal, seed dormancy, growth habit,

gigantism, earliness, photoperiod sensitivity and harvest index in dry beans (Table 7.4). John Doebley and his laboratory (Doebley *et al.*, 1990) have isolated three key QTL that regulate glume toughness, sex expression and the number and length of internodes in both lateral branches and inflorescences. One of these, *teosinte glume architecture 1*, probably played a particularly important role in the appearance of maize 5000 years ago, as it disrupts reproductive development in such a way that the kernels are naked, rather than encased in tough glumes (Doebley *et al.*, 1995; Dorweiler and Doebley, 1997; Lukens and Doebley, 1999).

Major QTL for domestication traits have also been described in pearl millet (Poncet *et al.*, 1998), sorghum (Paterson *et al.*, 1998) (Table 7.5), tomato (Grandillo and Tanksley, 1996, 1999), pea (Weeden, 2007), sunflower (Burke *et al.*, 2002) and rice (Xiong *et al.*, 1999; Bres-Patry *et al.*, 2001). Li *et al.* (2005) found QTL with large phenotypic effects for a reduction in seed shattering and seed dormancy and the synchronization of seed maturation in wild × domesticated rice. A cluster of genes on chromosome 7 and several other locations resulted in substantial improvements in plant architecture and panicle structure. Xiao *et al.* (1998) located two *Oryza rufipogan* alleles on two chromosomes that were associated with almost 20% increases each on grain yield per plant. The locations of QTL associated with domestication are sometimes conserved across related species (Doganlar *et al.*, 2002), but not always (Paran and van der Knaap, 2007; Weeden, 2007).

Not only are there commonly major genes for the individual traits associated with plant

Table 7.3. Genetics and number of loci governing seed non-shattering in cultivated crops. (Reprinted with permission from G. Ladizinsky, © 1985, *Economic Botany* 39, 191–199, the New York Botanical Garden.)

Crop	Number of loci	Genetics of domesticated form
Rice – *Oryza sativa*	1	Recessive
Oat – *Avena sativa*		
Spikelet non-shedding	1	Dominant
Floral non-disjunction	1	Recessive
Barley – *Hordeum vulgare*	2	Recessive at both loci
Sorghum – *Sorghum bicolor*	2	Recessive at both loci
Pearl millet – *Pennisetum glaucum*	2	Recessive at both loci
Lentil – *Lens culinaris*	1	Recessive
Wheat – *Triticum monococcum*	2	Recessive complementary

Table 7.4. Genetic factors influencing the domestication syndrome in common bean. (Used with permission from E.M. Koinange, S.P. Singh and P. Gepts, © 1996, *Crop Science* 36, 1037–1045.)

General attribute	Trait	Gene or marker	Magnitude of effect on trait determined by regression
Seed dispersal	Pod suture fibers	*St*	Single gene
	Pod wall fibers	*St* (?)[a]	Single gene
Seed dormancy	Germination	*PvPR-2-1*	18%
		D1132	52%
		D1009-2	19%
		D1066	12%
Growth habit	Determinacy	*fin*	Single gene
	Twining	*Tor* or *fin*	Single gene
	Number of nodes	*fin*	53%
		D1492-3	20%
		D1468-3	16%
	Number of pods	*fin*	32%
		D1468-3	21%
		D1009-2	14%
	Internode length	D1032	19%
Gigantism	Pod length	D1520	23%
		PvPR-2-1	20%
		*LegH*16	16%
	100-seed weight	D1492-3	18%
		Phs	27%
		Uri-2	16%
		D0252	15%
Earliness	Days to flowering	*fin*	38%
		Ppd (?)[a]	19%
		D1468-3	12%
	Days to maturity	*fin*	30%
		Ppd (?)[a]	18%
		D1468-3	14%
Photoperiod sensitivity	Delay under 16 h days	*Ppd*	44%
		D1479	44%
		Ppd (?)[a]	28%
		D1468-3	28%
Seed pigmentation	Presence vs absence	*P*	Single gene

[a]Pleiotropy and tight linkage could not be clearly distinguished for genes with question marks.

Table 7.5. Action of QTL for traits related to domestication in *Sorghum propinquum*. (Used by permission from A.H. Paterson, K.F. Schertz, Y. Lin and Z. Li, © 1998, *Case History in Plant Domestication: Sorghum, an Example of Cereal Evolution*, CRC Press, London.)

Trait	Number of QTL	Mode of gene action			
		Dominant	Additive	Recessive	Overdominance
Shattering	1	1	0	0	0
Height	6	4	1	0	1
Flowering date	3	2	0	1	0
Seed size	9	2	1	6	0
Tiller number	4	1	2	1	0
Rhizomatousness	8	2	4	1	1
Overall	31	12	8	9	2

domestication, but in many cases these genes have pleiotropic effects, where they affect a number of traits simultaneously. As a result, their selection would make change across the whole domestication syndrome much more rapid than if the traits were evolving separately. In maize, Doebley has discovered two QTL that coordinately regulate plant and inflorescence architecture and have much stronger effects together than separately (Doebley et al., 1995). Allard (1988) identified a number of marker loci for quantitative traits that had significant additive effects on several to many quantitative traits in barley. Koinange et al. (1996) found several cases where the large effect of individual genes in bean was further magnified by their having pleiotropic effects. For example, the gene fin influenced determinacy, node number, pod number days to flowering and days to maturity (Table 7.4). A major gene Q has been shown to regulate several traits associated with wheat domestication, including the tendency of the spike to shatter, how tightly the chaff surround-

ing the grain is held and whether the spike is compact or elongated (Simons et al., 2006). Grandillo and Tanksley (1996) also discovered several regions of the tomato genome that had effects on more than one trait.

Many of the QTL associated with the domestication syndromes are clustered close together on the same chromosome. Such close associations of genes would reduce the amount of segregation between these adaptively important genes and in a sense "fix" the crop type, again allowing for rapid change. In many crops like wheat and oats, self-pollination would help maintain these linkages, once the genes had reached homozygosity. Koinange et al. (1996) found the distribution of domestication syndrome genes to be concentrated in three genomic regions, one of which greatly affected growth habit and phenology, the other seed dispersal and dormancy, and a third the size of fruit and seed (Fig. 7.5). Doebley et al. (1990) found five of the QTL that distinguish maize and teosinte in a tight cluster on chromosome 8. These genes regulated:

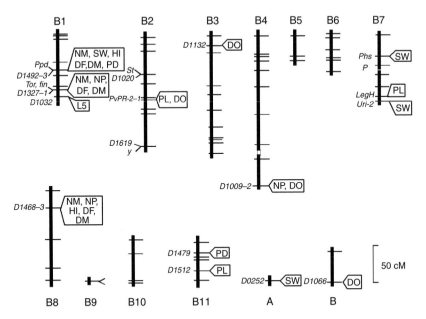

Fig. 7.5. Linkage map location of known genes and marker loci controlling the domestication syndrome in common bean. Symbols for the genes: *fin*, determinacy; *P*, anthocyanin pigmentation; *Ppd*, photoperiod induced delay in flowering; *St*, pod string; *y*, yellow pod color. Symbols for the marker loci: DF, days to flowering; DM, days to maturity; DO, seed dormancy; HI, harvest index; L5, length of the fifth internode; NM, number of nodes on the main stem; PL, pod length; NP, number of pods per plant; PD, photoperiod induced delay in flowering; SW, seed weight. (Used with permission from E.M. Koinange, S.P. Singh and P. Gepts, © 1996, *Crop Science* 36, 1037–1045.)

(i) the tendency of the ear to shatter; (ii) the percentage of male spikelets in the primary inflorescence; (iii) the average length of internodes on the primary lateral branch; (iv) the percentage of cupules lacking the pedicellate spikelet; and (v) the number of cupules in a single rank. Many of the genetic factors controlling domestication-related traits were also concentrated on a few chromosomal blocks in pearl millet (Poncet et al., 1998), tomato (Grandillo and Tanksley, 1996) and rice (Xiong et al., 1999).

Strong selection pressures acting on domestication loci have been shown to influence levels of sequence diversity, sometimes distant from the gene itself. These "selective sweeps" have been discovered by sequencing the DNA in proximity to the domestication genes and searching for reduced levels of nucleotide variation. Around the rice Waxy gene that played an important role in the emergence of non-glutinous japonica rice varieties, there is a selective sweep of greater than 250 kb that affects genetic diversity across at least 39 genes (Olsen et al., 2006). On chromosome 10 in maize, which contains several large-effect QTLs for local adaptation and domestication traits, there is a selective sweep of 1.1 Mb nucleotides encompassing > 15 genes (Tian et al., 2009). Around the maize Y1 gene that regulates endosperm color is a 600 kb selective sweep that carries ten genes (Palaisa et al., 2004).

Strong selection pressure on single domestication traits, in conjunction with genetic hitch-hiking, can in some instances "drag" other genes with apparently maladaptive phenotypic effects. In evaluating the genetic consequences of selection during the domestication of sunflower, Burke et al. (2002) found that numerous QTL with antagonistic effects were common in the cultivar parent that produced more wild-type phenotypes. For example, one of the ten QTL positively influencing achene size "w[as] embedded within a cluster of other apparently maladaptive QTL" (Burke et al., 2005). It turned out that a major, oil-related QTL cluster mapped in the linkage group, indicating that strong directional selection for that trait by breeders in concert with genetic hitchhiking had carried along more minor, negative QTL, as part of a selective sweep.

Comparisons of sequence diversity throughout genomes in wild versus cultivated germplasm have indicated that numerous genes have been the subject of selection since domestication.

By screening maize microsatellites for their levels of genetic diversity, Vigouroux et al. (2002) found about 5% of the genome was targeted by selection. Wright et al. (2005) examined 774 maize genes and found about 2–4% were under selection. Using these values, Tian et al. (2009) estimate that if the maize genome contains 59,000 genes, then a minimum of 1200 have been targets of selection during maize domestication.

The Causative Changes in Domestication Genes

Many of the domestication-related traits are conditioned by recessive, loss-of-function alleles, but certainly not all (Doebley et al., 2006; Burger et al., 2008). The kinds of sequence variability leading to domestication traits vary widely. In a review of the differences in 26 cloned genes that control domestication traits or varietal differences, Doebley et al. (2006) found eight regulatory changes, two regulatory plus amino acids changes, one regulatory plus an early stop codon, three early stop codons, three otherwise disrupted coding sequences, two transposon insertions, one intron splicing defect and six amino acid changes. About half of the mutations were in regulatory genes and half in structural ones. Among the domestication genes mentioned in the above discussions, fw2.2 in tomato is caused by a regulatory change, tga1 in maize by an amino acid change, waxy in rice by an intron splicing defect, hd1 in rice by a disrupted coding sequence and Q by a regulatory and amino acid change.

Rate of Domestication

Until recently, it was thought that once humans began cultivation, selection for the domestication genes would be rapid (Allaby, 2010). For example, Hillman and Davis (1990) suggested that farmers could have fixed the tough rachis trait in cereals in perhaps 20–30 years if they harvested with sickles or uprooted partially or near-ripe crops. The frequency of genotypes with a tough rachis would steadily increase each year since the farmer would recover a higher proportion of the seeds of the mutant genotype each harvest, assuming that most of the wild-type grains that fell to the ground were predated.

However, new archeological evidence indicates that pre-domestication cultivation may have actually taken thousands of years. Fuller (2007) provides evidence that grain size increases evolved over 500–1000 years and the fixation of non-shattering inflorescences took even longer. The fixation of non-shattering rachises of wheat, barley and rice appears to have taken 1000–2000 years (Tanno and Willcox, 2006a; Fuller et al., 2007, 2009; Willcox et al., 2008). The non-shattering phenotypes of oat and rye did not predominate at archeological sites for thousands of years (Weiss et al., 2006).

There are a number of reasons why full domestication may have taken so long (Allaby, 2010). As Hillman and Davis (1990) suggested, the rate of increase of the mutant phenotype would be greatly diminished if immature plants were harvested by beating the ripe spikelets into baskets, sickle-reaping unripe crops, uprooting unripe crops or hand-plucking or stripping fruit. This would also be the case if seed were selected from the ground (Kislev et al., 2004). With all these harvesting methods, there would be little selection for the tough rachis. The model of Hillman and Davis also assumes that most of the seeds that fell to the ground were predated. If a large proportion of seed escaped predation, it would dilute the frequency of the mutant type when the new planting was sown. The rate of change would be further slowed if fewer seeds were planted in some years than others because poor harvests meant less seed could be spared for sowing.

Gene flow from native genotypes could have also severely diluted the proportion of domestication alleles in the seed gathered each year. It has been suggested by some researchers that fixation of the domestication traits may actually have not occurred until the crop was sufficiently isolated from wild strands or planted outside its native range (Allaby et al., 2008; Jones and Liu, 2009).

Evolution of Weeds

In many cases, species that started out as weeds of crops were eventually domesticated themselves. Sauer (1952) states "The ancestors of most New World seed plants appear to have been attractive weeds. They were not tenacious intruders that the cultivator had difficulty getting rid of, nor are they such as grow on trodden ground. They were gentle, well-behaved weeds that liked the sunshine, loose earth and plant food of tilled species, and had no great root systems. Such volunteers were first tolerated, then protected, and finally planted." He mentioned that the cherry tomato in Mexico and Central America is not planted, but protected. Madia sativa and Bromus mango were initially weeds of root crops in Chile, but now are minor seed crops. In the northern periphery of root crops, amaranth, chenopod, squashes and beans may have started out as weeds, but "where climatic advantage shifted from the root plant to seed plant, the attention of the cultivator shifted from the former to the latter. Instead of selecting root variants to meet the local situation, he began to select the attractive weeds". Rye and oats are thought to have begun as weeds of the Near Eastern assemblage that were domesticated in more northern climates where their potential was more apparent. Hexaploid wheat may also have begun its existence as a weed in cultivated tetraploid fields, before its full benefit was noted.

As farming progressed, humans also began inadvertently to select for weedy crop mimics through tilling, weeding and harvesting (Harlan et al., 1973). As we discussed in Chapter 5, many of these weedy races arose after introgression with the crop type, but many non-crop species have also developed weedy races. Two major forms of crop mimics are generally recognized, vegetative and seed: vegetative means the weed is similar looking to crop seedlings and their vegetative stage; and seed means weed seeds have similar density and appearance to those of crops, making it difficult to separate them before planting.

Numerous examples of crop mimicry have been reported. Several species of wild rice and barnyard grass have evolved developmental and growth patterns that make them very difficult to distinguish from cultivated types (Barrett, 1983). In the case of barnyard grass, the crop mimic Echinochloa crus-galli var. oryzicola is actually more similar to rice in many attributes than to its own progenitors. Weedy races of grain chenopods, teosinte, amaranths, pearl millet and sorghum invade agricultural fields and look almost identical to their related crop species until their inflorescences shatter just before harvest (Sauer, 1967; Harlan *et al.*, 1973; Wilson and Heiser, 1979). Considerable differentiation is often observed in weedy races depending on the types of cultivars grown in a region and the natural diversity present. This is particularly apparent in weedy rice, where there are numerous different indica- and japonica-mimicking races, both where wild species are present and absent (LingHwa and Morishima, 1997; Xiong *et al.*, 1999). One of the most unusual adaptations has been described in maize fields in Mazatlan, Mexico, where fields are cropped one year and then fallowed the next. Here teosinte populations have arisen with an inhibitor that prevents germination for 1 year and therefore protects the plants from grazing in the fallow years (Wilkes, 1977).

Numerous examples of seed mimicry have also been described. One of the earliest cases involved races of *Camelina sativa*, whose seeds were so similar to flax that they could not be separated by winnowing, where heavy seeds are removed from chaff and lighter seeds by wind (Stebbins, 1950). Seeds of *Vicia sativa* are normally a different shape and size than lentils, but in central Europe, the seeds of the two species are very similar and as a result *V. sativa* can be a very serious weed (Rowlands, 1959). In many cases, vegetative and crop mimicry are combined in weeds, making it almost impossible to identify the invaders. The *Camelina* species have the same growth habit, branching pattern, flowering time and fruit characteristics as flax.

With the advent of herbicides, a third class of mimics has arisen, herbicide mimicry. Resistance to S-triazine herbicides, atrazine and simazine, has been found in several different weed species, including *Brassica campestris* and *Chenopodium alba* (Souza Machado *et al.*, 1977; Warwick and Black, 1980). The distribution is highly localized in most cases, but in *Senecio vulgaris*, a wide range of susceptibility was found to simazine among the fruit farms in England. A population's susceptibility was correlated with the number of years of continuous herbicide use, strongly implicating selection (Holliday and Putwain, 1980); however, continuous selection with herbicides has not always been necessary for resistance to emerge. Friesen *et al.* (2000) found native populations of *Avena fatua* in Manitoba, Canada that were resistant to imazamethabenz,

even though this herbicide had not been previously applied. Numerous factors influence the rate at which herbicide resistance evolves, including rates of genetic mutation, initial frequency of resistance genes, type of inheritance, mating system and gene flow (Jasieniuk and Maxwell, 1994; Mortensen *et al.*, 2000).

Genetic Diversity and Domestication

An important ramification of domestication was a reduction in levels of genetic variability in both plants and animals. Virtually all domesticants are substantially less diverse than their progenitors due to the bottleneck associated with selecting a few elite types and directional selection. However, several factors had an important influence on the amount of diversity captured during domestication, including geographical isolation, sexual structure and levels of diversity in the original species population.

The way crops were planted and their mode of reproduction had a substantial influence on the amount of diversity maintained in them. Those crops that were outcrossed and planted individually would have remained much more diverse than the broadcast, selfed seed crops. Ancient races of maize and chili peppers show astonishing levels of variability across South America (Fig. 7.6), in large part because they were outcrossed and planted individually in hills, so that individual diversity could be recognized and exploited. In the asexually propagated crops like potatoes, similar high levels of diversity have also been maintained, since specific genotypes can be selected for propagation without the necessity of gathering segregating populations of seeds.

Hybridizations between sexually propagated crops and their wild progenitors would have occasionally increased their variability and improved local adaptations, particularly as the crops diffused from their point of origin. As we discussed in Chapter 5, crop × wild progenitor hybridizations played an important early role in the evolution of numerous crops. Some of the most dramatic examples have been maize and kidney beans at a local level, and sorghum and apples across large geographical areas

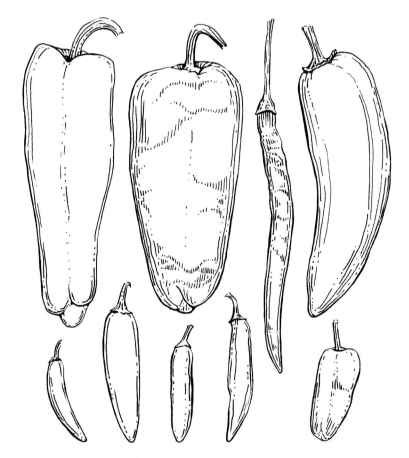

Fig. 7.6. Range of variation found in South American chili peppers.

(see later chapters on these crops). In some cases, hybrids may even have been directly accepted as new cultigens; possible examples being the grain amaranths, mangos and the allopolyploid bread wheat. In the selfed grains and legumes, crop × wild hybridizations would have been restricted, but even the strongest selfers occasionally outcross (Allard and Kahler, 1971).

Domestication and Native Diversity Patterns

In his landmark work on cultivated plants, N.I. Vavilov (1926) used the centers of diversity of native crop species to predict where they were initially domesticated. This system works well for a high percentage of crops, but a number such as wheat, sorghum, pearl millet and beans do not have a true center of diversity, and others such as barley and rice were domesticated far from their centers of diversity (Harlan, 1992). To adjust for this problem, Vavilov developed the concept of secondary centers to describe those cases in which centers of diversity and origin were not the same. This approach helps with several crops including wheat, barley and rice, but there are still a number of crop species that appear to have been domesticated completely outside their native ranges. These *transdomestications* might have resulted from long-range ocean drift (bottle gourd outside of Africa, sweet potato in Polynesia), dispersal by migratory birds (perhaps tomato in Mexico, arabica coffee in Arabia), human trade of wild material (cotton from Africa to India, perhaps *pepo* gourds in eastern North America), or original movement as a weed (rye and oats).

Table 7.6. Little-known plants domesticated by the Incas in South America. (Source: NRC, 1989.)

Type of crop	Common name	Species	Distinctive properties
Roots and tubers	Achira	*Canna edulis*	Staple with unusually large starch grains
	Ahipa	*Pachyrhizus ahipa*	Crisp like apples, addition to salads
	Arracacha	*Arracacia xanthorrhiza*	Has flavors of celery, cabbage and chestnut
	Maca	*Lepidium meyenii*	Sweet tangy flavor; can be stored for years
	Mashua	*Tropaeolum tuberosum*	Starchy staple; very easy to grow
	Mauka	*Mirabilis expansa*	"Cassava" of the highlands
	Oca	*Oxalis tuberosa*	Second most important staple in the highlands (to potatoes)
	Potatoes	Many other than *Solanum tuberosum*	Most important staple; huge diversity
	Ulluco	*Ullucus tuberosus*	Very brightly colored; staple in some areas
	Yacon	*Polymnia sonchifolia*	Sweet and juicy, but almost calorie free
Grains	Kaniwa	*Chenopodium pallidicaule*	Very nutritious and extremely hardy
	Kiwicha	*Amaranthus caudatus*	Protein quality is equivalent to milk
	Quinoa	*Chenopodium quinoa*	Excellent source of protein; extremely hardy
Vegetables	Basul	*Erythrina edulis*	Tree with edible seeds; staple in some areas
	Nuñas	*Phaseolus vulgaris*	Dropped into hot oil and popped
	Tarwi	*Lupinus mutabilis*	Extremely rich in protein and oil
	Peppers	Many *Capsicums*	Huge range in pungency and taste
	Squashes	Many *Cucurbita*	Unusually robust and productive
Fruits	Mora de Castilla	*Rubus glaucus*	Superior in flavor and size to other raspberries
	Ugni	*Myrtus ugni*	Spritely flavor; blueberry relative
	Capuli cherry	*Prunus capuli*	A large, sweet black cherry
	Cherimoya	*Annona cherimola*	Has flavors of papaya, pineapple and banana
	Goldenberry	*Physalis peruviana*	Yellow fruits are great in jam; very hardy
	Highland papayas	*Carica* species	Unusually cold adapted
	Lucuma	*Pouteria lucuma*	Used as both a staple crop and a fresh fruit; a tree can feed a family
	Naranjilla (Lulo)	*Solanum quitoense*	Like tomato; has particularly refreshing juice
	Pacay	*Inga* sp.	Long pods filled with soft white pulp (called ice-cream beans)
	Passionfruits	*Passiflora* sp.	High quality and huge variability
	Pepino	*Solanum muricatum*	Tastes like sweet melon
	Tamarillo	*Cyphomandra betacea*	A tree with a tomato-like fruit
Nuts	Quito palm	*Parajubaea cocoides*	Nut tastes like tiny coconuts

Harlan (1976) has classified domestication patterns into five classes: (i) *endemic* – the domesticant occupies a well-defined, small geographic region (guinea millet); (ii) *semi-endemic* – the domesticant occupies a small range with some dispersal out of it (African rice and tef); (iii) *monocentric* – the domesticant has a wide distribution with a discernible center of origin (the later plantation crops such as coffee, rubber, cacao and oil palm); (iv) *oligocentric* – domesticant has a wide distribution and two or more centers of diversity (our major food crops such as wheat, barley, pea, lentil, chickpea, flax, maize and lima bean); and (v) *non-centric* – domesticant has a wide distribution, but no discernible centers of diversity (American beans, radish, sorghum, pearl millet, cole crops and bottle gourd).

While people in developed countries are familiar with only a few dozen centric and non-centric crops, in reality hundreds of endemic and semi-endemic crops were domesticated. In his book *Crops and Man*, Harlan (1992a) provides what he calls a "short list" of world crops that encompass 11 pages of text. Most of these crops are unknown outside of their region of origin. In the publication *Lost Crops of the Incas* (National Research Council, 1989), over 30 species are described that were domesticated by the Andean peoples, but are little grown outside

of South America (Table 7.6); many of these are restricted to specific elevational gradients (Fig. 7.7). Ethiopia provided the world with coffee, but also a unique assemblage of endemic crops including the cereal tef, the oil crop noog, the mild narcotic chat, and enset, a relative of banana whose stem base rather than fruit is eaten (Harlan, 1992b). Thousands of plant species have been utilized by somebody somewhere, but very few of them have attracted widespread attention.

Summary

The transition from hunter-gatherer to farmer was a gradual one that took thousands of years. The earliest grains and legumes were not domesticated until about 10,000 years ago, and it took another 5000 years for the rest of our major crops to emerge. The earliest group of domesticants were all herbaceous annual species and most were selfing, which provided quick harvests after planting and true-to-type seed. Once plants were domesticated, they were dramatically altered by humans through conscious and unconscious selection. The simple act of harvesting resulted in the selection of non-shattering

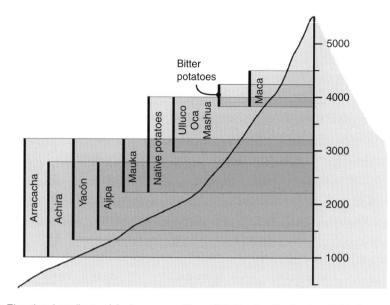

Fig. 7.7. Elevational gradients of Andean crops. (From CIP Circular, September 1994, Centro Internacional de Papas, Lima, Peru.)

traits, more determinant growth, more uniform ripening and higher seed production. The seedling competition that occurred after seed were scattered probably increased seedling vigor and rate of germination. Many of the traits associated with plant domestication are regulated by only a few genes, making rapid evolutionary change possible. Domestication often resulted in reduced levels of variability, but hybridization between crops and their wild progenitors occasionally increased their local adaptations and expanded their geographic range. Transfers of genes between wild and domesticated populations also led to the evolution of weeds that mimic the crop in such a way that their removal becomes difficult.

8

Cereal Grains

Introduction

The cereal grains represent a diverse group of species in the *Gramineae* (grasses) (Fig. 8.1). They are grown on all continents of the world except Antarctica and are the primary food source of most of the world. In fact, over two-thirds of our major food crops are cereals.

The quality of the information bearing on the evolutionary history of the grain species varies greatly. Some crops such as barley have relatively clear pasts due to good historical records and the presence of living ancestors. Others, such as rice and sorghum, are ambiguous due to spotty archeological remains and a confusing array of existing taxa. Still others like maize have missing or unclear progenitor species, making conjectures about their origins problematic. In this chapter, we review the available information on the evolutionary origin of each major grain species and tell as complete a story as possible.

Barley

Barley, *Hordeum vulgare* L., was certainly one of the earliest domesticants and it has a very simple evolutionary history (Zohary and Hopf, 2000). It was derived from what was traditionally considered a separate species, *Hordeum spontaneum* C. Koch, which is widely distributed in the initial area of cultivation (Fig. 8.2). The two taxa are so closely related that most investigators now consider them to be in the same species, *H. vulgare*, with the cultivated forms belonging to subsp. *vulgare* and the wild forms to subsp. *spontaneum*. The genes separating wild and cultivated forms are easily transferred, and natural hybrids can be found with combinations of wild and domesticated characteristics (Harlan, 1976; von Bothmer *et al.*, 1990). Some of the brittle, six-rowed hybrids have even been given their own name, *Hordeum agriocrithon* Åberg (Zohary, 1973; von Bothmer *et al.*, 1990).

Only a few recessive alleles separate cultivated and wild barley (Nilen, 1971; Salamini *et al.*, 2002): (i) the wild forms have spikes that shatter easily at maturity, while the cultigens have a non-brittle spike that results from a mutation in either of two tightly linked "brittle" genes, Bt_1 or Bt_2; (ii) the wild forms have tight glumes that are difficult to thresh, while the cultivated forms have naked grains that result from a recessive gene (n) that allows the husks to be easily removed by threshing; and (iii) the domesticated forms have six- rather than two-row glumes due to recessive mutations in the *Vrs1* locus.

Much evidence points towards a Near Eastern origin for barley cultivation. A number of molecular markers have been used to document this, including AFLPs (Badr *et al.*, 2000), chloroplast DNA (Clegg *et al.*, 1984; Neale *et al.*, 1988) and gene sequences (Kilian *et al.*, 2006; Morrel and Clegg, 2007). Strong arguments

Fig. 8.1. A comparison (left to right) of the inflorescences of rice, oats, barley, wheat and rye. The prominence of awns varies greatly in wheat and barley.

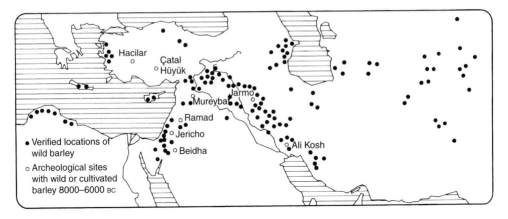

Fig. 8.2. Early archeological sites of *Hordeum vulgare* subsp. *vulgare* and the distribution of its wild progenitor *H. vulgare* subsp. *spontaneum*. (Used with permission from J.R. Harlan and D. Zohary, © 1966, *Science* 153, 1074–1080, American Association for the Advancement of Science.)

have also been made for additional origins on the eastern edge of the Iranian Plateau (Morrell and Clegg, 2007; Jones *et al.*, 2008) and the Horn of Africa (Orabi *et al.*, 2007). Saisho and Purugganan (2007) provide evidence that the barley that originated in the Fertile Crescent was moved west into Europe and North Africa, and that which originated at the eastern edge of the Iranian Plateau was moved towards Asia.

People first began gathering wild, two-rowed forms with easily shattered spikes about 19,000 years ago in the Near East and by ~10,500 years before present (BP) were cultivating non-brittle types. Barley was probably used first as a staple, but it did not take people very long before they were fermenting it into beer. Naked forms of two-row barley were being grown by about 8500 BP and by

6000–7000 BP six-rowed barley was dominant across the region (Zohary and Hopf, 2000). At least three independent mutations occurred in the wild-type *Vrs1* gene that resulted in a six-row spike (Komatsuda *et al.*, 2007). One of these is found worldwide, while the others are more restricted to East Asia and the western Mediterranean.

Barley was a common companion to einkorn and emmer wheat in the earliest farming communities, but its importance as a staple crop gradually diminished over the next 5000 years. It came to be considered a poor man's grain, being grown on marginal sites that did not adequately support wheat. Today, barley is grown all across temperate climates and is used primarily to make beer and feed livestock.

Maize

The history of maize has been difficult to trace as its fruiting cob and monoecious nature are very unique in the *Gramineae*. In other grasses the sexual parts are in close proximity rather than isolated at different locations in the plant, and the grains are protected individually by glumes rather than being naked.

The closest relatives of cultivated maize (*Zea mays* L. subsp. *mays*) are a few species of *Tripsacum* and a number of wild *Zea* subspecies (teosintes) (Table 8.1), which were for a long time considered in a separate genus, *Euchlaena*. The teosintes have a habit that resembles maize, but their tassels are not on a central spike like maize and their ears are much less complex (Figs 8.3 and 8.4). The *Tripsacum* species have pistillate and staminate flowers that are borne separately, but unlike maize, they are adjacent on the spike. Their seeds are embedded in segments of the rachis that scatter when ripe.

All the native *Zea* species have very restricted ranges in Mexico and Central America, and they carry the same chromosome number, $2n = 20$, except for *Zea perennis*, which is tetraploid. The *Tripsacum* species all have multiples of $x = 18$. Diploid teosinte crosses readily with maize, and reciprocal introgression may occasionally occur today in Mexico and Guatemala, where teosinte grows adjacent to cultivated maize, although evidence of modern gene flow is minimal (Doebley *et al.*, 1984; Doebley, 1990). *Tripsacum* can be crossed with maize (Eubanks, 2001a,b) and a natural hybrid species has been identified (Talbert *et al.*, 1990), but the sterility barriers

Table 8.1. *Zea* taxonomy according to Doebley and Iltis (1980) and Iltis and Doebley (1980).

Species and subspecies	Ploidy	Life history	Location and elevation
Z. diploperennis Iltis, Doebley & Guzman	Diploid	Perennial	Sierra de Manantlán in state of Jalisco; 1400–2400 m
Z. perennis (Hitchcock) Reeves & Mangelsdorf	Tetraploid	Perennial	Northern slopes of Volcán de Colima in state of Jalisco; 1500–2000 m
Z. luxurians (Durieu & Ascherson) Bird	Diploid	Annual	Southwest Guatemala, Honduras and Nicaragua; sea level to 1100 m
Z. mays L. subsp. *huehuetenangensis* (Iltis & Doebley) Doebley	Diploid	Annual	Guatemala; 900–1650 m
Z. mays L. subsp. *mexicana* (Schrader) Iltis	Diploid	Annual	Central and northern Mexico; 1700–2600 m
Z. mays L. subsp. *parviglumis* Iltis & Doebley	Diploid	Annual	Western escarpment of Mexico from Nayarit to Oaxaca; 400–1800 m
Z. mays L. subsp. *mays*	Diploid	Annual	Cultivated worldwide

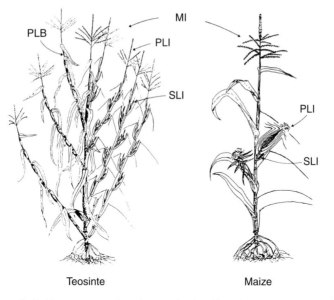

Teosinte   Maize

Fig. 8.3. Differences in fruiting structure of teosinte and maize. Note that tassels and ears are on the same fruiting stalk of teosinte, while on maize they are segregated on different spikes. MI, main inflorescence; PLI, primary lateral inflorescence; SLI, secondary lateral inflorescence; PLB, primary lateral branch. (Used with permission from J. Doebley *et al.*, © 1990, *Proceedings of the National Academy of Sciences* 87, 9888–9892.)

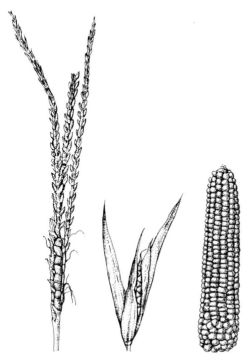

Fig. 8.4. Grain-bearing inflorescences of maize and its relatives. Left to right: *Tripsicum dactyloides, Zea mexicana* and *Zea mays.*

are sufficient to greatly limit natural introgression (Newell and DeWet, 1973; James, 1979; deWet *et al.*, 1984a).

The high chromosome number of maize suggests that it might be an ancient polyploid formed from two diploids with $2n = 10$. Several independent lines of evidence support this conjecture (Molina and Naranjo, 1987): (i) haploid maize shows many chromosome associations (Ting, 1985); (ii) normal taxa display secondary associations of bivalents (Vijendra Das, 1970); (iii) chromosomes of *Zea mays* form four subsets of five chromosomes in somatic metaphase cells rather than two sets of ten (Bennett, 1984); (iv) maize carries a high number of isozyme and RFLP duplications (Stuber and Goodman, 1983; Helentjaris *et al.*, 1988); (v) DNA sequence data have shown numerous duplications (Gaut and Doebley, 1997); and (vi) distant relatives *Coix* and *Sorghum* have a haploid number of five.

The lack of fossil evidence of a prototype maize plant has led to three major hypotheses concerning the origin of maize: (i) maize, teosinte and *Tripsacum* were separate lineages that evolved from a common, unknown ancestor (Weatherwax, 1954); (ii) an interspecific hybridization between two or more native grasses produced maize (Mangelsdorf and Reeves, 1939; Eubanks, 2001a,b); and (iii) maize evolved directly from teosinte (Beadle, 1939; Galinat, 1973; Dorweiler and Doebley, 1997; Iltis, 2000).

In a complex scenario, Mangelsdorf and Reeves (1939) initially suggested that modern maize arose through a series of interspecific crosses, and in fact was the progenitor of teosinte. In their "tripartite hypothesis" they envisioned that now extinct races of pod corn were introduced into Mexico and central America and subsequently hybridized with *Tripsacum* to form teosinte. This new teosinte then hybridized with maize to produce superior races. Mangelsdorf (1974) later altered this hypothesis and considered teosinte to be a mutant derivative of maize.

In support of Mangelsdorf's hypotheses, pod corn has been found among the fossils of ancient communities that lived in New Mexico 3000–4000 BP and even older ears have been located in the Tehuacán Valley, Mexico, that appear to have traits of both popcorn and pod corn. Mangelsdorf (1958) also crossed modern races of popcorn and pod corn and obtained a hybrid that had a combination of grass and maize characteristics. However, there is no hint of where the pod corn came from.

Eubanks (2001a,b) has suggested that maize was derived from a hybridization between *Tripsacum dactyloides* and *Zea diploperennis*. Support for this hypothesis has come from her generating recombinant progeny that have maize-like flowering spikes, and RFLP data showing that modern maize appears to carry a combination of fragments from a limited sample of native *Tripsacum* and *Zea*.

The most overwhelming support has been garnered for the teosinte hypothesis through molecular and isozyme studies (Bennetzen *et al.*, 2001; Smith, 2001). No intermediate forms between maize and teosinte have been found in the archeological record and there is little evidence that humans ever cultivated teosinte (Galinat, 1973; Mangelsdorf, 1974, 1986), but a punctuated change could have occurred in the ear or tassel that led to a dramatically different crop (Iltis, 1983, 2000). Longley (1941) showed that chromosome morphology of maize and teosinte was very similar.

In an extensive analysis of electrophoretic variation in native populations, one variety of Z. *mays* subsp. *parviglumis* was found to have a high genetic identity of 0.92 with maize, and the two were tightly grouped when the data were subjected to a principal component analysis (Doebley *et al.*, 1987). This suggests that they are directly related and are part of the same lineage. The close similarity between maize and Z. *mays* subsp. *parviglumis* has also been documented using cDNA restriction fragments (Doebley, 1990) and ribosomal ITS sequences (Buckler and Holtsford, 1996). Most recently, Matsuoka *et al.* (2002) were able to trace the origin of maize using SSR markers to a single domestication of subsp. *parviglumis* in the Rio Balsas region of southern Mexico about 9000 years ago.

Using starch grain and phytolith evidence, Piperno and co-workers found direct evidence of maize cultivation in the Central Balsas River Valley in Mexico at 8700 BP (Piperno *et al.*, 2009; Ranere *et al.*, 2009). They were also able to determine that the cobs rather than stems were being processed by the people in the region.

Even though maize and teosinte are separated by numerous polygenic traits, the punctuated change in maize morphology probably was facilitated by there only being a few major loci involved in the evolutionary change. Iltis (1983) suggested that the emergence of maize may have been due primarily to a feminization of the tassel; however, John Doebley's laboratory has found several key genes that separate teosinte from maize through regulation of glume toughness, naked kernels, sex expression and the number and length of internodes in both lateral branches and inflorescences of maize (reviewed in Chapter 7).

Maize became an integral part of a Mesoamerican crop assemblage that included beans and squash (Smith, 2001). From its early cultivation in southwestern Mexico before 8000 BP, maize spread to the Mexican Gulf coast and Panama by about 7000 BP (Dickau *et al.*, 2007; Pohl *et al.*, 2007). It is likely that subsequent hybridization with teosinte played a role in the early development of maize, as the modern crop appears to contain as much as 77% of the landraces' diversity (Eyre-Walker *et al.*, 1998; Tenaillon *et al.*, 2001). Maize arrived in eastern North America through the southwestern states about 2000 years ago (Smith, 1998). It appeared in South America by 6000 BP (Bush *et al.*, 1989) and was introduced from there into Florida via the Caribbean islands.

By the time the Europeans arrived in the Americas in the 1500s, maize was an important staple across a vast area from Argentina to Canada. Indians from all over North and South America had developed countless varieties of maize, many of which still exist in Mesoamerica (Doebley *et al.*, 1985; Bretting and Goodman, 1989). The primary forms that were developed were: (i) popcorn, which has extremely hard seeds that explode when heated; (ii) flint corns, which are composed of hard starch; (iii) flour corns, which have soft starch that can be ground into flour; (iv) dent corns, which have soft starch at the top of the kernel and hard starch below; and (v) sweetcorn, which is eaten as a sugary vegetable (Fig. 8.5). Within all these groups there exists tremendous variation for kernel color, ear size, maturation dates, and overall plant habit. The efforts of various primitive peoples represent a remarkable example of the changes possible under domestication. Heiser (1990) suggests that at least part of this diversity was produced by seeds being planted individually instead of being broadcast like the other grain species. This practice made people more aware of individual variation.

Millets

The millets represent a number of different genera in the grass tribe *Paniceae* (Fig. 8.6). *Panicum miliaceum* L. (proso or broomcorn millet), *Setaria italica* (L.) P. Beauv. (Italian or foxtail millet) and *Echinochloa utilis* Ohwi & Yabuno (Japanese barnyard millet) are known collectively as small millets and represent several million hectares across the world, particularly in India. Other important millets are *Pennisetum americanum* (L.) Schum. (bullrush, pearl or cat-tail millet), which is one of the principal grain crops in the driest areas of tropical Africa and India, and *Eleusine coracana* (finger millet), which is important in the highlands of Ethiopia and Uganda.

The evolutionary history of the small millets has not been extensively studied, but

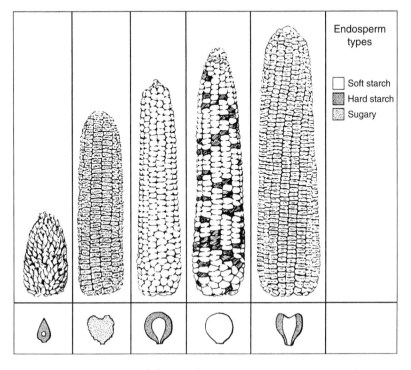

Fig. 8.5. Major types of modern maize (left to right): popcorn, sweetcorn, flint maize, flour maize and dent maize. Endosperm types are depicted below each ear. (Redrawn from H.G. Baker, © 1970, *Plants and Civilization*, Wadsworth Publishing Company, Belmont, California.)

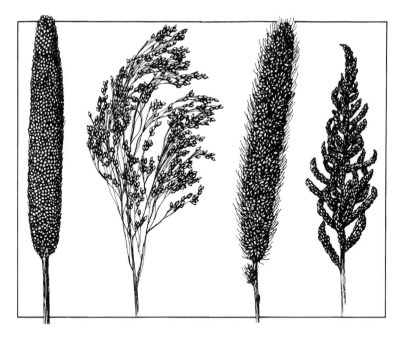

Fig. 8.6. Major types of millet (left to right): pearl, broomcorn, foxtail and barnyard.

a few conjectures can be made about their origins. Japanese barnyard millet is a hexaploid $(2n = 6x = 56)$ and is very similar to the widespread weed *Echinochloa crus-galli* (L.) Beauv., which is probably its progenitor (Yabuno, 1966; Hilu, 1994).

The crop *P. miliaceum* $(2n = 4x = 36)$ was selected from a widespread native species of the same name. It was domesticated in the Yellow River Valley of north China by about 10,000 BP, making it one of the earliest crop species (Lu *et al.*, 2009). Evidence of the domestication of *P. milaceum* has also been found in Europe from 8000 to 7500 BP (Hunt *et al.*, 2008), making it likely that common millet had at least two independent domestications.

S. italica $(2n = 2x = 18)$ probably arose from the common diploid weed *Setaria viridis* (L.) Beauv., which grows all over Eurasia (deWet *et al.*, 1979; Wang *et al.*, 1995; Fukunaga and Kato, 2003). The two are so similar that they can be considered subspecies of *S. viridis*, the cultivated forms being subsp. *italica* and the wild forms being subsp. *viridis* (Rao *et al.*, 1987). The earliest evidence of foxtail domestication is from about 8000 BP in north China, following the emergence of common millet by about

2000 years (Lu *et al.*, 2009; Bettinger *et al.*, 2010). Two additional domestication centers for foxtail millet have been suggested including Europe and an area ranging from Afghanistan to Lebanon (Li *et al.*, 1995; Benabdelmouna *et al.*, 2001; Fukunaga *et al.*, 2006). The evidence for foxtail millet domestication in Europe is about 1000 years later than in China (Hunt *et al.*, 2008).

Pearl millet $(2n = 2x = 14)$ belongs to a highly heterozygous group that contains both wild and cultivated forms. Stapf and Hubbard (1934) originally recognized 15 species of cultivated millets, but these taxa are now generally considered as one species because of high inter-fertility (Brunken *et al.*, 1977). There are some minor reproductive barriers that have arisen between cultivated and wild pearl millets during domestication (Amoukou and Marchais, 1993). Cultivation probably began in tropical West Africa by 3500 BP (D'Andrea *et al.*, 2001; Klee *et al.*, 2004; Oumar *et al.*, 2008) with the emergence of persistent spikelets and protruding grains instead of deciduous spikelets and enclosed grains. Pearl millet was being grown in India at about the same time (Fuller *et al.*, 2004).

Finger millet ($2n = 4x = 36$) probably arose from *Eleusine indica*, a native species that has both diploid (subsp. *indica*) and allotetraploid (subsp. *africana*) races. The diploid is located throughout the tropics and subtropics, while the tetraploid is largely confined to eastern and southern Africa. Genomic *in situ* hybridization (GISH) has identified diploid *E. indica* and *Eleusine floccifolia* as the "A" and "B" genome donors of *E. indica* subsp. *africana* (Bisht and Mukai, 2001).

There are two general types of finger millet cultivars: (i) an African highland race that has a strong *E. indica* subsp. *africana* influence; and (ii) an Afro-Asian type that has strong *E. indica* subsp. *indica* influence. Most authors feel that the cultivars originally arose in Uganda from subsp. *africana* and then diverged through introgression with native populations to produce the modern cultivated races (Kennedy-O'Byrne, 1957; Mehra, 1963a,b; Purseglove, 1972). The origin of the crop from subsp. *africana* has been documented using a diverse array of molecular markers (Hilu and Johnson, 1992; Salimath *et al.*, 1995). Finger millet was probably domesticated about 5000 BP in eastern Africa and arrived in India like the other millets by 3000 BP, where it underwent substantial differentiation (Hilu *et al.*, 1979; deWet *et al.*, 1984b; Fuller *et al.*, 2004).

Oat

Oats present a confusing array of morphological and cytogenetic groups. Baum (1977, 1985a,b) divided the genus into 29 species using morphological patterns, while Ladizinsky and Zohary (1971) recognized only seven species based on the inter-fertility of the various taxa. The genus is composed of three ploidy groups: diploids ($2n = 2x = 14$), tetraploids ($4n = 4x = 28$) and hexaploids ($2n = 6x = 42$) (Table 8.2). Species with similar genomes tend to cluster together in molecular analyses (Drossou *et al.*, 2004; Fu

Table 8.2. *Avena* species and their genomes. (Based on Baum, 1977; Loskutov, 2008.)

Species	Chromosome number ($2n$)	Genome
A. canariensis Baum.	14	AcAc
A. domascena Raj. et Baum		AdAd
A. hirtula Lag.		ApAp
A. longiglumis Dur.		AlAl
A. atlantica Baum		AsAs
A. brevis Roth		AsAs
A. hispanica Ard		AsAs
A. lustanica Baum		AsAs
A. strigosa Schreb.		AsAs
A. weistii Steud.		AsAs
A. macrostahya Balansa ex Coss et Durieu		CmCm
A. clauda Dur.		CpCp
A. eriantha Dur		CpCp
A. ventricosa Balan.		CvCv
A. abyssinica Hochst.	28	AABB
A. agadiriana Baum et Fed.		AABB ?
A. barbata Pott.		AABB
A. valviloviana (Malz.) Mordv.		AABB
A. insularis Ladiz.		AACC ?
A. maroccana Baum.		AACC
A. murphyi Ladiz.		AACC
A. fatua L.	42	AACCDD
A. occidentalis Dur.		AACCDD
A. sativa L.		AACCDD
A. sterilis L.		AACCDD

and Williams, 2008; Li *et al.*, 2009). Almost all the species are located in the Mediterranean basin and the Middle East. The cultivated species are represented by the widely grown hexaploid *Avena sativa* and the tetraploid *Avena abyssinica*, which is farmed regionally in Ethiopia.

There are four karyotypes (A, B, C, D) generally recognized in the group (Rajhathy and Thomas, 1974; Leggett and Thomas, 1995), although the difference between the A, B and D genomes may be minor enough to call the B-genome A′ and the D genome A″ (Loskutov, 2008; Nikoloudakis and Katsiotis, 2008). The existing karyotypic, crossibility and molecular data suggest that the diploid A-genome donor of the AC tetraploids and ACD hexaploids was *Avena canariensis* (Leggett, 1992; Leggett and Thomas, 1995; Li *et al.*, 2000). The C-genome donor was probably *Avena ventricosa* (Nikoloudakis and Katsiotis, 2008). All the allopolyploids cluster with the A-genome diploids, suggesting that there has been substantial alterations in the C genome post polyploidization. It is still unclear which diploid species donated the B genome to the polyploids as this genome is unknown in the diploids. However, Loskutov (2008) speculates that it is a derivative of the As genome, which is supported by Drossou *et al.* (2004) who found that the AB-genome tetraploids form a subcluster within the As-genome diploids. SSR variability patterns indicate that species with the As-genome are more variable than the species with the other genome contents (Li *et al.*, 2009).

The cultivated hexaploid oat, *A. sativa*, belongs to a genera of intergrading morphological and cytogenetic groups, but its immediate progenitor is known, as wild forms of *A. sativa* are found all over Western Asia and the Mediterranean region. These wild types have been variously referred to as *Avena sterilis* L. or *Avena fatua* L., depending on whether the spikelet has a single disarticulation point at the base of the floret. These taxa are inter-fertile with the cultivated forms and share the same karyotype, suggesting that they should be considered together as an *A. sativa* crop complex (Ladizinsky and Zohary, 1971).

The most widespread type, *A. sativa*, is separated from *Avena byzantina* only by the position at which the rachilla remains attached to the upper floret, while in *A. sativa* the rachilla breaks at the base of the upper floret. The other race, *Avena nuda*, is simply a free-threshing variant of *A. byzantina*.

The subtleties of these differences along with complete inter-fertility suggests these species should also be united under the umbrella *A. sativa*.

While oats were native to the ancient Near East, they are not thought to have been established as an independent crop until 3000 years ago in central Europe (Helbaek, 1959). Oats were probably carried along as a weed or secondary crop in the group of cultivated plants that spread out from the Near East (Zohary and Hopf, 2000). They were not exploited individually until they reached their full potential in the cooler, moist climates of central Europe. In fact, oats reached North America, Argentina and Australia during the colonization period, decades before they were grown as an individual crop in the Middle East.

Rice

The most commonly cultivated rice species, *Oryza sativa* (Asian or paddy rice), is grown primarily in the humid tropics and subtropics, with some cultivation on flooded upland sites such as central California. Another less important rice, *Oryza glaberrima* (African rice), is grown in East Africa, but is being replaced by *O. sativa*.

The cultivated rice and their ancestors are considered diploid ($2n = 24$), although their high chromosome number indicates that they could be ancient, diploidized polyploids. Ten genomes have been identified among the various sections of *Oryza*, based on chromosome pairing relationships, molecular markers and sequencing (Vaughan, 1994; Ge *et al.*, 1999, 2001). The cultivated species and their closest relatives carry the A genome and form what is referred to as the *sativa* complex. The A genome is further divided with superscripts to denote small pairing aberrations and partial sterility among the various diploid species (Table 8.3). There has been much debate on whether *Oryza rufipogon* and *Oryza nivara* should be considered as ecotypes rather than species, as they are separated primarily by life history and ecological differences and not reproductive barriers. Regardless, it is likely that annual *O. nivara* evolved from perennial *O. rufipogon* through an adaptive shift from wet to seasonally dry habitats (Grillo *et al.*, 2009; Zheng and Ge, 2010). The two cultivated species are also morphologically very similar, but they

Table 8.3. Genomes and geographical distribution of the species of *Oryza* in section *Oryza*. (Based on Vaughan, 1994; Chang, 1995: Ge *et al.*, 2001.)

Species and complex	Chromosome number (2*n*)	Genome designation	Geographical distribution
O. sativa complex	24		
O. sativa L.	24	AA	Worldwide (cultivated)
O. glaberrima Steud.	24	A^gA^g	West Africa (cultivated)
O. barthii A. Chev.	24	A^gA^g	Africa
O. glumaipatula	24	$A^{gp}A^{gp}$	South and Central America
O. longistaminata	24	A^lA^l	Africa
O. meridionalis	24	$A^?A^?$	Australia
O. nivara	24	AA	Tropical and subtropical Asia
O. rufipogon Griff.	24	AA	Tropical and subtropical Asia
O. officinalis complex	24		
O. punctata Kotschy	24, 48	BB, BBCC	Africa
O. eichingeri A. Peter	24	CC	Africa
O. officinalis Wall.	24	CC	Tropical and subtropical Asia
O. rhizomatis	24	CC	Sri Lanka
O. minuta J. S. Presl.	48	BBCC	Southeast Asia
O. alta Swallen	48	CCDD	South and Central America
O. grandiglumis Prod.	48	CCDD	South and Central America
O. latifolia Desv.	48	CCDD	South and Central America
O. australiensis Domin.	24	EE	Australia

show some chromosomal divergence and their hybrids are sterile (Oka, 1974, 1975).

O. glaberrima probably diverged from *Oryza barthi* (formerly known as *Oryza breviligulata*) (Linares, 2002), but the origin of *O. sativa* is much more cloudy. Strong cases have been made for both *O. rufipogon* and/or *O. nivara* being the progenitors. The two species are ecologically distinct with *O. nivara* being annual, photoperiod insensitive, self-fertilized and growing in drier habitats, while *O. rufipogon* is perennial, photoperiod sensitive, mostly cross-pollinated and adapted to wet areas. Molecular marker-based phylogenies have not been able to resolve this issue (Zhu and Ge, 2005), but studies on the genetics of rice domestication syndromes have shed some light. *O. sativa* shares the same life history characteristics as *O. nivara* and when an *indica* cultivar was crossed with an individual of *O. nivara*, there was no segregation for those life history traits. This suggests that their

phenotypic similarity is due to their carrying similar genes. However, it does not exclude the possibility that hybridization and introgression with *O. rufipogon* was not part of the evolutionary development of cultivated rice.

There has been considerable effort placed on finding the genes associated with the domestication syndrome in rice. Large QTL have been identified in a cross of *O. sativa* × *O. nivara* for genes associated with seed shattering, synchronization of seed maturity, seed dormancy, tiller number, panicle length, branching and spikelet number (Li *et al.*, 2005). QTL for many of these traits have also been discovered in crosses between *O. sativa* and *O. rufipogon* at similar chromosomal locations (Xiong *et al.*, 1999; Cai and Morishima, 2002). Two QTL regulating grain shattering have been cloned (*sh4* and *qSH1*), and have been shown to be the result of single nucleotide substitutions (Konishi *et al.*, 2006; Li *et al.*, 2006). Several other domestication-related genes have been cloned and sequenced, including *GN1a* (grain number), *Ghd7* (flowering time), *PROG1* (plant stature), *GIF1* (grain filling), *Rc* and *Rd* (seed pericarp color), *qSW5* (seed width) and *Wx* (texture of cooked rice) (Konishi *et al.*, 2008; Izawa *et al.*, 2009).

Two major races of *O. sativa* are recognized, subspp. *indica* and *japonica*. The *japonica* race is grown in colder, dryer environments and at higher elevation. Recent molecular studies have supported the independent domestication of these two subspecies, *indica* in a region south of the Himalayan mountains and *japonica* from southern China (Londo *et al.*, 2006). Of particular note is that while these two races appear to have separate origins, they both carry the same non-shattering allele, *sh4* (Zhang *et al.*, 2009). There are two possible hypotheses to explain how the same allele found its way into two highly divergent taxa (Fig. 8.7; Sang and Ge, 2007a,b). In one called the "snow-balling model", rice was first domesticated in a single region through the

accumulation of several domestication genes including *sh4*, and as these early domesticants were spread into other regions, they became introgressed with local wild populations of *Oryza* and the domestication genes including *sh4* became fixed in divergent backgrounds. This process ultimately produced the subspecies *indica* and *japonica*. In the other possibility called the "combination model", rice was originally domesticated in different regions through the accumulation of local domestication genes that made harvest more profitable, but *sh4* originated in only one region. Later hybridizations between the races containing *sh4* with those without it, resulted in rapid fixation of *sh4* in both gene pools. Tang *et al.* (2006) using gene sequence data found widespread, highly divergent linkage blocks in *indica* and *japonica* that may signal past mixing of the different *Oryza* taxa.

Rice domestication may have first occurred about 10,000 BP in the Yangtze region of China (Jiang and Liu, 2006), although full domestication including the non-shattering trait may have been closer to 6000 BP (Fuller *et al.*, 2009; Purugganan and Fuller, 2009). Rice was probably in Korea and Japan by 3000 BP. The cultivation of African rice probably began in the Niger Delta about 3500 years ago and it spread gradually across tropical East Africa. Asian rices arrived in Africa about 2000 BP (Chang, 1975). Rice found its way to the New World in 1647, when its cultivation was begun in the Carolinas.

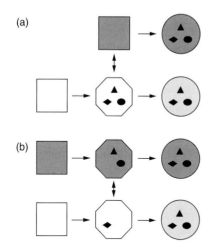

Fig. 8.7. Illustration of rice domestication models: (a) "snow-balling" model; (b) "combination" model. The squares represent wild populations from which rice was domesticated; hexagons represent the earliest domesticated rice; circles represent contemporary rice cultivars. Differences in shading represent genomic divergence. Shapes inside (triangle, diamond and ellipse) represent critical domestication genes that are now fixed in contemporary cultivars. (Used with permission from T. Sang and S. Ge, © 2007a, The puzzle of rice domestication. *Journal of Integrative Plant Biology* 49, 760–768.)

Rye

There are currently three groups of rye species recognized (Frederiksen and Peterson, 1998; Chikmawati *et al.*, 2005): (i) *Secale cereale* L., the cultivated species that also exists as highly diverse annual weeds including non-shattering, semi-shattering and shattering types in Iran, Afghanistan and Transcaspia; (ii) *Secale sylvestre* Host., another annual, inbreeder that is widely distributed from Hungary to the steppes of southern Russia; and (iii) *Secale strictum*, a perennial outcrossing species found in North Africa, West Asia and southern Europe. In the most recent taxonomies, *S. cereale* contains the previously recognized species *Secale vavilovii*, while *S. strictum* contains what used to be called

Secale montanum. Cultivated *S. cereale* probably evolved from individuals similar to *S. vavilovii* (Davies and Hillman, 1992; Zohary and Hopf, 2000). *S. montanum* is distinct from *S. cereale* by two reciprocal translocations involving three pairs of chromosomes (Riley, 1955; Khush and Stebbins, 1961).

Many have proposed that rye developed as a secondary crop, similar to oats. In this scenario, rye was picked up as a weed when the wheat–barley assemblage arrived in western Asia where the native species are widely distributed. However, wild forms of rye with a brittle rachis appear to have been cultivated by about 12,000 BP in northern Syria (Hillman, 2000). This makes it possible that wild rye was originally a weed in wheat and barley fields in the Near East and was domesticated there. Like the other grain species, agronomic traits such as rachis fragility, ear branching and growth habit are determined by only a few genes.

Rye arrived in Europe as a cultivated crop by 4000 BP (Khush, 1962). Because of its tough constitution, it may have performed better than wheat and barley in the cooler, nutrient-poor northern climates and therefore attracted human attention. In modern times, tetraploid and hexaploid wheats have been artificially hybridized with rye to form the new crop called *Triticale* (Larter, 1995).

Sorghum

The wild sorghums present a diverse array of morphological variability and have presented taxonomists with an interesting challenge. In the most complex taxonomy, Snowden (1936) recognized over 70 species, but most researchers now accept a much smaller number after multivariate analyses were employed to cluster groups via morphological traits (deWet, 1978; Doggett, 1988). The three main native species located in Asia and Africa are *Sorghum bicolor* subsp. *arundinaceum* (Desv.) deWet and Harlan, *Sorghum propinquum* (Kunth) Hitchc. and *Sorghum halepense* (L.) Pers. *S. bicolor* subsp. *arundinaceum* is an allotetraploid ($2n = 2x = 20$) that is found in tropical Africa, while *S. propinquum* shares the same chromosome number and is located in Asia, Indonesia and the Philippines. *S. halepense*,

or Johnsongrass, is an octoploid ($2n = 2x = 40$) and is located in an overlapping zone between the other two species. Molecular and cytogenetic data have strongly implicated *S. bicolor* as one of the progenitors of the major weed Johnsongrass, if not the only one (Hoang-Tang and Liang, 1988; Morden *et al.*, 1990; Hoang-Tang *et al.*, 1991).

Cultivated *S. bicolor* subsp. *bicolor* was probably derived from *S. bicolor* subsp. *arundinaceum* through human selection for non-shattering heads, large seeds and heads, threshability and suitable maturities (Doggett and Rao, 1995). These traits are regulated by relatively few QTL (Paterson *et al.*, 1998). Patterns of morphological and molecular variability suggest that the center of origin of sorghum is in the Ethiopia–Sudan region of Africa (Aldrich *et al.*, 1992; Deu *et al.*, 1994; Doggett and Rao, 1995; Perumal *et al.*, 2007). It is likely that sorghum cultivation spread from Africa to India through trade routes some 2000–3000 BP and it arrived in China about the beginning of the Christian era.

Harlan and de Wet (1972) recognized five main races of cultivated sorghum based on morphological patterns (Fig. 8.8): (i) Bicolor, which is widely distributed across the African savannah and Asia; (ii) Caudatum, found in central Sudan and the surrounding areas; (iii) Guinea, which is grown in eastern and western Africa; (iv) Durra, found primarily in Arabia and Asia Minor, but also grown in India, Burma, Ethiopia and along the Nile Valley; and (v) Kafir, which is cultivated primarily in southern Africa. AFLP and SSR data support this grouping (Perumal *et al.*, 2007), with the Guinea-race showing the most molecular divergence and eco-geographic patterning (Folkertsma *et al.*, 2005). The races overlap in many areas and freely hybridize amongst themselves and the wild species to produce intermediate races. A weedy race of sorghum, *S. bicolor* subsp. *drummondii* (Steud.) de Wet, exists wherever cultivated and wild sorghums are sympatric (deWet, 1978).

The various cultivated types of today could have been derived from several wild races of *S. bicolor* subsp. *arundinaceum* that are located in different parts of Africa (Snowden, 1936; deWet and Huckabay, 1967). The West African sorghums probably came from var. *arundinaceum*, while the eastern-central group were ennobled from var. *verticilliflorum*. The northern-eastern

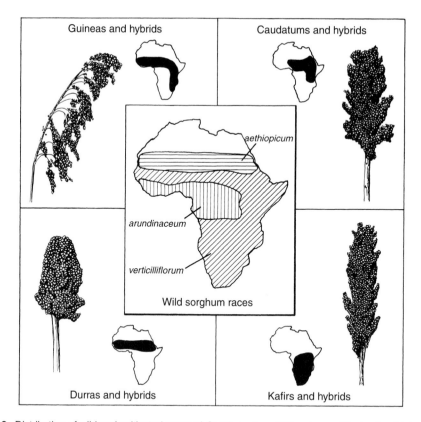

Fig. 8.8. Distribution of wild and cultivated races of *Sorghum*. An additional race, Bicolor, is widely scattered across the African savannah and in Asia from India to Japan and Indonesia.

sorghums were developed from var. *aethiopicum*. It is not known whether these races were domesticated separately or were stimulated by adjacent cultivations. There was undoubtedly much hybridization occurring between wild and cultivated forms as agriculture spread throughout Africa. In fact, recent molecular analyses have indicated that most variation is more closely associated with geographical than racial differentiation (Morden *et al.*, 1989; Menkir *et al.*, 1997; Djè *et al.*, 2000).

Wheat

The taxonomy of *Triticum* has undergone numerous reorganizations that vary according to how much importance is given to levels of inter-fertility, ecological ranges and specific morphological traits (for an excellent review

see http://www.ksu.edu/wgrc). We will use the recent taxonomy of van Slageren (1994) and cite other relevant synonyms where warranted (Table 8.4).

Three ploidy levels of wheat were domesticated, diploid, tetraploid and hexaploid. The diploids are represented by the single cultivated species *Triticum monococcum* or einkorn, meaning one seed per spikelet. Einkorn is grown for a dark bread in some parts of the Middle East and southern Europe, but its glumes fit tightly around its seeds, making them hard to remove. For this reason, it is more commonly used to feed cattle and horses.

The tetraploid group is represented by a complex of several subspecies, with the most important cultivated ones being *Triticum turgidum* subsp. *dicoccoides* (emmer) and *T. turgidum* subsp. *durum* (durum). Emmer was widely grown in the Mediterranean region until Greco-Roman times and makes good bread and pastry, but like einkorn has clinging glumes and delicate stems

Table 8.4. A classification of the wheat species *Triticum* L. (Based on Van Slageren, 1994.)

Species	Subspecies	Status	Chromosome number	Genome
T. monococcum	*aegilopoides*	Wild	2*n*=14	AA
	monococcum	Cultivated		
T. urartu		Wild		
T. turgidum	*carthlicum*	Cultivated	2*n*=28	AABB
	dicoccoides	Wild		
	dicoccum	Cultivated		
	durum	Cultivated		
	turgidum	Cultivated		
	paleocolchicum	Archeological		
	polonicum	Cultivated		
T. timopheevi	*armeniacum*	Wild	2*n*=28	AAGG
	timopheevi	Cultivated		
T. aestivum	*spelta*	Cultivated	2*n*=42	AABBDD
	macha	Cultivated		
	aestivum	Cultivated		
	compactum	Cultivated		
	sphaerococcum	Cultivated		

that make it difficult to harvest and thresh. Emmer is now locally important as livestock feed in a diverse region including the mountains of Europe and the Dakotas of the USA. Durum is grown widely in Italy, Spain and the USA. Its seeds are separated easily from its glumes, and it has a high gluten content that makes it sticky when wet. This makes it excellent for making spaghetti and macaroni. There is also another tetraploid species, *Triticum timopheevi*, but it is only cultivated in a restricted region of Russia.

The hexaploids are represented by the bread wheat, *Triticum aestivum*, which was the last domesticated wheat to appear, but is now the most widely planted. It is a highly variable group that is divided into numerous subspecies (Table 8.4). With the exception of subsp. *spelta*, the bread wheats have tough inflorescence stems that do not shatter when harvested, and the seeds are easily threshed after gathering. Some of the hexaploid subspecies must be hulled (*spelta* and *macha*), while others are free threshing (*aestivum*, *compactum* and *sphaerococcum*). All hexaploid wheats are high in the protein gluten, which makes a fluffy, leavened bread. *T. aestivum* is grown throughout the world, but is particularly important in the continental climates of Ukraine, central USA, Canada and Australia, and the cool temperate climates of northern Europe, China and New Zealand.

Cultivated einkorn was probably derived from the wild subspecies *T. monococcum* subsp. *aegilopoides* (*Triticum boeoticum* or *T. monococcum* subsp. *boeoticum*). The cultivated and wild subspecies look very similar and share numerous molecular markers (Asins and Carbonell, 1986; Hammer *et al.*, 2000). The native distribution of einkorns is in northern Syria, southern Turkey, northern Iraq, Iran and western Anatolia. The only substantial difference between the wild and cultivated forms is that the wild species has fragile spikes that allow its seeds to scatter freely, while the cultivated variety has a tough rachis that allows for more effective harvesting. In the cultivated forms, the mature ear breaks into individual spikelets through threshing, and the kernels are wider.

The earliest archeological evidence of people gathering brittle *T. monococcum* spp. *aegilopoides* is found in remains at Tell Abu Hureyra that are about 10,000 years old. The first cultivated einkorn is present by about 8000 BP at several sites in the Near East including Ali Kosh in Iran, Jarmo in Iraq and Çayönü and Can Hasan in Turkey (Zohary and Hopf, 2000). Both brittle and non-brittle forms are found together at these locations, suggesting that gathering and farming were being employed simultaneously. Studying patterns of AFLP and RFLP variation in wild and cultivated races, Heun *et al.* (1997)

and Luo *et al.* (2007) were able to target the first domestication of einkorn to southeastern Turkey. Kilian *et al.* (2007) found using AFLPs that there are three separate races of wild *T. monococcum* spp. *aegilopoides* and only one of them was domesticated.

The tetraploid wheat *T. turgidum* was derived from the natural hybridization of wild einkorn (AA) and another diploid species carrying the B genome (Fig. 8.9). Sequence data from plastid acetyl-CoA carboxylase and 3-phosphoglycerate (Huang *et al.*, 2002) and repeated nucleotide sequences (Dvořák *et al.*, 1993) show that

Triticum urartu is the diploid AA donor. No diploid with the B genome has been identified; however, a wide variety of molecular evidence points to the S-genome species, *Aegilops speltoides* (*Triticum speltoides*), as being the closest living relative (Talbert *et al.*, 1995; Blake *et al.*, 1999; Huang *et al.*, 2002; Kilian *et al.*, 2006). *A. speltoides* appears to be the plastid donor of all polyploid wheats (Wang *et al.*, 1997).

The cultivated forms of emmer wheat were probably derived from wild populations of *Triticum dicoccoides*. Wild emmer is most common in the catchment area of the upper Jordan

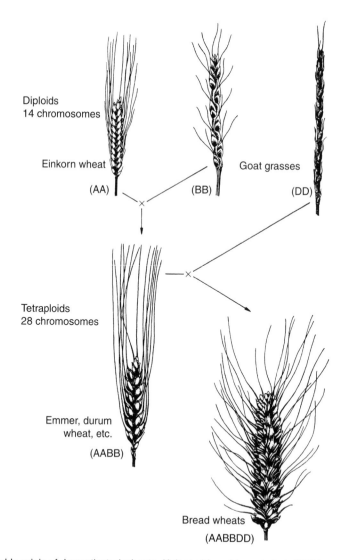

Fig. 8.9. Probable origin of domesticated wheats. (Adapted from Mangelsdorf, 1953.)

Valley, where it forms natural fields of grain with wild barley and oats. Further north it grows wild with barley and einkorn wheat. An analysis of AFLP data at 204 loci places the origin of the tetraploid wheats to southeast Turkey (Özkan et al., 2002). Wild forms of emmer wheat have been found at numerous prehistoric sites, with the oldest remains being at least 19,000 years old. The earliest cultivated forms have been discovered at Tell Aswad in Syria (van Zeist and Bakker-Heeres, 1985), Tell Abu Hureyra (Hillman, 1975) and Jerico (Hopf, 1983) in deposits over 9000 years old. Cultivated types have a non-brittle spike that is regulated by a single major gene, and their grains are wider and thicker, with a rounder cross section. Emmer wheat spread rapidly across the Near East, even to areas where einkorn was not cultivated, and became the principal wheat in Neolithic and Early Bronze ages.

Durum wheat probably arose soon after emmer cultivation began (Zohary and Hopf, 2000). The free-threshing durum types probably evolved under cultivation through a gradual accumulation of genes reducing the toughness of glumes. This shift from hulledness to nakedness is regulated by a polygenic system. Scattered wild tetraploids also have the Q factor found in *T. aestivum* (see below), but these may have been picked up through hybridization with the bread wheats.

The hexaploid *T. aestivum* (AABBDD) is a relatively recent product of hybridizations between the cultivated tetraploid wheats *T. turgidum* subsp. *dicoccoides* (AABB) and the wild diploid goat grass *Aegilops tauschii* (*Ae. squarrosa*) (Fig. 8.9) (McFadden and Sears, 1946; Kihara, 1954; Dvořák et al., 1998; Matsuoka and Nasuda, 2004). The geographic origin of the hexaploid wheats was probably outside the original Fertile Crescent

as the native range of *Ae. squarrosa* is in continental central Asia rather than the Mediterranean Near East. A southern Caspian Sea–Transcaucasus origin for hexaploid wheat has been confirmed by isozyme (Jaaska, 1983) and ribosomal RFLPs (Dvořák et al., 1998), although Talbert et al. (1998) provide evidence using sequence comparisons of low-copy DNA that hexaploid wheat formed more than once. The range of goat grass probably expanded as a weed in secondary, man-made habitats as farming spread.

The first evidence of hulled bread wheats (*T. aestivum* subsp. *spelta*) has been found at Arukhlo 1, Transcaucasia (Janushevich, 1984), dating about 7000 BP, but by this time hulled forms were probably grown all across the Caspian belt. As with the other wheats, the genetics of the free-threshing trait is relatively simple, being regulated by the *Q* gene on chromosome 5A, and *tg* (tenacious glumes) on chromosome 2D (Zohary and Hopf, 2000). The *Q* gene has been cloned and sequenced (Simons et al., 2006).

The addition of the D genome greatly expanded the range of wheats. The hexaploids spread rapidly and diverged into numerous races as new environments were met, mutations were accumulated and unique alleles were introgressed from native species. The extensive amount of interspecies hybridization now occurring naturally in the Middle East supports a polyphyletic origin for most of the polyploid species. Substantial chromosomal and genic modifications probably arose after the early hybridizations as the various taxa adapted to new environmental challenges, and their self-pollinating nature aided in the fixation of new coadapted complexes. This reorganization and diversification probably enhanced the adaptive potential of the genus, but has given great headaches to evolutionists trying to determine genomic origins.

9

Protein Plants

Introduction

Wheat and barley were probably the first crops to be cultivated by our ancestors, but the legumes were not far behind. Lentils and peas were being harvested in the wild at about the same time grain cultivation was beginning in the Near East, and it did not take people long to add the other pulses to their list of cultivated crops. Beans may even have been domesticated before maize in the New World.

In this chapter, we explore the origins of the most important legume crops. The evolutionary relationships of the legumes are often straightforward, because they have direct living relatives and simple ploidy relationships. They are also carbonized readily like the grains and are well represented in the archeological record.

Chickpea

Chickpea (*Cicer arietinum* L.) belongs to a genus of 40 species that is located primarily in central and western Asia. The majority of the *Cicer* species are perennial shrubs, but the group containing the cultivated forms is annual. Most species are diploid with $2n = 2x = 16$. Three crossability groups have been identified among the species (Ladizinsky, 1995); the cultivated chickpea can be crossed with *Cicer reticulatum* Davis and *Cicer*

echinospermum Ladiz. Of these two wild species, only *C. reticulatum* is fully inter-fertile with the cultivated forms and morphological, cytological and molecular studies have confirmed their close relationship (Ahmad, 2000; Irula *et al.*, 2002; Singh *et al.*, 2008). Therefore, *C. reticulatum* should probably be regarded as a subspecies, *C. arietinum* subsp. *reticulatum* (Ladiz) Cubero et Morreno. Crosses between the cultivated species and other wild species have generally failed (Ladazinsky and Adler, 1976a,b; Singh and Ocampo, 1993, 1997), although there are reports of successful hybridizations between cultivated chickpea and *Cicer pinnatifolium* Jaub et Spach, *Cicer judaicum* Boiss and *Cicer bijugum* Rech. (Singh *et al.*, 1994; Badami *et al.*, 1997).

Chickpea cultivation was probably associated with the emergence of the grain crops in the Near East. However, its origin is quite unique, as the distribution of its progenitor, *C. reticulatum*, is quite narrow compared to the other species domesticated in the region (Ladizinsky, 1995), its seeds are the smallest of its relatives in the Near East (Abbo *et al.*, 2008), and its pods are retained rather than dehisced as in lentil and pea (Ladizinsky *et al.*, 1984). Also, it is the only Near Eastern crop sown in the spring rather than the fall. It has been suggested that while the progenitor of chickpea was very small seeded, it attracted early attention because of its pod retention (Ladazinsky, 1979) and perhaps its high tryptophan content (Kerem *et al.*, 2007). There was probably a fungus present (Ascochyta blight) in the Near East that could cause total yield loss if chickpea was sown in the fall, forcing early farmers to plant in the spring to avoid infection (Abbo *et al.*, 2003). This would have resulted in a major shift in the physiology of domesticated chickpea, as vernalization-insensitive types would have had to be selected by early farmers. The arrival in the Near East of African spring-planted crops at about 6000 BP may have encouraged early farmers to make the shift in planting date. This possibility is supported by the observation that while chickpeas are present by 10,000 BP in the archeological record, they disappear for several thousand years before reemerging in about 6000 BP (Abbo *et al.*, 2003). In short, cowpea domestication probably required more conscious thought by early farmers than the other Near East crops.

The oldest chickpea remains have been found in the Near East at Tell el-Kerkh, Cayönü and Nevali Çori, dating to around 9300 BP (Tanno and Wilcox, 2006b). Chickpeas probably traveled to Europe along with the early grain crops. Seeds have been found in Greece from 8000 BP and southern France from 5000 BP. They arrived in India about 4000 BP, and in Ethiopia from the Mediterranean about 3000 BP.

Over the course of domestication, the size of the seeds increased dramatically from 3.5 to 6.0 mm (Fig. 9.1). The seed coat also became smoother and thinner. As the crop was dispersed, two major morphological types emerged (Hawtin *et al.*, 1980): (i) plants with large, owl-shaped, light colored and smooth seeds (Kabuli) with pale cream flowers; and (ii) plants with smaller, ram-shaped, dark and wrinkled seeds (Desi) with purple flowers. The Kabuli types are located primarily on the Mediterranean side of the chickpea distribution, while the Desi forms are on the eastern side. The Kabuli types are more distant from the wild types than the Desi ones, and this fits with their more distant geographical separation from their wild ancestor and their more variant ecological range (Rowewal *et al.*, 1969).

Cowpea

The genus of the cowpea, *Vigna*, is a relatively large pantropical genus with the majority of its species being found in Africa. All species have 22

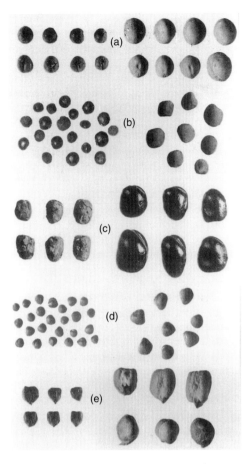

Fig. 9.1. Seeds of pulses from archeological sites (left) and modern varieties (right): (a) pea, (b) lentil, (c) broad bean, (d) bitter vetch and (e) chickpea. (Used with permission from M. Hopf, © 1986, *Archaeological evidence of the spread and use of some members of the Leguminosae family*. In: *The Origin and Domestication of Cultivated Crops*, Elsevier Science Publishers, New York.)

chromosomes with little cytogenetic divergence. The variability and complexity of morphological patterns within the group has led to the development of several different taxonomies (Ng, 1995). In one of the most recent, one annual subspecies is recognized, *Vigna unguiculata* subsp. *unguiculata*, and ten wild perennial subspecies (Pasquet, 1998). *V. unguiculata* subsp. *unguiculata* is divided into two varieties, *V. unguiculata* subsp. *unguiculata* var. *unguiculata* representing the cultivated forms, and *V. unguiculata* subsp. *unguiculata* var. *spontanea*, the likely wild progenitor. The cultivars are further subdivided into five groups based on seed and pod characteristics: (i) 'Unguiculata', grown as a seed crop; (ii) 'Biflora' (catjang), used as fodder; (iii) 'Sesquipedalis' (yardlong or asparagus bean), grown as a green pod vegetable; (iv) 'Textilis', used for its peduncle fibers; and (v) 'Melanophthalmus' (blackeyed pea), grown as a seed crop. Nigeria, Niger and Brazil are the leading cowpea producers in the world.

There is conflicting evidence on where the crop was first domesticated, probably due to the considerable amount of hybridization that has occurred between cultivated and wild races (Coulibaly et al., 2002). Pasquet (1999), using isozyme data, found no clear center of origin. Vaillancourt and Weeden (1992) found a loss in a BamHI restriction site in chloroplast DNA in all domesticated clones and some wild V. unguicultata subsp. unguiculata var. spontanea, which suggests a West African origin. AFLP data led Coulibaly et al. (2002) to propose a northeastern Africa origin. Cowpea was probably cultivated by 6000–7000 BP and arrived in India about 4000 BP with the grain species (Ng, 1995). Cowpea was probably introduced to Brazil from West Africa by European explorers and African slaves in the 16th and 17th centuries.

Considerable divergence occurred in cowpea soon after its cultivation began. The original forms were probably spreading, short-day plants that readily scattered their seeds. Pod dehiscence and seed dormancy were most likely lost quickly in conjunction with domestication. Upright, day-neutral types may have first emerged after introgression with local wild relatives in the rainforests of Africa (Alexander and Coursey, 1969; Lush and Evans, 1981). The fodder and green pod vegetable were probably developed after the crop arrived in Asia.

Pea

The genus Pisum has undergone numerous revisions of what constitutes species and subspecies. One of the more recent taxonomies suggest that the genus is represented by only two self-pollinating, diploid species ($2n = 2x = 14$), Pisum fulvum Sibth. & Sm. and Pisum sativum L. (Davies, 1995), while another recognizes P. sativum and P. fulvum as well as an additional

species, Pisum abyssinicum A. Br (Ambrose and Maxted, 2000). In general, molecular analyses support the three species model (Smýkal et al., 2010). P. fulvum is a wild species located in the eastern Mediterranean, while P. sativum (Fig. 9.2) is a complex aggregate of wild and cultivated races centered in the Mediterranean basin and Near East. P. abyssinicum is a cultivated species restricted to Yemen and Ethiopia. P. fulvum and P. sativum can only be crossed with difficulty, although the hybrids are largely sterile (Ben-Ze'ev and Zohary, 1973). The interspecies cross using P. sativum as the maternal parent is more successful than the reciprocal, but the F_1 hybrids are weak and have many quadrivalents, trivalents and univalents at metaphase I. Regardless, hybrids are often found wherever the two species overlap (Davies, 1995). P. abyssinicum differs from P. sativum by several chromosomal rearrangements making them difficult to intercross and the two species have a nuclear-cytoplasmic incompatibility (Bogdanova et al., 2009).

Within P. sativum are two subspecies, P. sativum subsp. sativum and P. sativum subsp. elatius. The wild forms of P. sativum fall into two general morphological-ecological classes: (i) a tall, climbing type that was formerly called Pisum elatius Bieb., which prefers mesic habitats; and (ii) a more xeric-loving, short type formerly called P. humile Boiss., which often invades cereal farms. The 'humile' form was thought to be the ancestor of the cultivated types (Ben-Ze'ev and Zohary, 1973), as 'elatius' populations differ from the cultivated forms by a single chromosomal translocation, and some populations of 'humile' in Turkey and Syria share the same chromosomal complement. However, molecular data indicate that the source of the domesticants could also have been 'elatius' in the Fertile Crescent (Jing et al., 2010; Kosterin et al., 2010; Smýkal et al., 2010). P. abyssinicum was domesticated separately from P. sativum, perhaps from a hybrid of P. fulvum and P. sativum subsp. elatius (Vershinin et al., 2003; Jing et al., 2010).

Wild and domesticated peas are difficult to distinguish in the archeological record; however, the seed coat appears to be diagnostic when it exists, as domesticated forms have a smooth seed coat, while the wild forms are rough coated (Zohary and Hopf, 2000).

Fig. 9.2. The pea plant (*Pisum sativum*). (Used with permission from R.H.M. Langer and C.D. Hill, © 1982, *Agricultural Plants*, Cambridge University Press, New York.)

Smooth-coated forms are found with the earliest domesticated barleys and wheat in the Near East. Peas probably followed the spread of Neolithic agriculture into Europe, and they were widespread in central Germany by 6000 BP. They arrived in the Nile Valley by 7000 BP and in India by 4000 BP.

As with most other seed crops, there was a general increase in size associated with pea cultivation, along with the development of non-dehiscence of the pod, seed retention and the elimination of seed dormancy. The group has diverged substantially over the years, due at least in part to their self-breeding mechanism. Tremendous variation can be found in height, habit, flower color and seed characteristics. Literally hundreds of landraces exist, and breeders have provided us with large numbers of varieties (Zohary and Hopf, 2000).

Lentil

Lentil belongs to the small genus *Lens*, with six annual diploid species ($2n = 2x = 14$) located in the Mediterranean basin and Southwest Asia. *Lens* contains the cultivated *Lens culinaris* subsp. *culinaris* (Fig. 9.3), its wild progenitor *L. culinaris* subsp. *orientalis* (= *L. orientalis*) and five other wild species. Three crossability groups have been identified among these species: (i) *L. culinaris* and *Lens odemensis* Ladiz.; (ii) *Lens ervoides* (Bring.) Grand., *Lens nigricans* (M. Bieb.) Grand and *Lens lamottei* Czefranova; and (iii) *Lens tomentosus* Ladiz. (Ladizinsky *et al.*, 1984; van Oss *et al.*, 1997; Mayer and Bagga, 2002). Two integrating forms of cultivated lentil are recognized: (i) subsp. *microsperma*, with small grains and seeds (3–6 mm in diameter); and (ii) subsp. *macrosperma*, with large pods and seeds (Zohary, 1995a). The species are primarily self-pollinated like most other legume crops.

0 1 2 3
⌐—⌐—⌐—⌐ cm

Fig. 9.3. The lentil plant (*Lens culinaris*). (Reprinted with permission from Zohary, © 1972, Israel Academy of Sciences and Humanities.)

L. culinaris subsp. *culinaris* and subsp. *orientalis* have overlapping ranges, look very similar and share a high percentage of their molecular markers (Havey and Muehlbauer, 1989; Muench *et al.*, 1991; Mayer and Soltis, 1994), although subsp. *orientalis* is much smaller and has pods that burst before harvest. There is considerable genetic divergence in subsp. *orientalis*, as some races share chromosome homology and crossability with the cultivated types, while others do not. Based on cpDNA restriction data, chromosome behavior and crossability, Ladizinsky (1999) believes that three lines collected from Turkey, northern Syria and southern Syria are part of the genetic stock from which lentil was domesticated. Further pinpointing of the origin of the lentil will take much more comprehensive collection and screening.

Lentil cultivation, like that of chickpea, was closely associated with the domestication of wheat and barley. It may even be one of the founder crops of Old World agriculture, since carbonized lentil seeds have been found in many of the ancient Near Eastern farming villages (Zohary and Hopf, 2000). Its pattern of migration across Asia and Europe closely matched that of the seminal grain species and other legume crops, and it arrived in Spain and Germany 6000–7000 years ago (Fig. 9.4). Lentils were grown in India by 4500 BP.

Phaseolus Beans

There are dozens of *Phaseolus* species, all of American origin (Delgado-Salinas *et al.*, 2006). Five of them are cultivated (Fig. 9.5): *P. vulgaris* L. (common, haricot, navy, French or snap bean), *P. coccineus* L. (runner or scarlet bean), *P. lunatus* L. (lima, sieva, butter or Madagascar bean), *P. dumosus* Macfady (= *P. polyanthus* Greenm.) (year bean) and *P. acutifolius* A. Gray (tepary bean). Until recently, *P. polyanthus* was

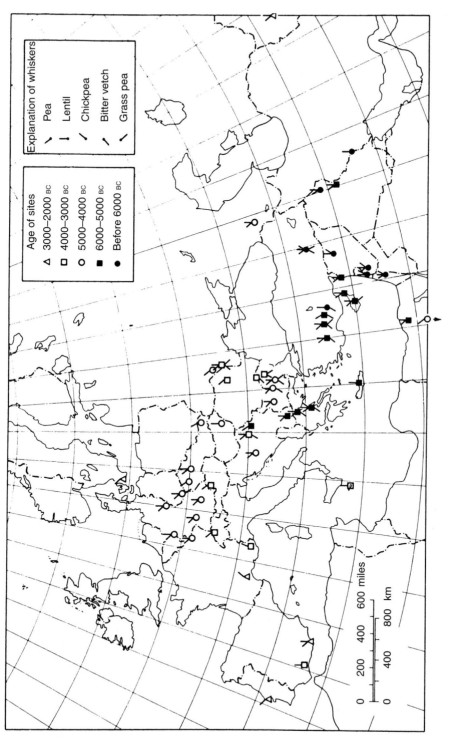

Fig. 9.4. Spread of pulses to Europe, west Asia and the Nile Valley. (Used with permission from D. Zohary and M. Hopf, © 2000, *Domestication of Plants in the Old World: the Origin and Spread of Cultivated Plants in West Asia, Europe and the Nile Valley*, 3rd edn, Clarendon Press, Oxford, UK.)

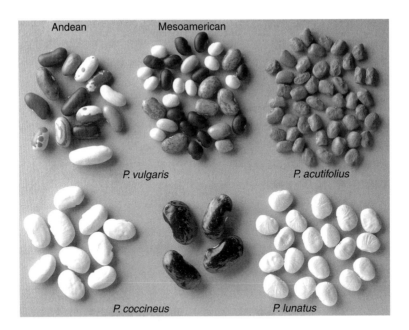

Andean   Mesoamerican

P. vulgaris   P. acutifolius

P. coccineus   P. lunatus

Fig. 9.5. Cultivated types of *Phaseolus* beans. (Jim Kelly at Michigan State University provided representatives of the bean types.)

considered a subspecies of *P. coccineus* (Schmit and Debouck, 1991).

All the cultivated species are diploid ($2n = 2x = 22$) and their direct progenitors are found in the wild. The wild and cultivated races of all species are fully to nearly fully compatible. The *Phaseolus* beans inhabit a wide range of habitats, including the cool, humid uplands of Guatemala (*P. coccineus* and *P. dumosus*), semi-arid regions in Guatemala, Arizona and Mexico (*P. acutifolius*), tropics and subtropics of Central and South America (*P. lunatus*), and warm, temperate areas in Mexico–Guatemala (*P. vulgaris*) (Debouck, 2000). *Phaseolus dumosus* and *P. coccineus* are outcrossed but can also self, while the other species are primarily self-pollinating.

The evolutionary relationships among the cultivated species have been studied at a number of different levels including crossability, seed proteins and molecular markers (Gepts, 1998; Debouck, 2000). *P. vulgaris*, *P. dumosus* and *P. coccineus* are very closely related, with *P. acutifolius* in the middle and *P. lunatus* being the most distant. A number of other wild species share varying levels of compatibility with the cultigens, but they are not thought to be in their direct ancestry (Table 9.1).

The common bean was domesticated independently in Mesoamerica and South America from two distinct wild taxa of *P. vulgaris*, var. *aborigineus* in South America and var. *mexicanus* in Mesoamerica (Gepts, 1998). Kwak *et al.* (2009) have pinpointed the Mesoamerican origin to the Rio Lerma–Rio Grande de Santiago basin in west-central Mexico using microsatellite markers. Ancient seeds have been found in both Puebla, Mexico (2300 BP) and the Peruvian Andes (4400 BP) (Kaplan and Lynch, 1999), and landraces from the two regions have distinct floral structures, seed sizes, phytopathology, phaseolin seed proteins, allozymes and DNA markers (Khairallah *et al.*, 1990; Gepts, 1998; Kwak *et al.*, 2009). In addition, hybrids between plants from the two areas are mostly infertile. The combination of the dwarf lethal genes, DL_1 from Middle America and DL_2 from the Andean region, produces very weak F_1s.

The lima bean appears to have a polycentric origin, with three cultigroups being described based on seed morphology (Baudet, 1977): the 'Sieva' group with smaller, flat kidney-shaped seeds (40–70 g/100 seeds), the 'Potato' group with globular to elliptical, smaller seeds (40 g/100), and the 'Big Lima' group with

Table 9.1. Gene pools of the various *Phaseolus* cultigens (adapted from Debouck, 2000).

Primary gene pool (GP1)	Secondary gene pool (GP2)			Tertiary gene pool (GP3)		
	Close	Intermediate	Distant	Close	Intermediate	Distant
P. vulgaris	*P. costaricensis*	*P. polyanthus*	*P. coccineus*		*P. acutifolius*	*P. filiformis*
P. dumosus	*P. coccineus*	*P. costaricensis*	*P. vulgaris*			
P. coccineus	*P. dumosus*		*P. vulgaris*			
P. acutifolius		*P. parvifolius*			*P. vulgaris*	
P. lunatus	*P. pachyrrhizoides*		*P. maculatus*	*P. jaliscanus*	*P. salicifolius*	

flat and large seeds (70 g/100). Evidence from chloroplast and nuclear DNA polymorphisms places the domestication of the large-seeded types to the Andes of southern Ecuador–northwestern Peru and the small seeded ones to central-western Mexico (Motta-Aldana *et al.*, 2010). This work did not support a separate origin for the two small-seeded types. The earliest physical evidence of domesticated lima beans comes from coastal Peru (5600 BP) (Kaplan and Lynch, 1999).

Many investigators feel that *P. acutifolius* was domesticated originally in Central America (Debouck and Smartt, 1995), although an origin in northwest Mexico–southwest USA has also been suggested (Pratt and Nabhan, 1988). Recent AFLP data indicate the likely origin to be one of two Mexican states, Sinaloa or Jalisco (Muños *et al.*, 2006). The earliest evidence of *P. acutifolius* in the archeological record is from about 2500 BP in the Tehuacán Valley. Sketchy data indicate that *P. coccineus* was first tamed in Mexico about 1000 BP (Kaplan and Lynch, 1999). *Phaseolus dumosus* was probably domesticated in Guatemala in pre-Columbian times (Schmit and Debouck, 1991).

The common bean spread widely across North and South America over several thousand years and arrived in the Ohio Valley of the central USA by about 1000 BP. The Spanish explorers and traders took lima beans from Peru to Asia and Madagascar. The lima and common bean arrived in Africa via the slave trade. Remarkably, much of the variation in seed coat, shape and growth habit found in South America has been maintained in the African landraces, even though the initial introductions were probably limited (Fig. 9.6; Martin and Adams, 1987a,b). *Phaseolus vulgaris* eventually reached Europe by the 16th century and was introduced

back into eastern North America in the late 19th century.

In the last 10,000 years, all of the *Phaseolus* beans have undergone very similar changes under domestication that are often regulated by only a few genes (Smartt, 1999). Some of the most dramatic changes associated with cultivation have been the shift from the perennial to annual habit, larger and softer seeds, a shift from short-day to day-neutral photoperiods, and the development of more persistent pods. Considerable variation has also been developed in plant architecture, from indeterminant climbers to determinant bush types (Adams, 1974).

Faba Beans

Vicia faba L. (field or broad bean) is diploid ($2n = 2x = 12$) and it is outcrossed, although it has varying degrees of self-fertility. Several wild vetches show strong morphological similarities to the faba bean, such as *Vicia narbonensis* L. and *Vicia galiliea* Plitm. Et Zoh. (Schäfer, 1973; Zohary and Hopf, 1973), but most of them are $2n = 2x = 14$ and interspecific hybridizations have failed (Ladizinsky, 1975). *V. faba* chromosomes are much larger than the wild species and contain much more DNA (Chooi, 1971), indicating that any evolutionary relationship with known species is remote (Bond, 1995). Numerous molecular comparisons have supported the "independence" of *V. faba* (Duc *et al.*, 2010).

It is likely that the faba bean was domesticated in the Near East. The oldest, strong evidence of faba bean cultivation comes from Tell el-Kerkh in north-west Syria from 9300 BP,

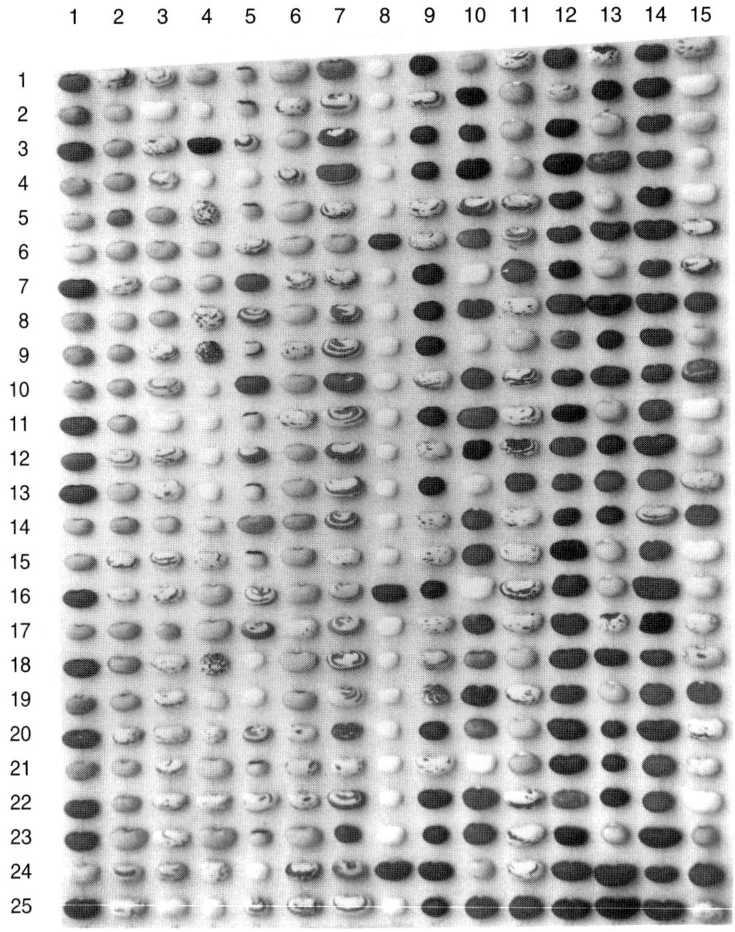

Fig. 9.6. *Phaseolus lunatus* seed types found in Malawi. Seeds are arranged in columns corresponding to 15 landraces. (Reprinted with permission from G.B. Martin and M.W. Adams, © 1987a, *Economic Botany* 41, 190–203, New York Botanical Garden, New York.)

although earlier evidence of a few seeds comes from Iraq ed-Dubb dated at 11,145 to 9950 BP (Tanno and Willcox, 2006b). The crop probably originated in the Near East, and spread west along the Mediterranean coasts until it reached Spain by 4000–5000 BP. Interestingly, faba beans did not arrive in China until the last millennium (presumably through the silk trade), even though China is now the leading producer in the world (Hanelt, 1972). The Spaniards took faba beans to Mexico and South America only a few hundred years ago.

Today, several different, intergrading subspecies of faba beans are recognized, based primarily on seed size: *minor, equina, faba* and

paucijuga (Cubero, 1974). The smallest type, *minor*, is considered the most primitive, but all the subspecies contain varieties that carry primitive traits such as shattering pods. *V. faba* subsp. *paucijuga* is the most self-fertile of all the subspecies and can set seed without bee activity.

Soybean

The soybean genus, *Glycine*, is divided into two subgenera, *Soya* and *Glycine*. The subgenus *Glycine* contains a group of wild perennial species whose center of distribution is Australia; all carry

$2n = 38$, 40 or 80 chromosomes (Table 9.2). The subgenus *Soya* is composed of the annual cultivated species *Glycine max* and the wild species *Glycine soya*, which both have $2n = 40$ chromosomes. *G. soya* is distributed naturally in China, Japan, Korea, Taiwan and the former USSR (Hymowitz, 1995). It is likely that cultivated *G. max* naturally hybridizes with wild *G. soya* as "big seed types" are commonly found in the wild (Wang *et al.*, 2008, 2010). These intermediate types are referred to as *Glycine gracilis*.

Much information has been developed on the evolutionary relationships within the subgenus *Glycine* (Doyle and Beachy, 1990; Doyle *et al.*, 1990, 2004). Numerous investigations on morphology, cytology, molecular markers and gene sequences have revealed a high number of nuclear genome groups and cytoplasms (Table 9.2, Fig. 9.7). As we discussed in Chapter 4, considerable diversity exists within each of these groups, and some of the polyploids have multiple origins. *Glycine tabacina* includes a diploid and two tetraploid types, which share one genome but differ at a second (Doyle *et al.*, 1999). *Glycine tomentella* represents a broad group of cytotypes ranging

Table 9.2. Species of *Glycine* and their genomes (based on Kollipara *et al.*, 1997; Hymowitz *et al.*, 1998).

Subgenus	Species	$2n$	Genome
Soya	*soya*	40	GG
	max	40	GG
Glycine	*albicans*	40	II
	arenaria	40	HH
	argyrea	40	A_2A_2
	canescens	40	AA
	cladestina	40	A_1A_1
	curvata	40	$C_1 C_1$
	cyrtoloba	40	CC
	falcata	40	FF
	hirticaulis	80	$H_1 H_1$
	lactovirens	40	$I_1 I_1$
	latifolia	40	$B_1 B_1$
	latrobeana	40	$A_3 A_3$
	microphyta	40	BB
	pindanica	40	$H_2 H_2$
	tabacina	40	$B_2 B_2$
		80	Complex
	tomentella	38	EE
		40	DD
		78	Complex
		80	Complex

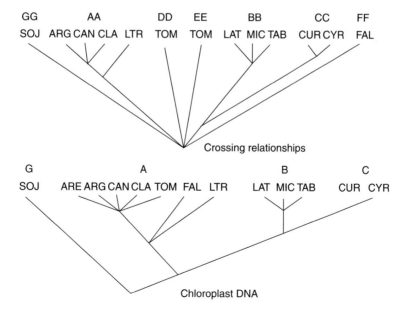

Fig. 9.7. Chloroplast DNA phylogeny of 13 of the *Glycine* species compared with their crossing relationships. Abbreviations are the first three letters of taxa listed in Table 9.2 (except: SOJ, *soya*; LTR, *latrobeana*). AA to GG are genome designations based on crossing studies; A to G are chloroplast clades. (Used with permission from J.J. Doyle, J.L. Doyle and A.H.D. Brown, © 1990, *Evolution* 44, 371–389, Society for the Study of Evolution.)

from $2n = 38$, 40, 78 and 80 and carrying two or more genomes (Doyle *et al.*, 2002). Many of the races have more than one origin, and inter-breeding between races has probably led to lineage recombination.

It is relatively clear that the cultivated soybean arose from *G. soya*. They are inter-fertile and have similar morphologies, distributions, isozyme banding patterns and DNA polymorphisms. What is still unclear is who their diploid ancestor was, because *G. soya* and *G. max* have chromosome numbers that are too high to be anything but ancient polyploids. Only a few remote relatives in the same subtribe have $2n = 20$, suggesting that the direct ancestor is either extinct or undiscovered (Kumar and Hymowitz, 1989). The duplication is estimated to have occurred about 15 Mya (Doyle *et al.*, 2004).

Most investigators feel that soybean cultivation began about 3000–4000 years ago (Hymowitz, 1995). Patterns of SSR and SNP diversity indicate that the origin of soybean was along the Yellow River of China (Guo *et al.*, 2010; Li, Y.-H. *et al.*, 2010). The major improvements instituted by people were the usual increased seed size, erect habit of growth and reduced seed shattering. Soybeans spread throughout China into Korea, Japan and Southeastern Asia during the expansion of the Ming dynasty 2000 to 3000 years ago. Europeans were aware of the soybean by 1712 and the Americans by 1765, but little planting was done before the 20th century (Hymowitz and Harlan, 1983). Soybeans were introduced into Brazil in 1882 to complete their entry into the New World (Hymowitz, 1995).

10

Starchy Staples and Sugar

Introduction

Cereals and legumes may feed the world, but numerous other crops are rich sources of carbohydrates and are regionally important. Cassava, taro, yam, sweet potato and white potato are the primary foodstuffs of people in many parts of the world where grains do not do well, particularly in the tropics. Sugarcane and sugarbeet are widely grown as a source of sugar for sweetening. The various starchy staples and sugars are not evolutionarily related, but most are propagated asexually and have far less protein than the legumes. In addition, most produce their edible parts under the ground.

Banana

There are two genera of wild bananas in the family *Musaceae*, *Musa* and *Ensete*. *Musa* species have been grouped into four or five sections using morphological patterns and chromosome numbers (Simmonds, 1962; Argent, 1976), and three sections using AFLP markers (Wong *et al.*, 2002). Most of our domesticated bananas were derived from the wild, diploid species ($2n = 2x = 22$) *Musa acuminata* Colla (A genome) and *Musa balbisiana* Colla (B genome) in the section *Eumusa*. The diploids are all located in Southeast Asia and the Pacific. As discussed in Chapter 4,

the majority of the cultivars are triploids with the genomic constitutions of AAA, AAB and ABB, although there are a few cultivated diploids (AB) and tetraploids (AAAB, AABB and ABBB) (Simmonds, 1995a). A wide array of nuclear and cytoplasmic molecular markers have been used to describe patterns of variability in wild bananas, trace evolutionary patterns and confirm the genomic identities of cultivated species (Heslop-Harrison and Schwarzacher, 2007; Perrier *et al.*, 2009; Nayar, 2010).

Considerable variability exists among the cultivated bananas, both within their center of origin in Southeast Asia and sub-Saharan Africa (Ortiz, 1997; Perrier *et al.*, 2009; Nayar, 2010). Humans have taken great advantage of the diversity produced in bananas through interspecific hybridization and the accumulation of somatic mutations. Numerous inflorescence types (Fig. 10.1) and flavors abound. While only a few cultivars of bananas are grown on a large scale for export, many distinct clones are cultivated on the local scale in Asia, Africa and South America. The AAA types produce the sweetest fruit (bananas) and are eaten fresh; the various AB hybrids are more starchy (plantains) and are cooked or used to make beer.

Two other bananas are of minor importance. Manila hemp, *Mus textilis* Née ($2n = 2x = 20$), was once popular in the Philippine islands and Borneo for its sheath fibers, which were made into rope. However, it is now little grown

Fig. 10.1. Extremes in bunch and fruit shapes of bananas found in Asia. From left: 'Klue Teparod' (AABB), 'Ripping' (ABB), 'Pisang Seribu' (AAB), 'Pitogo' (ABB) and a wild, non-pathenocarpic *Musa acuminata* subsp. *malaccensis* (AA). (Drawn from a photograph in P. Rowe and F.E. Rosales, © 1996, Bananas and plantains. In: J. Janick and J.N. Moore (eds) *Fruit Breeding*, John Wiley & Sons, New York.)

(Simmonds, 1995a). *Ensete ventricosum* (Welw.) Cheesm. $(2n = 2x = 18)$ is still grown in Ethiopia for its stems, which provide both starch and fiber. Both of these species have undergone minimal selection outside the wild.

All the varieties cultivated for fruit are parthenocarpic and seedless. Sexually produced bananas have large flinty seeds (Fig. 10.2), making edibility dependent on two characteristics: sterility and parthenocarpy. Seed set and zygotic viability vary greatly in both wild and cultivated populations of both *M. acuminata* and *M. balbisiana*, particularly among hybrids. Genetic female sterility is relatively common, and various types of meiotic errors lead to gametic

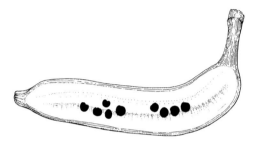

Fig. 10.2. Cross section of a mature, diploid banana showing some fully developed seeds. (Drawn from a photograph in P. Rowe and F.E. Rosales, © 1996, Bananas and plantains. In: J. Janick and J.N. Moore (eds) *Fruit Breeding*, John Wiley & Sons, New York.)

sterility, including chromosomal rearrangements and numerical imbalances. Parthenocarpy is regulated by several complementary dominant genes that are found in wild populations of *M. acuminata* (Simmonds, 1995a).

Diploids were probably first domesticated in the Malay region of Southeastern Asia, after their fruit had been harvested from the wild for tens of thousands of years. Bananas were probably being grown in the highlands of central New Guinea by about 7000 BP (Denham *et al.*, 2003, 2004). At a very early stage, human beings must have selected individuals with both parthenocarpy and seed sterility to obtain an edible fruit. Edibility is thought to have first appeared in wild *M. acuminata*. Hybridization with *M. balbisiana* occurred as AA cultivars were taken out of their center of origin, forming a classical polyploid complex (Fig. 10.3). The *balbisiana* types provided increased tolerance to cold and drought, while edibility in *M. balbisiana* was greatly enhanced by crosses with *M. acuminata* (Simmonds, 1995a).

The banana was probably introduced into Africa, India and the Polynesian islands from Indonesia–Malaysia. There is evidence of its cultivation in south Cameroon and Uganda, 2000–6000 years ago (Neumann and Hildebrand, 2009). The banana was first brought to the attention of the Europeans by Alexander the Great during his conquests. Portuguese travelers took the banana to the Canary Islands in the late 1400s and it reached Santo Domingo from there in 1516. Over the next century its cultivation spread throughout tropical America and the Caribbean region. Shipments to the USA began in the early 1900s from Central America and then Ecuador.

Cassava

Manihot esculenta Crantz (manioc, cassava or yuca) is a critical food source in tropical areas of South America, Africa, Asia and several oceanic islands (Fig. 10.4). It is grown primarily in wet lowlands, but produces well in somewhat arid environments and at elevations up to 2000 m. Cassava is a perennial shrub whose tuberous roots are higher in carbohydrate than rice or maize, although its roots are lower in protein (1–3%) (Heiser, 1990). New plantings are started by placing stem cuttings directly in the soil, and the plants often mature in less than 1 year. It is grown by mostly peasant farmers as a staple, but is also used to produce industrial starches, tapioca and animal feed. Cassava leaves are an excellent source of protein and are eaten in Africa as a pot herb.

Humans domesticated both bitter and sweet forms of cassava. The bitter types have

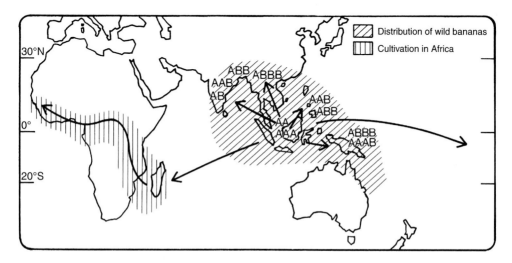

Fig. 10.3. Origin and migration of banana cultivation. (Used with permission from N.W. Simmonds, © 1985, *Principles of Crop Improvement*, Longman, London.)

Fig. 10.4. The manioc or cassava plant. Its total height can reach over 5 m.

high levels of cyanogenic glycosides that must be removed, which is done by peeling, grating, boiling and draining (Cock, 1982). The resulting pulp is then used directly to make a kind of bread, or it is dried and powdered for later consumption after frying. The grated cassava is also squeezed into porous tubes made of basketry and the resulting sap is used for sauces and alcoholic beverages. Tapioca is prepared by partially cooking small pellets of cassava starch.

The genus *Manihot* is a member of the *Euphorbiaceae*. There are about 100 species of *Manihot* and all have 36 chromosomes. The Brazilian species have been arranged into 16 groups (Allem *et al.*, 2001; Allem, 2002). Many of the species have been successfully intercrossed; most of the resulting hybrids have normal chromosome pairing at meiosis, with only a few univalents and tetravalents (Hahn *et al.*, 1990; Bai, 1992; Nassar *et al.*, 1995).

Even though cassava shows regular bivalent pairing, it is probably polyploid in origin due to its high chromosome number. Magoon *et al.* (1969) suggested that it was a segmental polyploid derived from two species with six similar and three dissimilar chromosomes. However, disomic inheritance has been documented across a wide range of RAPD, microsatellite and isozyme loci (Fregene *et al.*, 1997). Since all existing species have the same chromosome number, polyploidization must have been an ancient event

that occurred when the genera first emerged, long before domestication.

Until recently, manioc was believed to be hybrid derived. Rogers and Appan (1973) felt that cassava arose from the hybridization of two closely related species. Likely candidates were thought to be *Manihot aesulifolia*, *Manihot isoloba*, *Manihot rubricaulis* and *Manihot pringeli*, as all are erect with tuberous roots like *M. esculenta*. Using patterns of variability in chloroplast DNA, nuclear ribosomal DNA and AFLPs, *M. esculenta* subsp. *flabellifolia* (or *Manihot tristis*) was determined to be one of the progenitors, but these markers left the possibility of other relationships unresolved (Fregene *et al.*, 1994; Roa *et al.*, 1997). Most recently, Olsen and Schaal used DNA sequence and microsatellite data to show that cassava was derived solely from wild *M. esculenta* subsp. *flabellifolia* (Olsen and Schaal, 1999, 2001; Olsen, 2004). They found little evidence of recent hybridization, even with *Manihot pruinosa*, which is sympatric and interfertile (Olsen, 2002).

The location where cassava was first domesticated has also been long debated, with origins proposed in the dry scrublands of northeastern Brazil, the savannahs of Columbia and Venezuela, the rainforests of Amazonia, and the warm, moist lowlands of Mexico and Central America (Ugent *et al.*, 1986). The earliest archeological evidence of cassava cultivation has been found in Panama from 7000–7800 BP (Piperno *et al.*, 2000; Dickau *et al.*, 2007). The molecular data indicate that cassava was probably first domesticated along the southern border of the Amazon basin (Olsen and Schaal, 1999, 2001).

At the arrival of the Europeans, only sweet forms of cassava were cultivated on the Peruvian coast, Central America and Mexico. Cassava was used as an important garden crop, but not a staple (Sauer, 1993). In South America, sweet types were also grown as garden vegetables, but bitter forms were utilized as a staple crop. Cassava was restricted to the New World until the end of the 16th century, when sailors took it to the west coast of Africa. It had been used in the African slave trade as a staple provision. Once established on the west coast of Africa, it spread rapidly along the islands of Réunion, Madagascar and Zanzibar to the east coast, and finally moved into the interior from both coasts by the 19th century (Jennings, 1995). It found its way to India by 1800. Most of the introductions were by cuttings, but self-sown seedlings became established in most regions, resulting in high levels of local and regional diversity. Numerous landraces still exist, particularly among indigenous farmers who maintain tremendous amounts of diversity (Boster, 1985; Salick *et al.*, 1997).

Potato

The edible, tuber-bearing *Solanum* species (Fig. 10.5) are a small part of a very large genus. Potatoes belong to section *Petota* (*Tuberarium*), subsection *Potato* (*Hyperbasarthrum*) of *Solanum*. About 100 wild species and four cultivated species are now recognized (van den Berg *et al.*, 1998; Spooner *et al.*, 2007; Ovchinnikova *et al.*, 2011), down from a previous estimate of 217 wild species and seven cultivated ones (Hawkes, 1990). Their basic chromosome number is

$x = 12$, and they are found at several ploidy levels, including diploid, triploid, tetraploid, pentaploid and hexaploid. Both autopolyploids and allopolyploids are present among the higher ploidies. The diploids have an *S*-allele incompatibility system and are primarily outcrossed, while the polyploids are self-compatible and generally self-pollinated (Simmonds, 1995b). The most important cultivated species, *Solanum tuberosum*, is considered by most to be autopolyploid because it displays tetrasomic inheritance (Martinez-Zapater and Oliver, 1984; Douches and Quiros, 1988). All the diploid species produce relatively high proportions of unreduced gametes (Mok and Peloquin, 1975; Iwanaga and Peloquin, 1982).

Tetraploid *S. tuberosum* subsp. *tuberosum* is by far the most widely grown taxa, but diploid *Solanum ajanhuiri*, triploid *Solanum juzepczukii* and pentaploid *Solanum curtilobum*, are still grown by native South Americans. All three of these species are of hybrid origin (Rodríquez *et al.*, 2010).

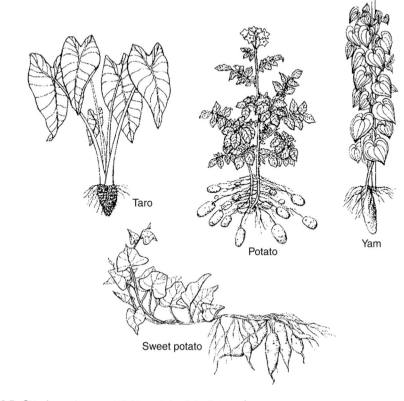

Fig. 10.5. Starchy root crops, potato, sweet potato, taro and yam.

The related tetraploid, *S. tuberosum* subsp. *andigena*, is also cultivated at some locations in cool temperate Chile. The origins of the two subspecies of *S. tuberosum* are unclear, although molecular data do suggest that they are distinct (Raker and Spooner, 2002).

The majority of native species are located in South America and 75% of them are diploid (Hawkes, 1990; Ochoa, 1990; Spooner and Hijmans, 2001). In the frost-free Andean valleys are found diploids in the *Solanum brevicaule* complex. Most cross easily, and natural hybrids have been reported in most combinations (Hawkes, 1990). Substantial interspecific hybridization is still going on within wild and cultivated populations, and many hybrid swarms have been described (Brush *et al.*, 1981; Johns and Keen, 1986). Because of this promiscuity and morphological continuity, van den Berg *et al.* (1998) have suggested that all the South American species in the *brevicaule* complex should be considered one species, *S. tuberosum*. Molecular data indicate there may be northern and southern populations that could be considered as species, but they cannot be readily distinguished from each other (Spooner *et al.*, 2003).

Probably the first step in South American potato cultivation was the collection of clones that were edible. Wild tubers are generally very bitter to the taste and can contain toxic amounts of alkaloids. The exact period when people began this activity is unknown, but the oldest evidence of potato use dates to 13,000 BP (Urgent *et al.*, 1982). The original area of domestication was probably in the high plateau of Bolivia–Peru and probably had a monophyletic origin (Spooner *et al.*, 2005a). Potato cultivation spread throughout highland South America as a complex of diploid, triploid and tetraploid forms.

The potato was first taken to Europe by Spanish invaders in 1567 and was spread throughout Europe by the end of the century. An Andean origin for the European material has long been accepted for those early potatoes (Salaman, 1949; Hawkes, 1967) and a recent comparison of herbarium specimens using a plastid deletion marker supports this view (Ames and Spooner, 2008). *S. tuberosum* subsp. *andigena* was an important crop until the fungus disease late blight almost eliminated it in the 1840s. It has been widely asserted that a clone of resistant *S. tuberosum* subsp. *tuberosum* (Rough Purple Chili) was introduced in the mid-1800s from Chile and filled the void left by subsp. *andigena* (Goodrich, 1863; Grun, 1990); however, data from Ames and Spooner (2008) indicate that the Chilean potato became prominent before the late blight epidemics. The Andean potato first appeared in Europe around 1700 and persisted until 1892, while the Chilean potato arrived in Europe in 1811, about 30 years before the late blight epidemic. Over 99% of the modern potato cultivars now possess Chilean cytoplasm (Hosaka, 1993, 1995; Powell *et al.*, 1993; Provan *et al.*, 1999).

Sugarbeet

The cultivated beet, *Beta vulgaris* L. spp. *vulgaris*, belongs to the goosefoot family, *Chenopodiaceae*. There are numerous types of domesticated beets: (i) leaf beets, which are used as a leafy vegetable and do not have a swollen hypocotyl; (ii) garden beets, whose swollen hypocotyl is eaten as a salad vegetable; (iii) mangels, whose swollen hypocotyl is used primarily as a fodder crop; (iv) sugarbeets, whose root is an important sugar source; and (v) fodder beets, which are used to feed livestock and are hybrids of mangels and sugarbeets (Fig. 10.6). By far the most widely grown type is the sugarbeet, which predominates in Europe, the former USSR and North America; it is the source of about half of the world's sugar production.

All the cultivated types were derived from the subsp. *maritima* (L.) Thell., which occurs naturally on seashores in cool temperate parts of Europe and Asia. Both morphological and molecular evidence are consistent with this possibility (Coons, 1975; Mikami *et al.*, 1984; Kishima *et al.*, 1987). *B. vulgaris* is mostly outcrossed and anemophilous (Dark, 1971). It is primarily a diploid ($2n = 2x = 18$), although autopolyploids arise periodically via unreduced gametes. Tetraploid cultivars have been produced artificially with little commercial success, but triploid types have become important in Europe. Cytoplasmic male sterility has been exploited to produce commercial hybrid seed. There is considerable gene flow between wild and cultivated forms, and the weed beets found in many production fields are introgressants of wild and cultivated beets (Bartsch, 2011). Also, the wild beets found in California

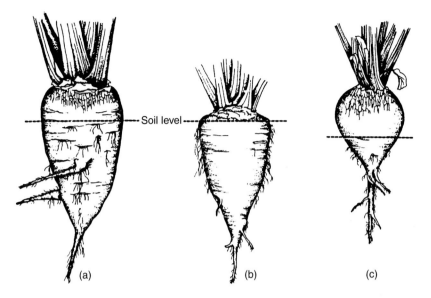

Fig. 10.6. Various forms of beets: (a) fodder beet, (b) sugarbeet and (c) garden beet. Note differences in soil level. (Used with permission from R.H.M. Langer and C.D. Hill, © 1982, *Agricultural Plants*, Cambridge University Press, New York.)

probably evolved from feralized populations of cultivated beet (Bartsch and Ellstrand, 1999).

Domestication probably began in the eastern Mediterranean, where the leaves were used as pot herbs and animal feed (Ford-Lloyd and Williams, 1975). Leaf beets were described in the ancient writings of Aristotle and Theophrastus. Romans used the beet rather extensively, and it found its way from Italy to Europe with the "barbarian" invaders. Beets were introduced into the USA in 1800.

Beet was first recognized to have high sugar levels in the 17th century (Deerr, 1949). Marggraf discovered sugar in the sap of the root in 1747, and his student Achard was given a grant by the King of Prussia to develop a commercial industry. The first sugar factory was built in 1801 at Kunern, Silesia. The sugarbeet industry in the USA was established at the end of the 19th century. The early sugarbeet cultivars contained approximately 6% sugar, while the modern types have over 20%.

Sugarcane

There are six commonly recognized species of *Saccharum*, the genus of sugarcane (Amalraj and Balasundaram, 2006; Table 10.1). They are all largely cross-pollinated by wind and suffer inbreeding depression. All are polyploids with very high chromosome numbers and considerable amounts of aneuploidy. Polysomic inheritance predominates within the group, although it is not complete (Al-Janabi *et al.*, 1993, 1994; Da Silva *et al.*, 1993). All sugarcane species can be successfully crossed, and they have limited interfertility with numerous other genera, including *Narenga*, *Erianthus*, *Sorghum*, *Schlerostachya*, *Imperata* and *Zea* (Stevenson, 1965).

A generally consistent phylogeny of *Saccharum* has emerged using morphological and cytological data (Daniels and Roach, 1987). *Saccharum spontaneum* is an autopolyploid, but the other species originated from interspecies hybridization. *Saccharum robustum* represents a very diverse population of introgressants and may have arisen as an intergeneric hybrid of *S. spontaneum* and several other species. *Saccharum edule* may be an allopolyploid of *Saccharum officinarum* and *S. robustum*. *Saccharum barberi* probably resulted from the natural hybridization of *S. spontaneum* and *S. officinarum*. *Saccharum sinense* is closely related to *S. barberi*, but can be distinguished from it based on its horticultural characteristics (Daniels *et al.*, 1991). Various molecular markers have

Table 10.1. Sugarcane species (*Saccharum*) and their chromosomal numbers.

Species	Status	Chromosome Number (2n)	Location
S. barberi Jeswiet	Cultivated	82–92	Northern India
S. edule Hassk.	Cultivated	60, 70 and 80	New Guinea
S. officinarum L.	Cultivated	80	Worldwide tropics
S. robustum Brandes & Jeswiet	Wild	60 and 80	Borneo to New Guinea
S. sinense Roxb.	Cultivated	82–124	Southeast Asia, China
S. spontaneum L.	Wild	40–180	Northern Africa to China

supported the separation of these species, as well as the separation of the genus *Saccharum* from *Miscanthus* and *Erianarum* (Jannoo et al., 1999; Takahashi *et al.*, 2005; Selvi *et al.*, 2006).

Sugarcane was probably first domesticated in or near New Guinea approximately 10,000 years ago. Because of its morphology and similar chromosome number, *S. robustum* is thought to be the progenitor of *S. officinarum*. Cultivation probably moved northward toward continental Asia, where *S. officinarum* ("noble" canes) hybridized with *S. spontaneum* to produce *Saccharum sinense* (hybrid "thin" canes). These hybrids were less sweet and less robust than the noble canes, but were hardier and could be grown successfully in subtropical mainlands (Sauer, 1993). This new type eventually became established in the India–China area. At the same time, *S. officinarum* was moving eastward across the Pacific. The other cultivated species, *S. barberi* and *S. edule*, also had early origins, but their spread was much more limited.

Sugar manufacture first began about 3000 years ago in India; before that time the canes were grown as garden plants for chewing (Sauer, 1993). Europeans did not become aware of the crop until the explorations of Alexander the Great in Asia. Columbus is credited with introducing sugarcane into the New World in the late 15th century. Today the various sugarcane species are found all across the tropics and are the source of approximately 50% of the world's sugar.

Sweet Potato

Ipomoea is a pantropical genus with more than 50 species recognized. It belongs to the family *Convolvulaceae* and displays a polyploid series of $2n = 30$, 60 and 90 (Austin, 1978; McDonald and Austin, 1990). The *Batatas* complex includes the sweet potato, 12 other species and an interspecific hybrid. A- and B-genome species are recognized (Rajapakse *et al.*, 2004). The domesticated sweet potato, *Ipomoea batatas* (L.) Lam, is the only hexaploid in the section.

There has been considerable debate about the phylogenetic organization of the sweet potato and its close allies. Based on morphological data, Ting and Kehr (1953) felt that the sweet potato was an allotetraploid derived from unknown species, while Nishiyama et al. (1975) suggested that it was an autohexaploid of the diploid *Ipomoea trifida* and was part of a polyploid complex (Fig. 10.7). Nishiyama and Temamura (1962) found what they thought was a wild hexaploid type in Mexico, but it was later shown to be a feral sweet potato (Jones, 1967). Jones (1967) and Martin and Jones (1972) felt that *I. trifida* and *I. batatas* evolved separately from an unknown, but common, ancestor. Recent molecular data have indicated that *I. batatas* is much more closely related to *I. trifida* than any other diploid species, strongly supporting an autopolyploid origin for the sweet potato (Huang and Sun, 2000; Srisuwan *et al.*, 2006). Diploid *I. trifida* produces $2n$ pollen (Jones, 1990; Orjeda *et al.*, 1990) and the chromosomes of $2x$ forms associate readily with $6x$ types (Shiotani and Kawase, 1989). *I. trifida* is the only diploid species that is self-incompatible, similar to the sweet potato (Diaz *et al.*, 1996).

The place where sweet potato was first domesticated is still unclear. The oldest archeological proofs of sweet potatoes are found in Peru (8000–10,000 BP), but these may be the remains of gathering (Ugent *et al.*, 1982) and a Middle

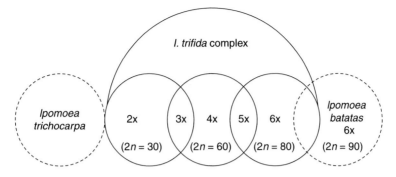

Fig. 10.7. Proposed polyploid complex of *Ipomoea trifida*. (Used with permission from K. Kobayashi, © 1984, *Proceedings of the Sixth Symposium of the International Society for Tropical Fruit Crops.*)

American origin cannot be excluded. Most likely, the sweet potato originated in either the Central or South American lowlands (O'Brien, 1972). Zhang *et al.* (2000) and Huang and Sun (2000) found the highest amount of molecular diversity in Central America, leading them to propose it as the center of origin. Other ancient remains have been found in Hawaii, New Zealand and across Polynesia (Yen, 1982). It is not clear how the crop reached Polynesia from America before the voyages of the Europeans; however, a number of explanations have been presented ranging from early transoceanic movements of humans to natural dispersal by drifting capsules (Sauer, 1993).

One thing is clear: Columbus discovered the sweet potato in the West Indies and was the first to take it back to Europe (Yen, 1982). It was then spread along the so-called '*batata* line' to Africa, Brazil and India by Portuguese explorers who used it to prevent scurvy (*batata* was the West Indian Arawak's name for sweet potato). Spanish trading galleons dispersed sweet potatoes along the 'camote line' between Mexico and the Philippines (the word camote was derived from the Mayan word *camotli*). Based on AFLP diversity patterns, Rossel *et al.* (2000) suggest that the major source of sweet potato germplasm into Oceania came from Mesoamerica. The sweet potato was introduced into southern China by 1594 and Japan by 1674.

Taro

Taro (*Colocasia esculenta*) belongs to the *Araceae* or aroid family. It is the primary starch source of millions of tropical peoples, most notably those located in the Pacific and Caribbean islands and West Africa. It is grown for its underground corms (Fig. 10.5), and is farmed in flooded or swampy land that will support little else that is edible. Wild populations of *C. esculenta* are found throughout south-central Asia and exist at two ploidy levels, $2n = 2x = 26$ and $2n = 3x = 42$ (Kuruvilla and Singh, 1981; Coates *et al.*, 1988). Two botanical varieties can be distinguished by their corm characteristics, var. *esculenta* (diploid, dasheen type) and var. *antiquorum* (triploid, eddoe type). Two major germplasm pools have been identified based on isozyme and AFLP variation patterns, Southeast Asian and Melanesian (Lebot and Aradhya, 1991; Kreike *et al.*, 2004). Taro is reproduced almost exclusively asexually, which has resulted in its landraces being extremely variable (Kuruvilla and Singh, 1981; Sreekumari and Mathew, 1991; Quero-Garcia *et al.*, 2004).

Many anthropologists believe that taro was the first irrigated crop (Plucknett, 1979) and its cultivation may be more ancient than rice at many locations (Formosa, the Philippines, Assam and Timor). Domestication probably began in Indo-Malaya about 7000 BP, where it spread east to Asia and the Pacific islands and west towards Arabia. Fullagar *et al.* (2006) have found evidence that taro was first exploited in the highlands of Papua New Guinea by 10,200 BP and was being cultivated by about 6500–7000 BP. Traders and explorers took taro to West Africa in approximately 2000 BP, and it arrived several hundred years ago in the Caribbean islands and tropical America via slave ships. Taro was brought to Japan in prehistory, either from Taiwan or the Yangtze River area of China (Sasaki, 1986).

A number of other aroids are important food sources, including the Asian types *Alocasia indica*, *Alocasia macrorrhiza* and *Cyrtosperma chamissonis*, and the South American types *Xanthosoma atrovirens*, *Xanthosoma sagittifolium* and *Xanthosoma violaceum* (Table 10.2). Based on RAPD and isozyme data, *Alocasia* appears to be more closely related to *Colocasia* than *Xanthosoma*, which fits with their overlapping geographic distributions (Ochiai *et al.*, 2001). The origin of these aroids is generally clouded, although *Xanthosoma* or "cocoyam" cultivation is known to pre-date Columbus in Hispaniola, Central and South America and the Caribbean (Sauer, 1993). Cocoyam was introduced into West Africa in the colonial period as slave provisions and has become a major crop in many tropical areas. In Nigeria, taro (*Colocasia*) is called "old cocoyam" and *Xanthosoma* is referred to as "new cocoyam" to distinguish the original African aroid from the imported one (Valenzuela and DeFrank, 1995).

Yam

Yams are represented by hundreds of Old and New World species of *Dioscorea* in the family Dioscoreaceae; overall, there are ten yam species cultivated (Table 10.3). Of the cultivated yams, by far the most important are *Dioscorea cayenensis* and *Dioscorea rotundata* (Africa), *Dioscorea alata* (Southeast Asia and the South Pacific) and *Dioscorea trifida* (South America) (Arnau *et al.*, 2010). *D. trifida* has $2n = 80$ chromosomes, while *D. cayenensis* and *D. rotundata* have numbers ranging from 54 to 80 (Baquar, 1980; Dansi *et al.*, 2001), and *D. alata* has $2n = 40, 60$ and 80 (Arnau *et al.*, 2009; Obidiegwu *et al.*, 2010). The most common number in *D. alata* is $2n = 40$. The base chromosome number of *Dioscorea* has long been considered to be $x = 9$ or 10, but recent microsatellite segregation analyses suggest that the base number may actually be $x = 20$ (Hochu *et al.*, 2006; Arnau *et al.*, 2009; Bousalem *et al.*, 2010).

The evolutionary relationships of the yam species are ambiguous. Bousalem *et al.* (2006)

Table 10.2. Major species of edible aroids (based on Plucknett, 1979).

Location	Species	Chromosome number (2n)
Asia	*Alocasia indica* Schott.	28
	Alocasia macrorrhiza Schott.	26, 28
	Colocasia esculenta (L.) Schott.	28, 42
South America	*Cyrtosperma chamissonis* (Schott.) Merr.	26
	Xanthosoma atrovirens Koch.	26
	Xanthosoma sagittifolium Schott.	26
	Xanthosoma violaceum Schott.	26

Table 10.3. Most important cultivated yam species (*Dioscorea*) (from Arnau *et al.*, 2010).

Species	Geographic origin	Common names
D. alata L.	Southeast Asia, Melanesia	Greater, water, winged yam
D. bulbifera L.	South America, Africa, Asia, Melanesia	Aerial, bulbil, bearing yam
D. cayenensis Lam.	West Africa	Yellow Guinea yam
D. esculenta (Lour.) Burk.	Southeast Asia, Melanesia	Lesser yam, Asiatic yam
D. nummularia Lam.	Melanesia	Spiny yam, wild yam
D. opposita-japonica Thunb.	Japan, China	Chinese, Japanese yam
D. pentaphylla L.	Southeast Asia, Melanesia	Five-leaved yam
D. rotundata Poir.	West Africa	White Guinea yam
D. transversa Br.	Australia, Melanesia	Marou, wael
D. trifida L.	South America	Aja, aje, cush-cush, yampi

have documented tetrasomic inheritance in cultivated *D. trifida*, implying that it is autopolyploid. However, no wild tetraploids of *D. trifida* have been found, only diploids in French Guiana (Bousalem *et al.*, 2010). Based on AFLP and chloroplast markers, the wild and cultivated accessions of *D. trifida* in French Guiana are distinct from any other wild yam species and are probably closely related (Bousalem *et al.*, 2010). Triploids are also found in the coastal region of Guiana, where wild and cultivated forms are in proximity.

Based on this information, the cultivated yam probably arose through the unification of unreduced gametes in northeastern Amazonia, where the wild progenitors are located. An Amazonian origin for *D. trifida* has been assumed for a long time (Alexander and Coursey, 1969), but a wild progenitor species was previously unknown. There is no archeological evidence of the early domestication of the yam in Amazonia but, similar to cassava, it was being grown in Central America by 5000–7000 BP (Piperno *et al.*, 2000; Dickau *et al.*, 2007).

The cultivated yams, *D. rotundata* and *D. cayenensis*, likely arose in the "yam belt", ranging from central Côte d'Ivoire to the mountains of Cameroon (Coursey, 1967; Langemann, 1977). Whether these two Guinea yams are separate species is debatable, as there are many intergrading, intermediate types (Hamon and Touré, 1990; Dansi *et al.*, 1999). A number of wild species have been identified as potential ancestors of the *D. rotundata–D. cayenensis* complex (Terauchi *et al.*, 1992; Mignouna and Dansi, 2003). Two wild species in savannah areas (*Dioscorea abyssinica* Hochst ex Kunth) and humid forests (*Dioscorea praehensilis* Benth) are thought to be closely related to *D. rotundata*, while a group of seven or eight rainforest species could have contributed to *D. cayenensis*. Yams were probably domesticated in West Africa about 5000–7000 BP (Alexander and Coursey, 1969; Dumont *et al.*, 2005).

The origin of *D. alata* is unclear, although Prain and Burkill (1936) suggested that it was first domesticated in Indochina from the wild species found there, *Dioscorea hamiltonii* and *Dioscorea persimilis*. Mignouna *et al.* (2002) suggested that $2n = 40$ *D. alata* was an allotetraploid with two genomes PPHH (P = *D. persimilis*, H = *D. hamiltonii*). Malpa *et al.* (2005, 2006), using AFLP markers, found *D. alata* to be closer to *Dioscorea nummularia* and *Dioscorea transversa* than *D. persimilis*. These species are restricted to the Southeast Asia islands and Oceania. Most recently, Arnau *et al.* (2009) has used SSR segregation data to propose that the $2n = 40$ *D. alata* are diploid rather than tetraploid. There are no direct data for when *D. alata* might have been domesticated, although Lebot (2009) has speculated around 6000 BP.

11

Fruits, Vegetables, Fibers and Oils

Introduction

Even though the grains, legumes and starchy roots are of primary importance in feeding the world, there are numerous other crops that play an important role in human diets. As was already mentioned, Harlan (1992) lists over 250 food crops and suggests there are hundreds more. These crops are sometimes a major source of calories, but they are more commonly a supplemental source of nutrients that break up the tedium of limited crop diets. They are also used as drugs, medicine and fiber. In this chapter, we will discuss a number of the most frequently consumed fruits, vegetables, fibers and oils.

Fruits

Apple

The genus of apple, *Malus*, belongs to the subfamily *Pomoideae* of the *Rosaceae* family. Another important fruit tree, pear (*Pyrus*), belongs to the same subfamily. There are over 30 primary species of apple and most can be readily hybridized (Korban, 1986; Way *et al.*, 1991). Its primary wild ancestor is *Malus sieversii*, whose range is centered at the border between western China and the former Soviet Union. Apples are the main forest tree there and display the full range

of colors, forms and tastes found in domesticated apples across the world (Forsline *et al.*, 1994; Hokanson *et al.*, 1997b). The domesticated apple has long been referred to as *Malus × domestica* (Korban and Skirvin, 1984), although Mabberley *et al.* (2001) suggest that *Malus pumila* should be used to refer to the domesticated apple and its presumed wild relative *M. sieversii*. Other species of *Malus* that contributed to the genetic background of the apple likely include: *M. orientalis* of Caucasia, *M. sylvestris* from Europe, *M. baccata* from Siberia, *M. mandshurica* from Manchuria and *M. prunifolia* from China. It is likely that these species hybridized with domesticated apples as they were spread by humans (Harris *et al.*, 2002).

The bulk of the apple species are $2n = 2x = 34$ (Table 11.1), although higher somatic numbers of 51, 68 and 85 exist; several of the cultivated types are triploid (Chyi and Weeden, 1984). Apples display disomic inheritance (Weeden and Lamb, 1987), are largely self-incompatible and many are apomictic. They are propagated vegetatively, usually as composites with a separate rootstock and scion.

It is likely that the high chromosome number of apple represents an ancient genomic duplication, since there are several other rosaceous fruit species with lower haploid chromosome numbers of $n = 8$ and 9. The genome sequence of apple indicates that there was a genome-wide duplication from nine ancestral chromosomes to

Table 11.1. Distribution of selected apple species (*Malus*) in subsection *Pumilae* and their chromosome numbers (adapted from Way *et al.*, 1991).

Species	Chromosome number (2*n*)	Distribution
M. asiatica Nakai	34	North and northeast China, Korea
M. baccata (L.) Borkh.	34, 68	North and northeast China
M. × *domestica*	34, 51, 68	Worldwide
M. floribunda (Siebold) ex Van Houtte	34	Japan
M. halliana Koehne	34	Japan
M. hupehensis (Pamp.) Rehder	51	Central China
M. mandshurica (Maxim.) Kom. ex Skvortsov	34	Manchuria
M. micromalus Makino	34	Southeast China, Korea
M. orientalis Uglitzk.	?	Caucasia
M. prunifolia (Willd.) Borkh.	34	North and northeast China, Korea
M. pumila Mill.	34	Europe
M. sieversii (Ledeb.) M. Roem.	?	Northwest China
M. spectabilis (Aiton) Borkh.	34, 68	China
M. sikkimensis (Wenz.) Koehne ex C. K. Schneid.	51	Himalayas
M. sylvestris (L.) Mill.	34	Europe

17 (Velasco *et al.*, 2010). Isozyme studies in *Malus* support an allopolyploid origin based on the presence of duplicated gene systems, allele segregations and fixed heterozygosities (Chevreau *et al.*, 1985; Weeden and Lamb, 1987; Dickson *et al.*, 1991). An allotetraploid origin involving ancestral *Spiroideae* (mostly *x* = 9) and *Amygdaloideae* (*x* = 7) was proposed by Sax (1931, 1933) and is supported by flavonoid chemistry (Challice, 1974; Challice and Kovanda, 1981) and morphological traits (Phipps *et al.*, 1991).

Apples were certainly one of the earliest fruits to be gathered by people, and their domestication was probably preceded by a long period of unintentional planting via garbage disposal. It is difficult to determine exactly when the apple was first domesticated, but the Greeks and Romans were growing apples at least 2500 years ago. They actively selected superior seedlings and were budding and grafting 2000 years ago (Janick *et al.*, 1996). The most likely beginning of cultivation was in the region between the Caspian and Black seas (Vavilov, 1949–1950); apple cultivation had reached the Near East by 3000 BP (Zohary and Hopf, 2000). The Romans spread the apple across Europe during their invasions and it was dispersed to the New World by European settlers during the 16th century.

The passage of trade routes from China to the Middle East and Europe through central Asia probably facilitated repeated short and long distance dispersal to the east and west, either intentionally or unintentionally, of *M. sieversii* and its hybrid derivatives. The *M.* × *domestica* Borkh complex then may have arisen through hybridization to the east with species native to China including *M. prunifolia* (Willd.) Borkh., *M. baccata* (L.) Borkh., *M. mandshurica* (Maxim.) Kom. ex Skvortsov and *M. sieboldii* (Regel) Rehder. To the west, hybridization with the local species *M. sylvestris* (L) Mill. and *M. orientalis* Uglitzk has been conjectured (Ponomarenko, 1983; Morgan and Richards, 1993; Hokanson *et al.*, 1997b; Juniper *et al.*, 1999).

In southern and eastern Asia, Nai, or the Chinese soft apple, *M.* × *asiatica* Nakai, was the primary cultivated apple for over 2000 years until *M.* × *domestica* was introduced in the late 19th and early 20th centuries (Morgan and Richards, 1993; Zhang *et al.*, 1993; Watkins, 1995; Zhou, 1999). *M.* × *asiatica* is probably a hybrid complex derived primarily from *M. sieversii* with *M. prunifolia* and perhaps other species.

Citrus

The genus *Citrus* contains a confusing array of diversity (Fig. 11.1), and anywhere from three

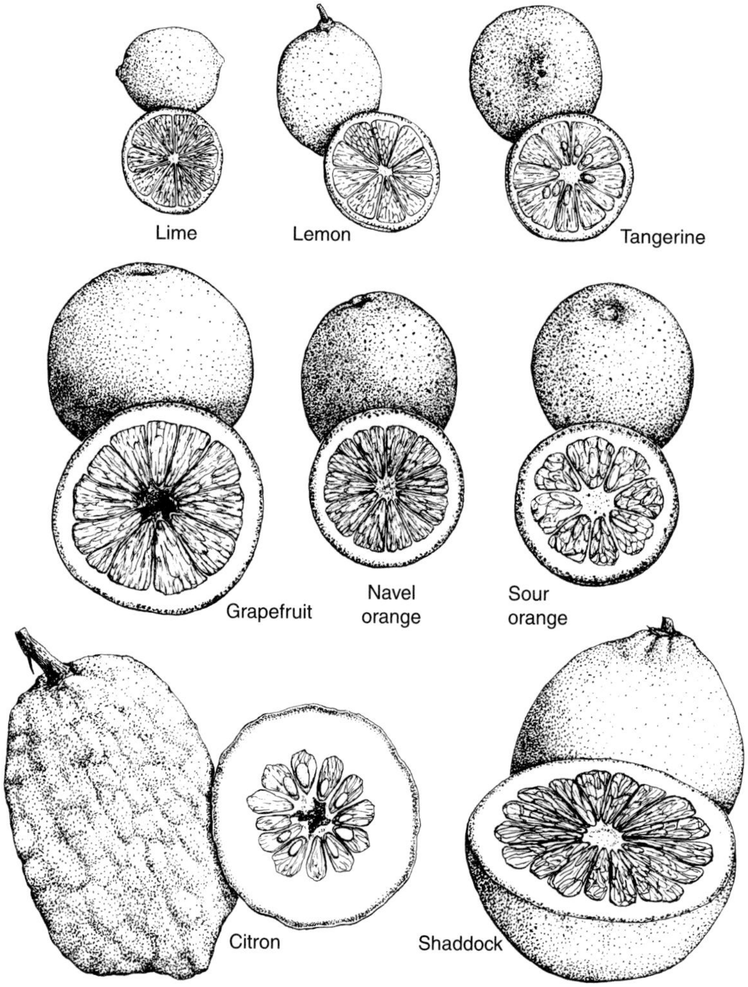

Fig. 11.1. Different types of citrus crops.

to 145 species have been recognized, depending on the authority (Tanaka, 1954; Hodgson, 1961; Swingle, 1967). Probably the most comprehensive attempt at citrus taxonomy was undertaken by Barrett and Rhodes (1976). They studied variation patterns in 147 morphological characters using numerical methods and came to the conclusion that there are only three clearly distinct taxa: (i) *Citrus medica* (citron); (ii) *Citrus gradis* (pummelo); and (iii) *Citrus reticulata* (mandarin). The difficulty in assorting more groups stems from the fact that many of the commercial species are morphologically similar and highly inter-fertile. Apomixis is also rampant in the genus, and many of the types prob-

ably arose from a common ancestor via asexual means (Roose *et al.*, 1995). The only simple part about *Citrus* taxonomy is that polyploidy is rare and most species share a basic chromosome number of $x = 9$.

Recent molecular data have provided information about the origin of many citrus taxa (Nicolosi *et al.*, 2000; Barkley *et al.*, 2006; Bayer *et al.*, 2009; Li, X. *et al.*, 2010). Early types of pummelo and mandarin probably hybridized to produce the sweet and sour oranges, and pummelo provided the cytoplasm. Lemon was derived from citron and sour orange. The lime was a hybrid of citron and one or more other taxa. Mandarin was the donor of at least some lime cytoplasm. The grapefruit was a likely product of pummelo and sweet orange. A number of different karyotypes are found in grapefruit, probably resulting from independent crosses between pummelos of variant karyotype constitutions and sweet oranges (de Moraes *et al.*, 2007).

The origins of most citrus taxa are clouded, but some conjectures can be made (Sauer, 1993). Citron may have originated in India and was spread prehistorically to the Near East and China. It arrived in Greece by 2500 BP. Pummelo and mandarin cultivation probably began in tropical Southeast Asia and spread to China by 2500 BP. Sweet and sour oranges are thought to have been generated repeatedly wherever pummelos and mandarins were grown together, although the complex and stable karyotype of sweet orange cultivars suggests a monophyletic origin (Pedrosa *et al.*, 2000). Oranges were first recorded in Chinese writings about the same time as pummelo and mandarin. The lime probably had a Southeastern Asian origin. The origin of the lemon is unknown, but it was most likely generated repeatedly wherever citron and lime cultivation overlapped. The grapefruit is of very modern origin, having emerged as a hybrid of pummelo and sweet orange on the island of Barbados in 1750.

The citrus species found their way to Europe and ultimately to the New World through a variety of routes. The pummelo moved through India to North Africa and arrived in Spain about 800 years ago. Lemon may have been spread from India to Rome via the Near East. Mandarin did not find its way into Europe before the 19th century and the lime did not reach temperate Asia or Europe until almost modern times. Columbus introduced a variety of citrus species to the New World in 1497, and these were spread all over the world by the Portuguese and Spanish during their expeditions in the 16th century.

Grape

As with citrus, the traditional grape taxonomy is confusing and somewhat artificial (Alleweldt *et al.*, 1991). About 60 species of *Vitis* are now recognized across the temperate world and all of them can be successfully intercrossed (Table 11.2). There is considerable geographical

overlap among species and interspecific hybrids are common (Olmo, 1995). In fact, interspecific hybridization played an important role in the spread of grape cultivation. The domesticated grape, *Vitis vinifera* L., was derived directly from native populations of *V. vinifera* subsp. *sylvestris*, but as grape cultivation entered new regions, the cultigens hybridized with local types to produce better adapted material. In spite of this genic mixing, wild and cultivated grapes still show considerable amounts of geographic and ecological divergence (Lamboy and Alpha, 1998; Sefc *et al.*, 2000).

Geographical origins and morphological characteristics have been used to subdivide *V. vinifera* into three morphotypes using morphology and geographical origins: *occidentalis*, *pontica* and *orientalis* (Negrul, 1938; Owens, 2008). The *occidentalis* group has the smallest berries and clusters, very fruitful shoots, and is of western European origin. The *orientalis* group has the largest berries, loose clusters and is from central Asia. The *pontica* group from Eastern Europe and the Black Sea Basin has intermediate characteristics. An important source of cultivar variation in *V. vinifera* has been through bud sports or somatic mutations (called "clones"), particularly those mutations altering berry pigmentation (e.g. 'Pinot noir', 'Pinot gris' and 'Pinot blanc' or 'Cabernet Sauvignon', 'Malian' and 'Shalistan') (Walker *et al.*, 2006; Owens, 2008).

The genus belongs to the family *Vitaceae*. It is primarily dioecious, although most cultivated types are hermaphroditic. *Vitis* is unique in the family in that most of its species have 38 chromosomes, rather than the more common $2n = 40$, and its chromosomes are much smaller. The muscadine grape, *Vitis rotundifolia* Michaux, is one of the few 40-chromosome species, and some have put it in a separate genus, *Muscadinia*. The high chromosome number of *Vitis* may represent ancient polyploidy, but the group has regular bivalents at meiosis (Patel and Olmo, 1955) and appears to have undergone considerable gene silencing (Weeden *et al.*, 1988; Parfitt and Arulsekar, 1989).

Cultivation of the grape had certainly begun in Middle Asia by 6000 BP, where wild populations of *V. vinifera* still exist (Levadoux, 1946; Zohary and Hopf, 2000; McGovern, 2003) (Fig. 11.2). Data utilizing chloroplast

Table 11.2. Grape species (*Vitis*) of the world (adapted from Owens, 2008).

Species	Distribution[a]
V. acerifolia Raf.	USA (NM, CO, KS, OK, TX)
V. aestivalis Michx.	USA (eastern, north-central and southern states)
V. amurensis Rupr.	China
V. arizonica Egelm.	Southwestern USA from AZ to west TX
V. balanseana Planch.	China, Southeast Asia
V. bashanica P.C. He	China (Shanxi)
V. bellula (Rehd.) W.T. Wang	China
V. betulifolia Diels & Gilg	China
V. blancoi Munson	Mexico
V. biformis Rose	Mexico
V. bloodworthiana Comeaux	Mexico
V. bourgaeanna Planch.	Mexico
V. bryoniifolia Bge.	China
V. californica Benth.	Central CA to southern OR
V. × champinii Planch.	South-central TX on and adjacent to the Edwards Plateau
V. chungii Metcalf	China (Fujian, Guangdong, Guangxi, Jiangxi)
V. chunganensis Hu	China
V. cinerea (Engelm.) Engelm. ex Millardet	USA (southern states)
V. cinerea var. *baileyana* (Munson) Comeaux	Interior regions of the southeastern USA from AL and GA in the south, north to OH, WV and PA
V. cinerea var. *cinerea*	Mississippi River Basin – from Gulf of Mexico to KS/NE/IA
V. cinerea var. *floridana* Munson	Coastal regions of southeastern USA, stretching from LA to MD
V. cinerea var. *helleri* (Bailey) M.O. Moore	South-central TX and extreme northern Mexico
V. coignetiae Pulliat ex Planch.	Japan, Korea, East Asia
V. × doaniana Munson ex Viala	North-central TX and adjacent OK
V. davidii (Roman. Du Caill.) Föex	China
V. erythrophylla W.T. Wang	China (Jiangxi, Zhejiang)
V. fengqinensis C.L. Li	China (Yunnan)
V. flexuosa Thunb.	China
V. girdiana Munson	Southern CA, Baja California
V. hancockii Hance	China
V. heyneana Roem. & Schult	China
V. hui Cheng	China (Jiangxi, Zhexi)
V. jacquemontii R. Parker	Central Asia, Pakistan, Afghanistan
V. jaegeriana Comeaux	Mexico
V. jinggangensis W.T. Wang	China
V. labrusca L.	East coast of USA from ME to SC, west to OH, MI down to LA
V. lanceolatifoliosa C.L. Li	China (Guangdong, Hunan, Jiangxi)
V. longquanensis P.L. Qiu	China (Fujian, Jiangxi, Zhejiang)
V. luochengensis W.T. Wang	China (Guangdong, Guangxi)
V. menghaiensis C.L. Li	China (Yunnan)
V. mengziensis C.L. Li	China (Yunnan)
V. monticola Buckley	USA (south-central TX)
V. mustangensis Buckley	USA (AL, AR, LA, OK, TX)
V. nesbittiana Comeaux	Mexico

Continued

Table 11.2. Continued.

Species	Distribution[a]
V. × novae-angliae Fernald	Northeastern USA
V. palmata Vahl	USA (southeastern USA, west to LA, and north to OK, TN, MO, IN, IL)
V. peninsularis M.E. Jones	Baja California
V. piasezkii Maxim	China
V. piloso-nerva Metcalf	China
V. popenoei J.H. Fennell	Mexico/Central America
V. pseudoreticulata W.T. Wang	China
V. retordii Roman. Du Cail. Ex Planch.	China
V. riparia Michaux	Great plains of USA into Canada through all of northeastern USA, south to northern LA and VA
V. romanetii Roman du Caill. ex Planch.	China
V. rotundifolia Michx.	Southeastern USA
V. rotundifolia var. *munsoniana* (J. Simpson ex Munson) M.O. Moore	USA (FL, AL, GA)
V. rupestris Scheele	USA (mostly southern MO, northern AR)
V. ruyuanensis C.L. Li	China (Guangdong)
V. shuttleworthii House	USA (FL)
V. shenxiensis C.L. Li	China (Shaanxi)
V. silvestrii Pamp.	China (western Hubei, southern Shaanxi)
V. × slavinii Rehder	Natural hybrid of *V. aestivalis* × *V. riparia*
V. sinocinerea W.T. Wang	China
V. tiliifolia Humb. & Bonpl. ex Schult.	Mexico, Central America, Caribbean
V. tsoii Merr.	China
V. vinifera L.	Western and central Europe, North Africa, Near East, Caucasus
V. vulpina L.	Southeastern USA
V. wenchowensis C. Ling ex W.T. Wang	China (Zhejiang)
V. wuhanensis C.L. Li	China
V. wilsonae Veitch	China
V. yunnanensis C.L. Li	China (Yunnan)
V. zhejiang-adstricta P.L. Qiu	China (Zhejiang)

[a]AL, Alabama; AR, Arkansas; AZ, Arizona; CA, California; CO, Colorado; FL, Florida; GA, Georgia; IA, Iowa; IL, Illinois; IN, Indiana; KS, Kansas; LA, Louisiana; MD, Maryland; ME, Maine; MI, Michigan; MO, Missouri; NE, Nebraska; NM, New Mexico; OH, Ohio; OK, Oklahoma; OR, Oregon; PA, Pennsylvania; SC, South Carolina; TN, Tennessee; TX, Texas; VA, Virginia; WV, West Virginia.

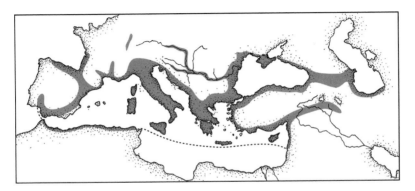

Fig. 11.2. Distribution of the wild grape, *Vitis vinifera* subsp. *sylvestris* (adapted from Zohary and Hopf, 2000).

molecular markers support the presence of at least two major domestication centers, which correspond to Negrul's *occidentalis* and *orientalis* groups (Arroyo-Garcia *et al.*, 2006). Using additional molecular markers, secondary domestication centers have also been proposed (Aradhya *et al.*, 2003; Grassi *et al.*, 2003).

There is ample archeological evidence that wine grapes were being grown in a large area spanning the Aegean, Mesopotamia and Egypt by 4000 BP (McGovern and Michel, 1995; Zohary and Hopf, 2000). From the very beginning, grapes were propagated by cuttings or layering. Grape growing began to spread out from Asia Minor and Greece about 3000 years ago, and by 2500 BP it had reached France. Its later distribution was strongly associated with the use of wine in the consecration of the Christian Mass.

Grapes were initially introduced to the New World by Portuguese and Spanish explorers (Olmo, 1995). European wine grapes were established in the Mediterranean-like climates of California in the middle 1800s. The first successful grape grown widely in the USA outside of California was 'Concord', which appeared in eastern North America as either a mutant of a native bunch grape (*Vitis labrusca*) or a

hybrid with *V. vinifera*. It was used primarily for fresh fruit and juice, and is still grown today. *V. rotundifolia* was domesticated for both wine and eating by the colonists in the Carolinas in the late 1700s. Native North American species found their way back to Europe in the late 1800s when it was discovered that they could be used as rootstocks to combat the devastation wrought by the root aphid, *Phylloxera* (Alleweldt *et al.*, 1991; Olmo, 1995). Recent problems with mildews and *Phylloxera* have further stimulated the use of wild species in breeding programs.

Peach

The peach (*Prunus persica* (L.) Batsch) is the most widely grown species in a very important genus containing plum (*Prunus domestica* L.), apricot (*Prunus armeniaca* (L.) Kostina), almond (*Prunus amygdalus* Batsch), sweet cherry (*Prunus avium* L.) and sour cherry (*Prunus cerasus* L.) (Table 11.3). Peach belongs to the family *Rosaceae* and the subgenus *Amygdalus*. It is largely outcrossed due to self-incompatibility. There are at least 77 wild species of *Prunus* and

Table 11.3. Native peach species (*Prunus*) and their domesticated relatives (adapted from Scorza and Okie, 1990).

Relatives	Common name	Chromosome number (2*n*)	Distribution
Wild			
P. davidiana (Carr.) Franch	Mountain peach, shan tao	16	North China
P. ferganensis (Kost. & Rjab) Kov. & Kost.	Xinjiang tao	16	Northeast China
P. kansuensis Rehd.	Wild peach, kansu tao	16	Northwest China
P. mira Koehne	Tibetan peach, xizang-tao	16	West China and Himalayas
P. persica (L.) Batsch	Peach, maotao	16	China
Domesticated			
P. domestica L.	European plum	48	West Asia, Europe
P. salicina Lindl.	Japanese plum	16	China
P. armeniaca (L.) Kostina	Apricot	16	Asia
P. amygdalus Batsch	Almond	16	Southwest Asia
P. avium L.	Sweet cherry	16	West Asia, Southeast Europe
P. cerasus L.	Sour cherry	32	West Asia, Southeast Europe

most of them are found in central Asia. While polyploidy is common in the genus *Prunus*, the cultivated peach is diploid and has a chromosome number of $2n = 2x = 16$.

Five species of peach are generally recognized: *P. persica*, *Prunus davidiana* (Carr.) Franch, *Prunus mira* Koehne, *Prunus kansuensis* Rehd. and *Prunus ferganensis* (Kost. & Rjab) Kov. & Kost. All are found in China. The domesticated peach can be readily hybridized with native populations of *P. persica* and all the other wild species. Successful hybrids have also been produced between peach and almond, apricot, plum and sour cherry (Parfitt *et al.*, 1985; Scorza and Okie, 1990). In most cases, these wide hybrids are largely sterile, although F_1s of almond and peach are highly fertile (Armstrong, 1957) and can be employed as rootstocks for both peach and almond.

Peach cultivation probably originated in western China from wild populations of *P. persica* (Hedrick, 1917; Scorza and Okie, 1990). The peach is mentioned in 4000-year-old Chinese writings, and most of the known variation in cultivated peaches is found in Chinese landraces. Peaches arrived in Greece through Persia about 2500 BP and in Rome 500 years later. The Romans spread the peach throughout their empire. The peach came to Florida, Mexico and South America in the mid-1500s via Spanish and Portuguese explorers. It became feral in the southeastern USA and Mexico, and was further spread throughout North America by native North Americans.

Strawberry

The genus of the strawberry is *Fragaria*, which belongs to the *Rosaceae* and the subfamily *Rosoideae*. The major cultivated species, *F.* × *ananassa*, is an octoploid ($2n = 8x = 56$) of interspecific origin. It originally appeared as an accidental hybrid in Europe about 1750 when plants of *Fragaria chiloensis* from Chile were planted next to *Fragaria virginiana* from the Atlantic seaboard of North America (Darrow, 1966). Its closest relatives are *Duchesnea* and *Potentilla*. There are five basic fertility groups in *Fragaria* that are associated primarily with their ploidy level or chromosome number (Table 11.4).

Diploid, tetraploid and hexaploid species are found in Europe and Asia (Table 11.4), but octoploids are restricted to the New World. Only one diploid species, *Fragaria vesca*, is located in North America. The genomic complement of the octoploids is probably AAA'A'BBB'B' (Bringhurst, 1990), with *F. vesca* probably being the A-genome donor and Japanese *Fragaria iinumae* being the B-genome donor (Rousseau-Gueutin *et al.*, 2009). In comparative mapping studies, the octoploid and diploid *Fragaria* show a high level of colinearity between their genomes, suggesting little chromosomal rearrangement has occurred subsequent to polyploidization (Rousseau-Gueutin *et al.*, 2008).

It is likely that the octoploids originated in northeastern Asia, where the A- and B-genome species are centered, and the polyploid derivatives then migrated across the Bering Strait and dispersed across North America (Hancock, 1999). It is possible that *F. chiloensis* and *F. virginiana* are extreme forms of the same biological species, separated during the Pleistocene, which subsequently evolved differential adaptations to coastal and mountain habitats. The two species are completely inter-fertile, carry similar cpDNA restriction fragment mutations (Harrison *et al.*, 1997a) and have very similar nuclear internal transcribed spacer (ITS) regions (Potter *et al.*, 2000).

Polyploidy in *Fragaria* probably arose through the unification of $2n$ gametes, as several investigators have noted that unreduced gametes are relatively common in *Fragaria* (Hancock, 1999). Staudt (1984) observed restitution in microsporogenesis of an F_1 hybrid of *F. virginiana* × *F. chiloensis*. In a study of native populations of *F. chiloensis* and *F. vesca*, Bringhurst and Senanayake (1966) found frequencies of giant pollen grains to be approximately 1% of the total. Over 10% of the natural hybrids generated between these two species were the result of unreduced gametes.

The inheritance patterns of the octoploids are in some dispute. Lerceteau-Köhler *et al.* (2003) concluded that *F.* × *ananassa* has mixed segregation ratios using AFLP markers, as they found the ratio of coupling versus repulsion markers fell between the fully disomic and polysomic expectations. However, two other studies evaluating isozyme, SSR and RFLP segregation observed predominantly disomic

Table 11.4. Native strawberry species (*Fragaria*) and their distribution (after Hancock, 1999; Hummer *et al.*, 2009; Rousseau-Gueutin *et al.*, 2009).

Species	Chromosome number (2*n*)	Distribution
F. bucarica Losinsk.	14	Western Himalayas
F. daltoniana J. Gay	14	Himalayas
F. iinumae Makino	14	Japan
F. mandshurica Staudt	14	North China
F. nilgerrensis Schlect.	14	Southeast Asia
F. nipponica Lindl.	14	Japan
F. nubicola Lindl.	14	Himalayas
F. pentaphylla Lozinsk	14	North China
F. vesca L.	14	North America, north Asia and Europe
F. viridis Duch.	14	Europe and Asia
F. yesoensis Hara.	14	Japan
F. corymbosa Losinsk.	28	North China
F. gracilis Losinsk.	28	Northwest China
F. moupinensis (French.) Card	28	South China
F. orientalis Losinsk	28	North Asia
F. tibetica spec. nov. Staudt	28	China
F. × bringhurstii Staudt	35, 42	California
F. moschata Duch.	42	North and central Europe
F. chiloensis (L.) Miller	56	Pacific Coast North America, Hawaii and Chile
F. virginiana Miller	56	Central and eastern North America
F. × ananassa Duch. ex Lamarck.	56	Cultivated worldwide
F. iturupensis Staudt	70	Iturup Island, Kurile Islands

ratios, indicating that the octoploid strawberry is completely diploidized (Arulsekar and Bringhurst, 1981; Ashley *et al.*, 2003).

Vegetables

Cole crops

The amazingly diverse genus *Brassica* has given us the oilseed rape or canola (*B. napus* L. and hybrids with *B. campestris*), the turnip and Chinese cabbage (*B. campestris* L.), rutabaga or swede (*B. napus* L.), black mustard (*B. nigra* Koch), brown mustard (*B. juncea* (L.) Czern) and Ethiopian mustard (*B. carinata*). The single *Cruciferae* species *Brassica oleracea* has, by itself, yielded a remarkable array of vegetables including cabbage, kale, Brussels sprout, cauliflower, broccoli and kohlrabi (Fig. 11.3).

An overall diagram of genomic relationships in the genus *Brassica* was first presented by U in 1935 (Fig. 11.4). He suggested that brown mustard (leaf mustard) was a hybrid of black mustard and turnip, rutabaga and oilseed rape were derived from turnip and kale, and Ethiopian mustard was a hybrid of black mustard and kale. This scheme has been supported by many different types of evidence including artificial synthesis of the hybrids, electrophoretic studies, molecular analysis of chloroplast and nuclear DNA, and genomic *in situ* hybridization (Song *et al.*, 1990; Lagercrantz and Lydiate, 1996; Snowdon *et al.*, 2002). The progenitor of the A, B and C genomes is thought to be an unknown hexaploid that was reorganized into the 8, 9 and 10 chromosome sets of *B. nigra*, *B. oleracea* and *B. rapa* through chromosomal fusions and rearrangements.

The first cultivated *Brassica* species was probably *B. campestris*, which was grown initially for its oilseed (Thompson, 1979). Another species, *B. napus*, was also domesticated for its oilseed, but not until the Middle Ages. There is only meager evidence of *B. campestris*' beginnings as an oil crop, but it is likely to have been

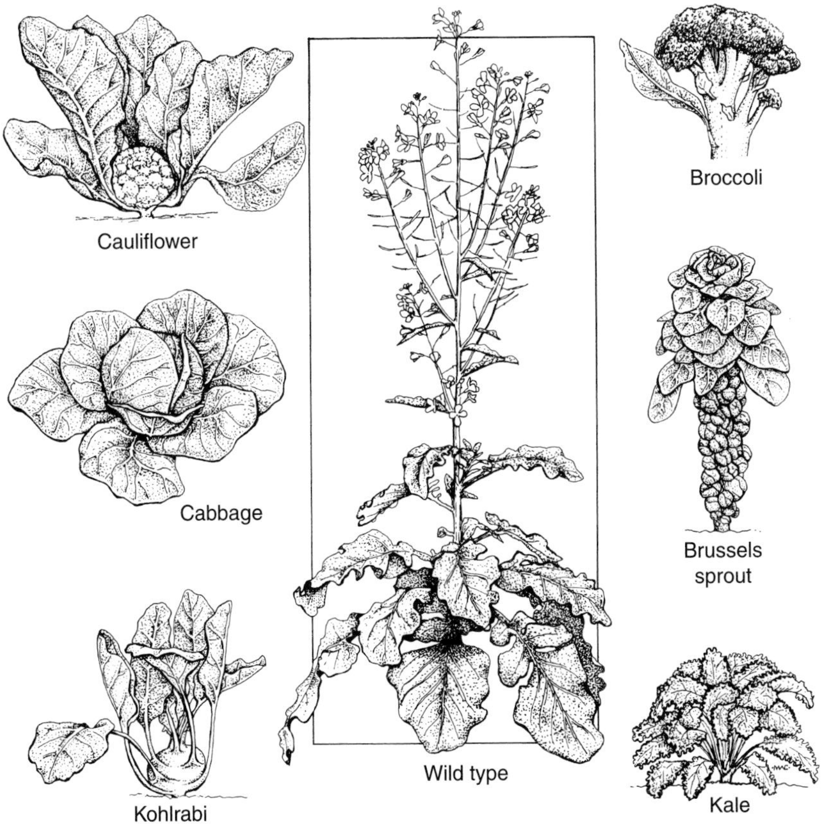

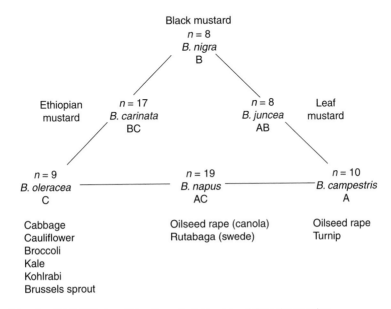

Fig. 11.3. Crops originating from *Brassica oleracea* subsp. *oleracea*.

Black mustard
n = 8
B. nigra
B

Ethiopian   *n* = 17   *n* = 8   Leaf
mustard   *B. carinata*   *B. juncea*   mustard
                  BC   AB

n = 9   *n* = 19   *n* = 10
B. oleracea ———————— *B. napus* ———————— *B. campestris*
   C   AC   A

Cabbage   Oilseed rape (canola)   Oilseed rape
Cauliflower   Rutabaga (swede)   Turnip
Broccoli
Kale
Kohlrabi
Brussels sprout

Fig. 11.4. Triangle of U (1935) describing the polyploid origin of *Brassica* species.

domesticated repeatedly about 4000 years ago from wild populations in an area spanning the Mediterranean to India. The turnip form of *B. campestris* emerged 3000 years later in northern Europe. Both *B. campestris* and *B. napus* probably first caught farmers' attention as weeds of wheat and other crops. These two species freely intercross and their hybrids are now our most important oilseed crops.

The leafy kales also have an early history dating back at least 2500 years, with linguistic, literary and historical data being compatible with domestication in the ancient Greek-speaking area of central and east Mediterranean (Maggioni *et al.*, 2010). The kales were derived from *B. oleracea* subsp. *oleracea*, found naturally across the coast of the Mediterranean from Greece to England. An early stage of their domestication must have been a reduction of the bitter-tasting glucosinolates, which are found in high levels in the wild species (Josefsson, 1967). The other types of *B. oleracea* emerged much later, as humans began to actively select for the enlargement of different plant parts (Gray, 1982). Early types of cabbages were first grown in ancient Rome and Germany over 1000 years ago. Broccoli, cauliflower and Brussels sprout are a more recent development, appearing within the last 500 years, cauliflower from Europe and broccoli from the eastern Mediterranean. Smith and King (2000) provide molecular evidence that cauliflower arose in southern Italy from a heading Calabrese broccoli. Brussels sprout first appeared as a spontaneous mutation in France in 1750.

The various types of *B. oleracea* were developed primarily by disruptive selection on the polymorphisms already available in wild *B. oleracea*, except for Brussels sprout. Buckman demonstrated this in 1860 at the Royal Agricultural College in southern England by selecting broccoli-like cultigens from wild plants in only a few generations. Cauliflower is most closely related to broccoli, while cabbage is most closely related to the kales (Song *et al.*, 1990). As each new *Brassica* type emerged, it spread rapidly throughout Europe and the Mediterranean countries, and hybridization with wild congeners undoubtedly played a role in the development of crop types. Palmer *et al.* (1983) found chloroplast DNA in two populations of *B. napus* that was likely the result of recent introgressive hybridization with *B. oleracea* and *B. campestris*.

Black, brown (Indian) and Ethiopian mustards were also very early domesticants, but probably followed oilseed rape and the leafy kales. All the mustards are used primarily as spices, with the exception of brown mustard, which is also used for its oilseed and as greens (Hemingway, 1995). Black mustard is referred to in the early written literature of both Babylonia and India and is found native in Asia Minor and Iran. Brown mustard is found naturally in central Asia to the Himalayas and was probably domesticated separately in India, China and the Caucasus. Ethiopian mustard arose in Ethiopia where the range of its parents overlapped, weedy *B. nigra* and cultivated *B. oleracea*.

Chili peppers

The peppers belong to the family *Solanaceae*. There are five cultivated and over 25 wild species that share the chromosome number $2n = 2x = 24$ (Eshbaugh, 1980, 1993). Molecular data have in general supported the traditional morphometric and cytogenetic classifications (Prince *et al.*, 1992, 1995; Rodriguez *et al.*, 1999; Walsh and Hoot, 2001). The most widely cultivated species is *Capsicum annuum* L., which contains a diverse array of types representing both sweet and hot peppers. It is grown worldwide. The rest of the cultigens, *Capsicum baccatum* L., *Capsicum frutescens* L., *Capsicum chinense* Joeg. and *Capsicum pubescens* Ruiz et Pav, are used primarily as spices, and their culture is largely confined to South America (Fig. 11.5) and, to a limited extent, Africa.

A major portion of *Capsicum* evolution appears to have occurred in south-central Bolivia (McLeod *et al.*, 1982). Primitive types probably migrated out of this region into the Andes and lowland Amazonia and speciated along the way. Cytogenetic studies indicate that some of the species differentiation was associated with differences in chromosomal size, morphology and rearrangements (Kumar *et al.*, 1987; Moscone *et al.*, 2007), although hybrids with varying levels of fertility can be obtained among most of the cultivated species and with many of the wild types (Fig. 11.6). *C. pubescens* is the species most strongly isolated from the rest. Separating *C. annuum*, *C. chinense* and *C. frutescens* into

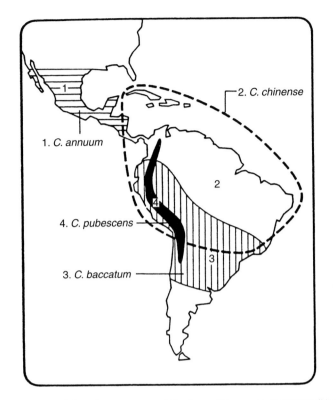

Fig. 11.5. Range of cultivated *Capsicum* peppers at the time of European discovery. (Used with permission from N.E. Simmonds, © 1985, *Principles of Crop Improvement*, Longman, (London.)

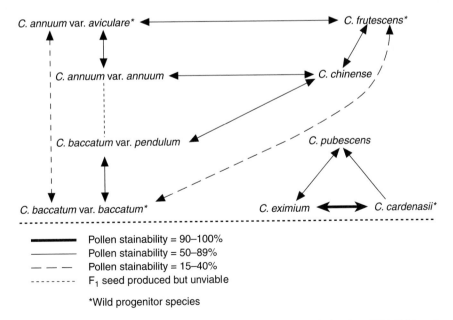

Pollen stainability = 90–100%
Pollen stainability = 50–89%
Pollen stainability = 15–40%
F$_1$ seed produced but unviable

*Wild progenitor species

Fig. 11.6. Crossing relationships of *Capsicum* species. (Used with permission from W.H. Eshbaugh, © 1975, *Bulletin of the Torrey Botanical Club* 102, 396–403.)

individual species may not be warranted as they inter-grade greatly and are difficult to distinguish at any level (Eshbaugh, 1993; Sudré et al., 2010); however, the separate names remain in force. All the cultigens are self-compatible, even though many of the wild species are outcrossed either through heterostyly or self-incompatibility (Heiser, 1995b).

Chili peppers were originally cultivated at several independent locations in the Americas (Pickersgill, 1989; Eshbaugh, 1993). Starch microfossils of *Capsicum* spp. have been found at seven sites from Panama to Peru that date to over 6000 BP (Perry et al., 2007). *C. baccatum* was probably domesticated in Bolivia, with the oldest remains being found along coastal Peru dating from 4000–5000 BP. *C. pubescens* probably originated in Bolivia. *C. chinense* originated in lowland Brazil and was spread to Peru by 2000–3000 BP. Pre-Columbian remains of *C. annuum* or *C. frutescens* have been located in the valley of Oaxaca in Mexico (Perry and Flannery, 2007). It is not known if each of these domestications was totally independent or stimulated by other domestications.

At one time, the five cultivated forms were thought to have originated from one progenitor species, but now most researchers feel there were at least three evolutionary lines leading to the cultivated taxa (Eshbaugh et al., 1983; Moscone et al., 2007). The *C. annuum* complex,

C. baccatum and *C. pubescens* are much too different to have emerged from the same progenitor in the last 2000–3000 years (McLeod et al., 1979, 1982). There is a wild form of *C. baccatum* subsp. *baccatum* that is very similar to cultivated *C. baccatum* var. *pendulum*. Cultivated *C. pubescens* shows a close relationship to the wild species *Capsicum eximum, Capsicum cardenasii* and *Capsicum tovari* (Eshbaugh, 1993; Heiser, 1995b). The cultigens *C. annuum* var. *annuum, C. chinense* and *C. frutescens* all appear to have a common ancestor, which may be *C. annuum* var. *aviculare*. Most of the changes associated with domestication involved fruit size and color, along with a shift from outcrossing to selfing.

Like so many other New World crops, Europeans became aware of chili peppers after the voyage of Columbus and they gained almost immediate acceptance (Heiser, 1995b). Peppers came to the widespread attention of people in North America and Africa in the last couple of centuries.

Cucurbits

The cucurbit family (*Cucurbitaceae*) has contributed a large number of crops to humankind (Table 11.5). Mesoamerica and South

Table 11.5. Major cucurbit crops and their center of origin.

Species	Common name	Chromosome number (2n)	Center of origin
Cucumis sativus L.	Cucumber	14	India
Cucumis melo L.	Muskmelon, cantaloupe	24	Africa, India
Citrullus lanatus (Thunb.) Mansf.	Watermelon	22	South Africa
Lagenaria siceraria Standl.	White-flowered gourd	22	Africa, the Americas
Cucurbita pepo L.	Summer squash, pumpkin, marrow	40	Mexico
Cucurbita argyrosperma Huber (= *Cucurbita mixta*)	Winter squash	40	Mexico, Central America
Cucurbita moschata Duchesne	Winter squash	40	Mexico, Central America
Cucurbita mixima Duchesne	Winter squash, pumpkin	40	Northern South America, Central America
Cucurbita ficifolia Bouché	Fig-leaf gourd	40	Mexico, Central America, northern South America

America are the origins of pumpkins and squashes (*Cucurbita pepo, Cucurbita argyrosperma* (= *Cucurbita mixta*), *Cucurbita moschata* and *Cucurbita maxima*). Cucumbers (*Cucumis sativa*) probably were domesticated in India, while muskmelons (*Cucumis melo* L.) and watermelons (*Citrullus lanatus*) originated in Africa. Gourd species were domesticated in Africa (*Lagenaria siceraria*) and Mesoamerica to northern South America (*Cucurbita ficifolia*).

In general, these taxa are reproductively isolated and are distinct for a number of molecular markers (den Nijs and Visser, 1985; Perl-Treves *et al.*, 1985; Jobst *et al.*, 1998). The cucumber is $2n=14$, while all the other species are $2n=22$ or 40. The high chromosome number of many of the cucurbit species suggests that they are of polyploid origin and this has been confirmed by examining isozyme segregation data (Weeden and Robinson, 1990). A comparison of the cucumber and melon genetic maps indicated that five of cucumber's seven chromosomes arose from fusions of ten ancestral chromosomes after divergence from melon (Huang *et al.*, 2009).

Recent studies indicate that each cultivated *Cucurbita* arose from a different wild taxon in the New World, but all their ancestors are not clear (Decker-Walters, 1990; Nee, 1990; Sanjur *et al.*, 2002). South American *Cucurbita andeana* was probably the ancestor of *C. maxima*, and *Cucurbita sororia* Bailey from Mexico was the likely progenitor of *C. argyrosperma*. Domesticated *C. pepo* appears to contain two lineages with separate origins, subsp. *pepo* and *ovifera*. *C. pepo* subsp. *ovifera* shares the same mtDNA with several wild taxa, including *C. pepo* subsp. *fraterna*, *C. pepo* subsp. *ovifera* var. *texana* and *C. pepo* subsp. *ovifera* var. *ozarkana*. *C. pepo* subsp. *pepo* carries a unique mtDNA haplotype that is not associated with any other known wild species. *C. moschata* shares some sequence homology with *C. sororia* but it is an unlikely progenitor, while *C. ficifolia* is distinct from all known species.

The *Cucurbita* species were among the earliest domesticants. Evidence for *C. pepo* cultivation comes from about 9000–10,000 BP in Mesoamerica, followed by *C. moschata* in Southern Mexico (7600–9200 BP), *C. argyrosperma* in southern Mexico (8700 BP), *C. ficifolia* in Peru (5000 BP) and *C. maxima* also in Peru (4000 BP) (Merrick, 1995; Smith, 1997; Dillehay

et al., 2007; Piperno *et al.*, 2009; Ranere *et al.*, 2009). There is also evidence of the domestication of *Cucurbita ecuadorensis* in Ecuador about 10,000–12,000 BP (Piperno and Stothert, 2003). Many of the species were first domesticated for their edible seeds and were gradually changed into thick-fleshed containers through artificial selection. The squashes and gourds were important components of the diets of the Aztecs, Incas and Mayas and the native North Americans carried them throughout North America. While most of the cucurbits were introduced into what is now the USA, an independent origin of domesticated *C. pepo* in eastern North America is likely (Decker, 1988; Smith, 2006). The various *Cucurbita* species were taken to Europe and Asia after the discovery of the New World.

There is evidence that the cucumber, *C. sativus* ($2n=14$), was domesticated in India and China 3000–4000 years ago, although its progenitors are unclear (Bates and Robinson, 1995). One possible candidate, *C. sativa* var. *hardwickii*, is found in the wild in Asia, but it might be a feral derivative. Another Chinese species, *Cucumis hystrix* Chakr., is morphologically similar to cucumber but has different chromosome numbers and distinct isozyme frequencies (Chen *et al.*, 1997). The muskmelon, *C. melo* ($2n=24$), came from eastern tropical Africa, and the watermelon, *C. lanatus* ($2n=22$), originated in central Africa. Both were domesticated in recent history from existing wild species of the same name. The bottle gourd, *L. siceraria*, does not appear to have a direct wild relative, but was presumably domesticated first in Africa where all the other wild *Lagenaria* species are found. Interestingly, the bottle gourd was being cultivated in the Americas by 9000–15,000 BP and by 6000–5000 BP in Asia, even though its roots lay in Africa (Heiser, 1990). Oceanic drifting is the most likely explanation for this disjunct distribution (Decker-Walters *et al.*, 2001).

Tomato

While some still refer to the genus of tomato as *Lycopersicon*, Spooner and his group have provided strong evidence that tomatoes are "deeply nested" in the genus *Solanum* and have placed them in a subsection, *Lycopersicon*

(Spooner *et al.*, 2005a; Rodriquez *et al.*, 2009). There are ten wild species distributed throughout Central and South America (Table 11.6). Wild *Solanum lycopersicon cerasiforme*, found in Mexico, Central America and South America, is probably the progenitor of the domesticated species (Jenkins, 1948; Rick, 1995). There do not appear to be any chromosomal structural differences separating the various species, and *S. lycopersicon* can be hybridized with all the other species with varying levels of success. There are both self-compatible and self-incompatible species; the self-incompatibility system in tomato is gametophytic and regulated by a single, multiallelic *S* locus (Rick, 1982). Changes in mating system have occurred in wild tomatoes from self-incompatibility to self-compatibility, and some mostly self-incompatible species have compatible relatives (Table 11.6). The self-compatible species can show substantial variation in rates of outcrossing (Rick *et al.*, 1977).

Mexico is generally accepted as the place where the tomato was first domesticated (Jenkins, 1948), although the evidence is far from conclusive (Rick, 1995; Peralta and Spooner, 2007). The other proposed origin is Peru (DeCandolle, 1886). In general, there are few linguistic and historical data on early tomato cultivation, and a molecular study would be complicated by not knowing which landraces are native. Isozyme data indicate that the domesticated tomato could have originated in the area of Ecuador–Peru and then spread to Mexico, or the Andes may have been a secondary area of domestication (Rick and Fobes, 1975; Rick and Holle, 1990). There is also evidence that the early Mexican domesticants may have found their way to South America, where they introgressed with native races of *Solanum pimpinellifolium* (Rick *et al.*, 1977).

The initial date of domestication is unknown, but tomato cultivation was well established when the European explorers arrived in Mexico and the Americas. The North American and European tomato cultivars probably came from Mesoamerica, as they are more similar to the Mexican types than the South American ones. Fears of toxicity initially slowed the spread of the tomato in Europe because it belongs to the deadly nightshade family, and it was not until the early 1800s that it was widely used for food. Tomato became established in North America a few decades later.

Table 11.6. Native species of tomato (*Solanum*) and their distribution. All species are $2n=2x=24$. (Based on Spooner *et al.*, 2005a.)

Species	Breeding system	Native distribution
S. cheesmaniae (L. Riley)	Self-compatible; exclusively autogamous	Galápagos Islands
S. chilense (Dunal) Reiche	Self-incompatible	Southern Peru to northern Chile
S. chmielewskii (C.M. Rick, Kesicki, Fobes & M. Holle), D.M. Spooner, G.L. Anderson & R.K. Jansen	Self-compatible; facultative allogamous	South-central Peru to northern Bolivia
S. galapagense S. Darwin & Peralta	Self-compatible; exclusively autogamous	Galápagos Islands
S. habrochaites S. Knapp & D.M. Spooner	Mostly self-incompatible	Southwest Ecuador to south-central Peru
S. lycopersicum L.	Self-compatible; facultative allogamous	Ecuador and Peru
S. neorickii D.M. Spooner, G.L. Anderson & R.K. Jansen	Self-compatible; highly autogamous	Southern Ecuador to south-central Peru
S. pennellii Correll	Mostly self-incompatible	Coastal Peru
S. peruvianum L.	Mostly self-incompatible	Northern Peru to northern Chile
S. pimpinellifolium L.	Self-compatible; varies from autogamous to facultative allogamous	Coastal Peru and Ecuador

Fibers and Oils

Cotton

The genus *Gossypium* belongs to the family *Malvaceae*. It contains about 43 diploid species ($2n = 2x = 26$) and five allotetraploids ($2n = 4n = 52$) (Table 11.7; Fryxell *et al.*, 1992). Four species of cotton are cultivated: the diploids *Gossypium herbaceum* and *Gossypium arboreum*, which carry the A genome and are cultivated to a limited extent in India, Pakistan and parts of Southeast Asia, and the tetraploids *Gossypium hirsutum* and *Gossypium barbadense*, which carry the A and D genomes. *G. arboreum* does not have a known wild progenitor and as a result has long been thought to be a derivative of *G. herbaceum*, but the two species are so different genetically that it is more likely that they were domesticated from different progenitors (Wendel, 1989). The ancestor of cultivated *G. herbaceum* is likely *G. herbaceum* subsp. *africanum*, an endemic found in South Africa. The high ploidy number of the so-called diploids, along with substantial levels of enzyme multiplicity and DNA duplication, suggest an ancient polyploid beginning (Endrizzi *et al.*, 1985; Wendel, 2000).

Extensive cytological and genetic data indicate that the progenitors of the tetraploids (AADD) are closely related to *Gossypium raimondii* (DD) and *G. herbaceum* (AA) (Phillips, 1966; Wendel, 1989; Small and Wendel, 1998). However, these two species are located on different continents, leaving much speculation about how their ancestors could have hybridized. There could have been a Cretaceous origin before the South American and African continents separated, but molecular data indicate a much more recent origin. Two hypotheses have been proposed on how the two genomes got together from different continents: (i) people carried an Old World form to America or vice versa and the hybridization occurred; or (ii) the polyploids arose in South America after an A-genome propagule drifted across the ocean from Africa or Asia. The latter hypothesis seems most plausible as the divergence pattern fits a time period between Pangea and the emergence of farming. Wendel and Albert (1992) suggest that the A genome originally evolved in Africa and Asia, and then dispersed to the Americas across the Pacific, rather than the Atlantic. This makes the most sense as the D-genome species are located on the western side of the New World. The tetraploids probably arose in northwestern Mexico and then spread to Peru. Molecular data suggest all the allopolyploids have a common, monophyletic origin (Wendel *et al.*, 2010).

The diploid species have a long history of cultivation that goes back at least 5000 years (Phillips, 1979). *G. herbaceum* was first domesticated in Arabia and Syria, while *G. arboreum* cultivation probably began in India. *G. arboreum* became dominant all across Africa and Asia, until it was replaced by the tetraploids in the last 100 years. Currently, *G. arboreum* is grown only to a limited extent in India, and *G. herbaceum* cultivation has only a spotty distribution in Africa and Asia.

Evidence of tetraploid cultivation also dates to antiquity. Cotton bolls and fibers of domesticated *G. barbadense* have been found in central-coastal Peru dating to 4500 BP. The oldest cultivated remains of *G. hirsutum* have been located in Mexico from 5500 BP, although they are not thought to represent the earliest

Table 11.7. Geographic distribution of the major lineages of *Gossypium* (Wendel *et al.*, 2010).

Genome group	Number of species	Geographic distribution
A	2	Africa, possibly Asia
B	3	Africa, Cape Verde Islands
C	2	Australia
D	13	Primarily Mexico; also Peru, Galápagos Islands, Arizona
E	7+	Arabian Peninsula, northeast Africa, southwest Asia
F	1	East Africa
G	3	Australia
K	12	Northwest Australia
AD	5	New World tropics and subtropics, including Hawaii

domestications. RFLP diversity patterns support the Yucatan peninsula as the primary site for *G. hirsutum* domestication (Brubaker and Wendel, 1994).

The Spanish and Portuguese colonists and traders spread the tetraploids to Spain, Africa and India via Cuba and Brazil (Phillips, 1979). *G. hirsutum* (upland cotton) was introduced into the southeastern USA in the mid-1700s from Mexico. *G. barbadense* (sea island cotton) arrived in the Carolinas and Georgia from South America in the late 1700s and flourished in the south for several decades until the boll weevil prevented its successful cultivation. *G. hirsutum* now comprises about 95% of the world's crop, while *G. barbadense* accounts for all but 1% of the rest.

Olive

The olive (*Olea europaea* L.) is one of the most ancient cultivated crops in the Mediterranean basin, and it was one of the first trees to be domesticated (Zohary and Spiegel-Roy, 1975; Zohary and Hopf, 2000). It is evergreen, outcrossing and vegetatively propagated (cuttings and grafts). The trees can live for more than 1000 years and there are well over 1000 cultivars worldwide. Cultivars are broken into two intergrading groups – oil varieties with up to 28% oil, and table olives that are pickled (Zohary, 1995b). Many varieties are used for both.

Olive is part of the *O. europaea* complex, which consists of six subspecies (Green, 2002): (i) *O. europaea* subsp. *europaea* (Miller) Lehr., which is the widespread Mediterranean olive; (ii) *O. europaea* subsp. *maroccana*, which is endemic to Morocco; (iii) *O. europaea* subsp. *laperrinei* in the Saharan mountains; (iv) *O. europaea* subsp. *cerasiformis* in Madeira; (v) *O. europaea* subsp. *guanchica* on the Canary Islands; and (vi) *O. europaea* subsp. *cuspidata* distributed from South Africa to China. There are two forms of the Mediterranean olive – wild olive or oleaster (*O. europaea* subsp. *europaea* var. *sylvestris*) and the cultivated olive (*O. europaea* subsp. *europaea* var. *europaea*). Many Mediterranean forests contain combinations of wild and feral forms, but some indigenous populations still exist (Lumaret and Ouazzani, 2001; Lumaret *et al.*, 2004).

All the olive taxa are inter-fertile (Besnard *et al.*, 2001) and gene exchange between wild and cultivated olive could have contributed to the evolution of the Mediterranean olive (Quézel, 1995; Besnard *et al.*, 2001). All of the olives have been reported to be $2n = 2x = 46$, but recent data from flow cytometry and microsatellite analysis indicate that *O. europaea* subsp. *cerasifolius* is tetraploid and *O. europaea* subsp. *maroccana* is hexaploid (Besnard *et al.*, 2008). The olive is most closely related to *O. europaea* subsp. *laperrinei* (Besnard *et al.*, 2001, 2002). Seven main cpDNA lineages have been identified in the *O. europaea* complex (Besnard *et al.*, 2007).

The first archeological evidence of olive cultivation has been found in the Near East from about 5000–5500 years ago, but it did not take long for olives to be grown all across the eastern shore of the Mediterranean, Crete, mainland Greece, and soon after in the western Mediterranean. Even though the earliest evidence of olive cultivation was in the Near East, the olive was likely domesticated a number of times at different locations across oleaster's range. When Breton *et al.* (2009) evaluated diversity patterns in wild and cultivated olive using nuclear and cytoplasmic markers, they identified nine genetic lineages of olive that are rooted in Turkey, Palestine, Tunisia, Algeria, Cyprus, Morocco, Corsica, Spain and France. Some of these lineages became widely dispersed, as the ancestry of some cultivars with French, Italian or Spanish names trace to the Near East or North Africa.

Groundnut

The genus *Arachis* (legume family) is restricted to South America and contains dozens of species in seven sections. It is grown throughout the world in warm temperate and tropical regions. The various species are mostly self-pollinated, but a high proportion of them can be successfully intercrossed. Their center of diversity falls in Bolivia, like the peppers. Most of the species are diploids with $2n = 2x = 20$, but there are at least two allotetraploids with $2n = 2x = 40$, including cultivated *Arachis hypogea* L. and wild *Arachis monticola* Krop.

et Rig. (Smartt *et al.*, 1978). Two distinct sets of chromosomes are found in groundnut, one marked by a pair of chromosomes significantly smaller than the others (A chromosomes, Fig. 11.7) and a set that carries a pair with a secondary constriction (B chromosomes) (Husted, 1936). Some mutivalents are formed during meiosis, suggesting segmental allopolyploidy, but the tetraploid species are mostly diploidized. Linkage maps of A and B donors of the groundnut have revealed a high degree of synteny between the genomes (Moretzsohn *et al.*, 2009).

The cultivated groundnut was probably derived from the wild tetraploid, *A monticola*, which is native to northwestern Argentina and freely hybridizes with it. For a long time, the most likely diploid progenitors were thought to be *Arachis cardenasii* and *Arachis batizocoi*, but more recent molecular and cytological data indicate that the progenitors are more closely related to *Arachis duranensis* and *Arachis ipaensis* (Seijo *et al.*, 2007; Burow *et al.*, 2009). *A. duranensis* carries the A genome, while *A. ipaensis* possesses the B genome. All subspecies and botanical varieties of the peanut have identical patterns of genomic hybridization using double GISH, which argues for a single allopolyploid origin (Seijo *et al.*, 2007).

The oldest archeological records of groundnuts are from northern Peru at around 8000 BP (Dillehay *et al.*, 2007; Piperno and Dillehay, 2008). However, their ancestry is probably much older, as there are no wild groundnut species found in Peru. Their cultivation more likely began in Bolivia when more robust types with less fragile pods were selected from *A. monticola*. By 1500, groundnuts were widely distributed throughout South America, the Caribbean and Mexico. Spanish travelers took them from Mexico to eastern Asia, and Portuguese sailors distributed them to Africa

from Brazil. India and North America received their first plants from Africa in the 1600s.

Sunflower

Two *Helianthus* species are cultivated, the sunflower, *H. annuus* L., and the Jerusalem artichoke, *H. tuberosum* L. The sunflower is widely grown as an oil crop, while the Jerusalem artichoke has a more limited use as a tuber crop. The genus *Helianthus* is a member of the *Asteraceae* family, which is divided into four sections (Schilling and Heiser, 1981). *H. annuus* belongs to the section *Annui*, whose 13 species are located in the western USA. They are all described as diploids, but cytological evidence suggests they are probably ancient polyploids; pre-meiotic treatment of microspores with colchicine induces quadrivalent formation (Jackson and Murry, 1983). Wild and cultivated *Helianthus* show considerable variability in the size of their ribosomal genes (Choumane and Heizmann, 1988) and their total DNA content (Sims and Price, 1985). Sunflowers can be crossed with most other species in their section, although hybrids generally display reduced fertility. *H. tuberosum* belongs to a different section than the sunflower (*Diaricati*) and is a hexaploid.

Sunflower was probably gathered by the earliest people inhabiting the western USA. It is a camp-following weed and was introduced by nomadic tribes into the central USA where it was first fully domesticated (Heiser, 1976). Biochemical and molecular evidence suggest that our cultivars arose from relatively few genotypes (Rieseberg and Seiler, 1990; Burke *et al.*, 2002a,b), and only a few major QTLs are associated with the major morphological differences between wild and domesticated forms (Cronn *et al.*, 1997). The first types selected were

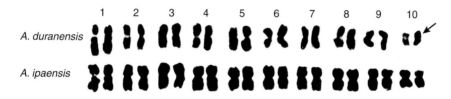

Fig. 11.7. Chromosomes of the two progenitors of the groundnut (*Arachis*). (Used with permission from G. Kochert *et al.*, © 1996, *American Journal of Botany* 83, 1282–1291.)

probably rare single-headed or monocephalic plants (Fig. 11.8).

The domestication of sunflower has generally been considered to have been in east-central USA (Smith, 2006), although recent archeological evidence has also suggested that a separate origin could have occurred in Mexico (Lentz *et al.*, 2008). Carbonized achenes found at the Hayes site in middle Tennessee and the San Andrés site in the Gulf Coast region of Tabasco, Mexico both date to about 4000–4500 BP (Crites, 1993; Lentz *et al.*, 2001). While the Mexican site could be of separate origin, it is more likely that the sunflowers in Mexico were disseminated from eastern North America. Two Mexican indigenous cultivars examined by Harter *et al.* (2004) and Wills and Burke (2006) using microsatellite data were shown to be closely aligned with wild populations from the eastern USA and "had a chloroplast DNA haplotype that is characteristic of domesticates from the United States" (Rieseberg and Burke, 2008). A substantial genetic bottleneck was shown to have occurred in response to that domestication (Harter *et al.*, 2004). Heiser (2008) suggests that there is little (if any) historical reference to the sunflower in Mexico, making a separate origin implausible and he concludes that the specimen recovered from San Andrés is actually a bottle gourd seed. Smith (2006) also questions the validity of this sunflower sample.

By the time Europeans came to the New World, the sunflower was being raised in a broad band from Mexico to southern Canada (Heiser, 1995a). The first European introductions were from Mexico to Spain in the 16th century, although sunflowers were carried to Europe from all over North America. Sunflowers' popularity as a food source grew very slowly in Europe until it arrived in Russia, where it quickly became an important oil crop. Its popularity may have exploded because it was not on the list of oily foods forbidden on certain Holy days (Heiser, 1995a).

Fig. 11.8. Wild *Helianthus annuus* next to a monocephalic cultivar.

12

Postscript: Germplasm Resources

Introduction

During the course of this book, we have described the evolutionary forces that produced crop species and have followed the changes associated with domestication. Genetic variation is the foundation on which all the crops were shaped, both before and after human intervention. Our species, *Homo sapiens*, was presented with a plentitude of potential crops and all we had to do was decide on how to most effectively exploit them.

As our population has continued to grow, we have found ourselves with reduced space and natural resources (Fig. 12.1). There are now almost 7 billion people in the world, and they all must find something to eat and a place to live. Associated with this crush of people is the loss of wild plant populations. These are being rapidly eliminated by the increasing sprawl of human activities and home sites. Vast areas of native vegetation have been eliminated, and many more are threatened by the plow, slash and burn farming, logging trucks and developers.

Modern breeding methods have greatly expanded the productivity of our major crops, but continued high yields hang on a thread of genetic vulnerability. We have developed most of our cultivars from a very narrow genetic base (Tanksley and McCouch, 1997) and, as a result, they are very susceptible to environmental perturbations. The decimation of huge acreages by the potato late blight famine in Ireland in 1845 and the corn blight in the central USA in 1970 serve as vivid reminders that genetic variability is the key to continued productivity (Plucknett *et al.*, 1987). Biotic and abiotic stresses can at any time decimate a crop with too narrow a base of variability.

Continued yield and quality improvements also depend on the acquisition of new genes (Maxted *et al.*, 2010). The Green Revolution that provided some of the less developed countries with new, high-yielding types of rice and wheat was based on the identification and transfer of novel genes for dwarfing and other desirable traits (Doyle, 1985). In addition, most breeders have found the solution to yield plateaus or new pest challenges to be the incorporation of unique germplasm.

This all means that the genetic variability necessary to feed our growing numbers is being eliminated by the expansion of that same exploding population. As natural populations are destroyed, we are losing a storehouse of allelic diversity. Wild populations carry a vast array of genetic variability that is available to breeders who want to improve crop quality, yield, and disease or stress tolerance. This makes the preservation of our native germplasm sources absolutely critical to our future breeding successes and perhaps the continued survival of our race.

Years ago	Cultural stage	Area populated	Assumed density per square kilometer	Total population (millions)
1,000,000	Lower Palaeolithic		 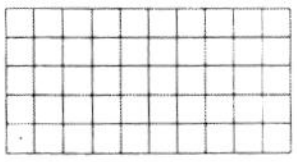 0.00425	0.125
300,000	Middle Palaeolithic		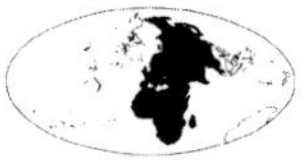 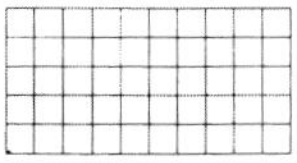 0.012	1
25,000	Upper Palaeolithic		 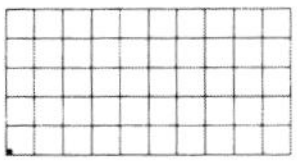 0.04	3.34
10,000	Mesolithic		 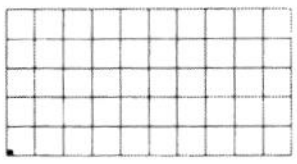 0.04	5.32
6000	Village farming and early urban		 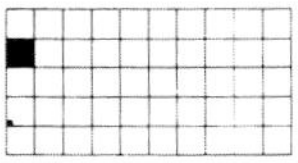 1.0[a] 0.04	86.5
2000	Village farming and urban		 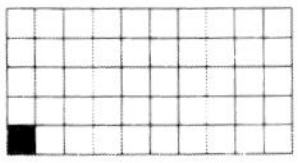 1.0	133

Year AD	Cultural stage	Area populated	Assumed density per square kilometer	Total population (millions)
1650	Farming and industrial		 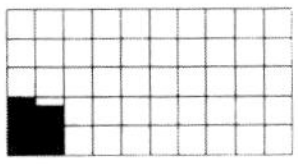 3.7	545
1750	Farming and industrial		 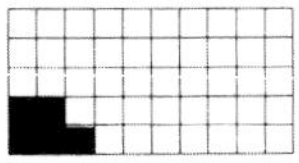 4.9	728
1800	Farming and industrial		 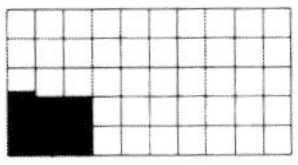 6.2	906
1900	Farming and industrial		 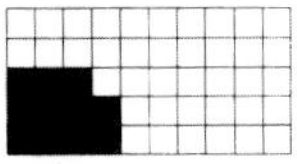 11.0	1610
1950	Farming and industrial		 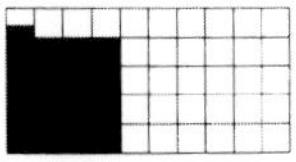 16.4	2400
2000	Farming and industrial		 46.0	6079

[a] The higher density refers to the area where farming has replaced the earlier type of economy (hunting and gathering).

Fig. 12.1. Changes in world population numbers and density over the last million years. (Used with permission from L.L. Cavalli-Sforza and W.F. Bodmer, © 1971, *The Genetics of Human Populations*, W.H. Freeman, San Francisco, California.)

Ex Situ Conservation

There has been a growing effort to collect wild germplasm and store it in repositories (Shands, 1990; Guarino *et al.*, 1995; Engles *et al.*, 2002). This *ex situ* conservation allows genotypes to be catalogued and made readily available to interested individuals. Probably the most extensive collections are those coordinated by the International Board for Plant Genetic Resources (headquartered in Rome, Italy) and the US National Germplasm System (headquartered in Beltsville, Maryland).

Such centralized collections are extremely critical as the march of civilization proceeds and the number of public breeders maintaining collections in industrialized countries declines (Brooks and Vest, 1988; Knight, 2003; Li and Pritchard, 2009). However, limits to storage space have demanded that collection sizes be kept to a minimum (Goodman, 1990). There is only so much material that can be stored in a viable state, particularly when it is clonally propagated and must be maintained as whole plants in the field or greenhouse. This limitation leaves a high likelihood that important germplasm will be left uncollected or poorly maintained.

The space problem has stimulated much research on improving methods of plant storage (Reed *et al.*, 2011). Seeds and pollen of many species have been found to survive years or even decades under proper conditions and require less space than whole plants, although they can only be used when cultivar purity is not important (Breese, 1989). Preservation of plants in tissue culture or as frozen buds has shown high promise as a way to save space and maintain specific genotypes (Forsline *et al.*, 1998; Reed, 2001), but these methods are costly and require specialized equipment. Clearly, no matter how efficient we become at storage, someone will still have to make a decision as to what stays and what goes.

Deciding what germplasm should be collected and stored is a great challenge. The general goal in any sampling strategy is to obtain as much of the genetic diversity as possible based on space and time limitations. High priority must be given to genes that solve existing problems, but other genes with unknown benefits should also be collected for future unforeseen circumstances. There is simply no reliable way to predict which alleles will be useful in future varieties until a need arises or they are incorporated.

Several different strategic considerations must be taken into consideration before collecting germplasm. A decision must be made on what constitutes the crop and where it is located (Brown and Marshall, 1995). Important related questions are: (i) What is the natural range of the species? (ii) Are there one or more relatives that should be collected? (iii) Is the crop sexually or asexually propagated? (iv) What is the best maintenance strategy? Decisions on these parameters are critical before effective sampling can even begin.

The germplasm collector must also face the big question of how many individuals to collect. Marshall and Brown (1975) and Brown and Marshall (1995) have devised equations that predict the sample size needed to adequately sample the common alleles in a population under different evolutionary models. In general, it takes 15–80 samples to adequately represent the common alleles. They conclude that localized common alleles should be the major target of a sampling strategy, since rare alleles will be acquired essentially at random except in the most comprehensive sampling efforts. Based on this assumption, the best strategy to use with no a priori knowledge of variation patterns is to collect 50–100 random individuals from as many diverse sites as possible.

Of course, this model assumes that each population and each species has the same level of genetic variation, which we know is not true. Each species has its own peculiar evolutionary history that must be taken into consideration in developing the most effective sampling strategy. For this reason, thorough studies of the ecological genetics of species are warranted before extensive collections are made. Important questions must be faced: (i) Are the species broken into environmentally and genetically distinct populations? (ii) Are the populations significantly substructured or even in their distribution? (iii) Are they outcrossed or inbred? Isozyme and DNA marker variability has proven instructive in determining variation patterns (Brown, 1988; Clegg, 1990; Hamrick and Godt, 1997), but it does not always accurately represent morphological patterns (Harrison *et al.*, 1997b). Nothing can replace extensive field experience.

In Situ Conservation

Probably the simplest solution to the storage problem is to maintain the wild populations themselves as genetic storehouses (Meilleur and Hodgkin, 2004). This is called *in situ* conservation. We may not know exactly what genes are contained in the native populations, but at least they will be there for later searching when a need arises. This approach depends less on the whims of individual breeders and repository directors and makes the loss of potentially valuable genes due to accident less likely. Also, the populations are still subject to the sieve of natural evolution, making it possible that new adaptations may arise.

A blanket preservation of all plant populations is, of course, impossible but efforts should be made to at least maintain large representative populations of the ecological range found in the progenitors of each crop species and its allies. A much larger sample of naturally segregating variability can be maintained in nature than can ever be held in even the largest repository. Effective *in situ* conservation requires knowledge of the variability patterns of the crop species and their ranges of adaptation, but this information is not unique to this type of conservation strategy. As we previously discussed, extensive evolutionary information is also needed to effectively collect native material for repositories.

While such surveys may seem like an impossible task, many of the crop species and their environments have already been subjected to census for their variation patterns and this information is available (Auricht *et al.*, 1995; Pendergast, 1995). It is hoped that the growing awareness of the fragility of our natural environment will not only result in the initiation of new ecological studies, but also the extensive cataloging of natural populations. The germplasm repositories could then concentrate on maintaining cultivars and natural materials that are in danger of eroding *in situ*. The repositories will still need to maintain a base collection of natural variability for ready deployment, but the emphasis would be placed on exploration and information transmittal, rather than the storage of huge numbers of genotypes.

References

Abbo, S., Lev-Yadun, S. and Ladizinsky, G. (2001) Tracing the wild genetic stocks of crop plants. *Genome* 44, 309–310.

Abbo, S., Shtienberg, D., Lichtenzveig, Lev-Yadun, S. and Gopher, A. (2003) The chickpea, summer cropping, and a new model for pulse domestication in the ancient Near East. *Quarterly Review of Botany* 78, 435–448.

Abbo, S., Zezak, I., Schwartz, E., Lev-Yadun, S., Karem, Z. and Gopher, A. (2008) Wild lentil and chickpea harvest in Israel: bearing on the origin of Near East farming. *Journal of Archeological Science* 35, 3172–3177.

Abbott, R.J. (1992) Plant invasions, interspecific hybridization and the evolution of new plant taxa. *TREE* 7, 401–405.

Abbott, R.J., Hegarthy, M.J., Hiscock, S.J. and Brennan, A.C. (2010) Homoploid hybrid speciation in action. *Taxon* 59, 1375–1386.

Adams, K.L. (2007) Evolution of duplicate gene expression in polyploidy and hybrid plants. *Journal of Heredity* 98, 136–141.

Adams, K.L. and Wendel, J.F. (2005) Polyploidy and genome evolution in plants. *Current Opinion in Plant Biology* 8, 135–141.

Adams, K.L., Cronn, R., Percifield, R. and Wendel, J.F. (2003) Genes duplicated by polyploidy show unequal contributions to the transcriptome and organ specific reciprocal silencing. *Proceedings of the National Academy of Sciences USA* 100, 4649–4654.

Adams, M.W. (1974) Plant architecture and physiological efficiency in the field bean. *Potential of Field Beans and Other Food Legumes in Latin America*. CIAT, Cali, Colombia, pp. 266–278.

Adams, W.T. and Allard, R.W. (1977) Effect of polyploidy on phosphoglucose isomerse diversity in *Festuca microstachys*. *Proceedings of the National Academy of Sciences USA* 74, 1652–1656.

Ahmad, F. (2000) A comparative study of chromosome morphology among nine annual species of *Cicer* L. *Cytobios* 101, 37–53.

Ahn, S. and Tanksley, S.D. (1993) Comparative linkage maps of rice and maize genomes. *Proceedings of the National Academy of Sciences USA* 90, 7980–7984.

Ainsworth, C. (2000) Boys and girls come out and play: the molecular biology of dioecious plants. *Annals of Botany* 86, 211–221.

Albrigio, A., Spettoli, P. and Cacco, G. (1978) Changes in gene expression from diploid to autotetraploid status of *Lycopersicon esculentum*. *Plant Physiology* 44, 77–80.

Aldrich, P.R., Doebley, J., Schertz, K.F. and Stec, A. (1992) Patterns of allozyme variation in cultivated and wild *Sorghum bicolor*. *Theoretical and Applied Genetics* 85, 451–460.

Alexander, D.E. (1960) Performance of genetically induced corn tetraploids. *American Seed Trade Association* 15, 68–74.

Alexander, J. and Coursey, D.G. (1969) The origins of yam cultivation. In: Ucho, P.J. and Dimbleby, G.H. (eds) *The Domestication and Exploitation of Plants and Animals*. Duckworth Publishers, London, pp. 405–425.

Al-Janabi, S.M., Honeycutt, R.J., McClelland, M. and Sobral, B.W.S. (1993) A genetic linkage map of *Saccharum spontaneum* L. *Genetics* 134, 1249–1260.

Al-Janabi, S.M., Honeycutt, R.J. and Sobral, B.W.S. (1994) Chromosomal assortment in *Saccharum*. *Theoretical and Applied Genetics* 89, 959–963.

Allaby, R. (2010) Integrating the processes in the evolutionary system of domestication. *Journal of Experimental Botany* 61, 935–944.

Allaby, R.G., Fuller, D.Q. and Brown, T.A. (2008) The genetic expectations of a protracted model for the origins of domesticated crops. *Proceedings of the National Academy of Sciences USA* 105, 13982–13986.

Allard, R.W. (1960) *Principles of Plant Breeding*. John Wiley & Sons, New York.

Allard, R.W. (1988) Genetic changes associated with the evolution of adaptedness in cultivated plants and their wild progenitors. *Journal of Heredity* 79, 225–238.

Allard, R.W. (1990) The genetics of host-pathogen coevolution: implications for genetic resource conservation. *Journal of Heredity* 81, 1–6.

Allard, R.W. and Kahler, A.L. (1971) Allozyme polymorphisms in plant populations. *Stadler Symposia* 3, 9–24.

Allard, R.W., Babble, E., Clegg, M. and Kahler, A. (1972) Evidence for coadaptation in *Avena barbata*. *Proceedings of the National Academy of Sciences USA* 69, 3043–3048.

Allem, A.C. (2002) The origin and taxonomy of cassava. In: Hillocks, R.J., Thresh, J.M. and Belloti, A.C. (eds) *Cassava, Biology, Production and Utilization*. CAB International, Wallingford, UK, pp. 1–16.

Allem, A.C., Mendes, R.A., Salomão, A.N. and Burle, M.L. (2001) The primary gene pool of cassava (*Manihot esculenta* Crantz subspecies *esculenta*, Euphorbiaceae). *Euphytica* 120, 127–132.

Alleweldt, G., Speigle-Roy, P. and Reisch, B. (1991) Grapes (*Vitus*). In: Moore, J.R. and Ballington, J.R. (eds) *Genetic Resources of Temperate Fruit and Nut Crops*. International Society of Horticultural Science, Wageningen, the Netherlands, pp. 289–328.

Amalraj, V.A. and Balasundaram (2006) On the taxonomy of the members of '*Saccharum* complex'. *Genetic Resources and Crop Evolution* 53, 35–41.

Ambrose, M. and Maxted, N. (2000) Peas (*Pisum* L.). In: Maxted, N. and Bennett, S.J. (eds) *Plant Genetic Resources of Legumes in the Mediterranean*. Kluwer Academic Publishers, Dordrecht, the Netherlands, pp.181–190.

Ames, M. and Spooner, D.M. (2008) DNA from herbarium specimens settles a controversy about the origins of the European potato. *American Journal of Botany* 95, 252–257.

Ammerman, A.J. and Cavalli-Sforza, L.L. (1984) *The Neolithic Transition and the Genetics of Populations in Europe*. Princeton University Press, Princeton, New Jersey.

Amoukou, A.L. and Marchais, L. (1993) Evidence of a partial reproductive barrier between wild and cultivated pearl millets (*Pennisetum glaucum*). *Euphytica* 67, 19–26.

Anderson, E. (1949) *Introgressive Hybridization*. John Wiley & Sons, New York.

Anderson, E. (1954) *Plants, Man and Life*. Melrose, London.

Anderson, E. (1961) The analysis of variation in cultivated plants with special reference to introgression. *Euphytica* 10, 79–86.

Anderson, E. and Hubricht, L. (1938) Hybridization in *Tradescantia* III. The evidence from introgressive hybridization. *American Journal of Botany* 25, 396–402.

Antonovics, J. (1971) The effects of a heterogeneous environment on the genetics of natural populations. *American Naturalist* 59, 593–599.

Antonovics, J. and Bradshaw, A.D. (1970) Evolution in closely adjacent plant populations. VIII. Clinal patterns at a mine boundary. *Heredity* 27, 349–362.

Aradhya, M.K., Dangl, G.S., Prins, B.H., Boursiquot, J.M., Walker, M.A., Meredith, C.P. and Simon, C.J. (2003) Genetic structure and differentiation in cultivated grape, *Vitis vinifera* L. *Genetical Research* 81, 179–182.

Argent, G. (1976) The wild bananas of Papua New Guinea. *Notes of the Royal Botanical Garden Edinburgh* 35, 77–114.

Argoncillo, C., Rodriguez-Loperen, M.A., Salcedo, G., Carbonero, P. and Garcia-Olmedo, F. (1978) Influence of homologous chromosomes on gene-dosage effects in allohexaploid wheat (*Triticum aestivum* L.). *Proceedings of the National Academy of Sciences USA* 75, 1446–1450.

Armstrong, D.L. (1957) Cytogenetic study of some derivatives of the F₁ hybrid *Prunus amygdalus* x *P. persica*. PhD thesis, University of California, Davis, California.

Armstrong, J.A., Powell, J.M. and Richards, A.J. (1982) *Pollinator and Evolution*. Royal Botanical Gardens, Sydney.

Arnau, G., Nemorin, A., Maledon, E. and Abraham, K. (2009) Revision of ploidy status of *Dioscorea alata* L. (Dioscoreaceae) by cytogenetic and microsatellite segregation analysis. *Theoretical and Applied Genetics* 118, 1239–1249.

Arnau, G., Abraham, K., Sheela, M.N., Chair, H., Sartie, A. and Asiedu, R. (2010) Yams. In: Bradshaw, J.E. (ed.) *Root and Tuber Crops, Handbook of Plant Breeding* 7. Springer Science+Business Media, New York, pp. 127–148.

Arnold, M.L. (1993) *Iris nelsonii* (Iridaceae): origin and genetic composition of a homolploid hybrid species. *American Journal of Botany* 80, 577–583.

Arriola, P.E. and Ellstrand, N.C. (1996) Fitness of interspecific hybrids in the genus *Sorghum*: Persistance of crop genes in wild populations. *Ecological Applications* 7, 512–518.

Arroyo-García, R., Ruiz-García, L., Bolling, L., Ocete, R., López, M.A., Arnold, C., Ergul, A., Söylemezoğlu, G., Uzun, H.I., Cabello, F., Ibáñez, J., Aradhya, M.K., Atanassov, A., Atanassov, I., Balint, S., Cenis, J.L., Costantini, L., Goris-Lavets, S., Grando, M.S., Klein, B.Y., McGovern, P.E., Merdinoglu, D., Pejic, I., Pelsy, F., Primikirios, N., Risovannaya, V., Roubelakis-Angelakis, K.A., Snoussi, H., Sotiri, P., Tamhankar, S., This, P., Troshin, L., Malpica, J.M., Lefort, F. and Martinez-Zapater, J.M. (2006) Multiple origins of cultivated grapevine (*Vitis vinifera* L. ssp. *sativa*) based on chloroplast DNA polymorphisms. *Molecular Ecology* 15, 3707–3714.

Arulsekar, S. and Bringhurst, R.S. (1981) Genetic model for the enzyme marker PGI in diploid California *Fragaria vesca*. *Journal of Heredity* 73, 117–120.

Ashley, M., Wilk, J., Styan, S., Craft, K., Jones, K., Feldman, K., Lewers, K. and Ashman, T. (2003) High variability and disomic segregation of microsatellites in the octoploid *Fragaria virginiana* Mill. (Rosaceae). *Theoretical and Applied Genetics* 107, 1201–1207.

Asins, M.J. and Carbonell, E.A. (1986) A comparative study on variability and phylogeny of *Triticum* species. 2. Interspecific relationships. *Theoretical and Applied Genetics* 72, 559–568.

Auricht, G.C., Reid, R. and Guarino, L. (1995) Published information on the natural and human environment. In: Guarino, L., Rao, V.R. and Reid, R. (eds) *Collecting Plant Genetic Diversity*. CAB International, Wallingford, UK.

Austin, D.F. (1978) The *Ipomoea batatas* complex I. Taxonomy. *Bulletin of the Torrey Botanical Club* 105, 114–129.

Ayala, F.J. (1975) Genetic differentiation during the speciation process. *Evolutionary Biology* 8, 1–78.

Ayala, F.J. (1982) *Population and Evolutionary Genetics: A Primer*. Benjamin/Cummings, Menlo Park, California.

Babcock, E.B. and Hall, H.M. (1924) *Hemizonia congeata*, a genetic, ecologic, and taxonomic study of the hay-field tarweeds. *University of California Publications in Botany* 13, 15–100.

Babcock, E.B. and Stebbins, G.L. (1938) *The American Species of Crepis. Their Interrelationships and Distribution as Affected by Polyploidy and Apomixis*. Publication No. 504, Carnegie Institute of Washington, Washington, DC.

Badami, P.S., Mallikarjuna, N. and Moss, J. (1997) Interspecific hybridization between *Cicer arietinum* and *C. pinnatifidum*. *Plant Breeding* 116, 393–395.

Badr, A., Müller, K., Schäfer-Pregl, R., El Rabey, H., Effgen, S., Ibrahim, H.H., Pozzi, C., Rohde, W. and Salamini, F. (2000) On the origin and domestication history of barley (*Hordeum vulgare*). *Molecular Biology and Evolution* 17, 499–510.

Bai, K.V. (1992) Cytogenetics of *Manihot* species and interspecific hybrids. In: Roca, W. and Thro, A.M. (eds) *Proceedings of the 1st International Scientific Meeting of the Cassava Biotechnology Network*. Centro Internacional de Agricultura Tropical, Cali, Colombia, pp. 51–55.

Bailey, C.H. and Hough, L.F. (1975) Apricots. In: Janick, J. and Moore, J.N. (eds) *Advances in Fruit Breeding*. Purdue University Press, West Lafayette, Indiana, pp. 367–384.

Bailey, D.C. (1983) Isozyme variation and plant breeders rights. In: Tanksley, S.D. and Orton, T.J. (eds) *Isozymes in Plant Genetics and Breeding*. Elsevier, New York.

Baker, H.G. (1970) *Plants and Civilization*. Wadsworth Publishing Company, Belmont, California.

Ballington, J.R. and Galletta, G.J. (1978) Comparative crossibility of 4 diploid *Vaccinium* species. *Journal of the American Society for Horticultural Science* 103, 554–560.

Balter, M. (1998) Why settle down? The mystery of communities. *Science* 282, 1442–1445.

Baquar, S.R. (1980) Chromosome behavior in Nigerian yams (*Dioscorea*). *Genetica* 54, 109.

Barakat, A., Matassi, G. and Bernardi, G. (1998) Distribution of genes in the genome of *Arabidopsis thaliana* and its implications for the genome reorganization of plants. *Proceedings of the National Academy of Sciences USA* 95, 10044–10049.

Barber, H.N. (1970) Hybridization and the evolution of plants. *Taxon* 19, 154–160.

Barigozzi, C. (ed.) (1986) *The Origin and Domestication of Cultivated Plants.* Elsevier, Amsterdam.

Barkley, N.A., Roose, M.L., Krueger, R.R. and Federici, C.T. (2006) Assessing genetic diversity and population structure in a citrus germplasm collection utilizing simple sequence repeat markers. *Theoretical and Applied Genetics* 112, 1519–1531.

Barrett, H.C. and Rhodes, A.M. (1976) A numerical taxonomy study of affinity relationships in cultivated citrus and close relatives. *Systematic Botany* 1, 105–136.

Barrett, S.C.H. (1983) Crop mimicry in weeds. *Economic Botany* 37, 255–282.

Bartsch, D. (2011) Gene flow in sugar beet. *Sugar Technology* DOI 10.1007/s12355-010-0053-1.

Bartsch, D. and Ellstrand, N.C. (1999) Genetic evidence for the origin of California wild beets (genus *Beta*). *Theoretical and Applied Genetics* 99, 1120–1130.

Bates, D.M. and Robinson, R.W. (1995) Cucumbers, melons and water-melons: *Cucumis* and *Citrullus* (Cucurbitaceae). In: Smartt, J. and Simmonds, N.W. (eds) *Evolution of Crop Plants.* Longman Scientific and Technical, Harlow, UK.

Baudet, J.C. (1977) Origine et classification des espèces cultivées du genre *Phaseolus. Bulletin de las Societe Botanique de Belgique* 10, 65–76.

Baum, B.R. (1977) *Oats: Wild and Cultivated: A Monograph of the Genus* Avena L. *(Poaceae).* Biosytematics Research Institute, Canada Department of Agriculture, Ottawa.

Baum, B.R. (1985a) *Avena atlantica*: a new diploid species of the oat genus from Morocco. *Canadian Journal of Botany* 63, 1057–1060.

Baum, B.R. (1985b) A new tetraploid species of oat. *Canadian Journal of Botany* 63, 1379–1385.

Baumgartner, B.J., Rapp, J.C. and Mullet, J.E. (1989) Plastid transcription activity and DNA copy number increase in early barley chloroplast development. *Plant Physiology* 89, 1011–1018.

Baur, E. (1924) Untersuchungen Ober das Wesen, die Entstehung und die Vererbung von Rassenunterscheiden bei *Antirrhinum majus. Bibliotheca Genetica* 4, 1–170.

Bayer, R.J., Mabberley, D.J., Morton, C., Miller, C.H., Sharma, I.K., Pfeil, B.E., Rich, S., Hitchcock, R. and Sykes, S. (2009) A molecular phylogeny of the orange subfamily (Rutaceae: Aurantioideae) using nine cpDNA sequences. *American Journal of Botany* 96, 668–685.

Bazzaz, F.A., Levin, D.A., Levy, M. and Schmierbach, M.R. (1982) The effect of chromosome doubling on photosynthetic rates in *Phlox. Photosynthetica* 17, 89–92.

Beadle, G.W. (1939) Teosinte and the origin of maize. *Journal of Heredity* 30, 245–247.

Beck, C.B. (1976) *Origin and Early Evolution of Angiosperms.* Columbia University Press, New York.

Bellon, M.R. and Berthaud, J. (2004) Transgenic maize and the evolution of landrace diversity in Mexico: the importance of farmers' behavior. *Plant Physiology* 134, 883–888.

Benabdelmouna, A., Abirached-Darmency, M. and Darmency, H. (2001) Phylogentic and genomic relationships in *Setaria italica* and its close relatives based on the molecular diversity and chromosomal organization of 5S and 18S-5.8S-25S rDNA genes. *Theoretical and Applied Genetics* 103, 668–677.

Bennett, M.D. (1972) Nuclear DNA content and minimum generation time in herbaceous plants. *Proceedings of the Royal Society, London, Series B* 181, 109–135.

Bennett, M.D. (1976) DNA amount, latitude and crop plant distribution. *Environmental and Experimental Botany* 16, 93–108.

Bennett, M.D. (1984) The genome, the natural karyotype and biosystematics. In: Grant, W.F. (ed.) *Plant Biosystematics.* Academic Press, New York, pp .41–66.

Bennett, M.D. (1987) Variation in genomic form in plants and its ecological implications. *New Phytologist* 106, 177–200.

Bennetzen, J.L. (2000a) Transposable element contributions to plant gene and genome evolution. *Plant Molecular Biology* 42, 251–269.

Bennetzen, J.L. (2000b) Comparative sequence analysis of plant nuclear genomes: microcolinearity and its many exceptions. *Plant Cell* 12, 1021–1029.

Bennetzen, J., Buckler, E., Chandler, V., Doebley, J., Dorweiler, J., Gaut, B., Freeling, M., Hake, S., Kellogg, E., Poethig, R.S., Walbot, V. and Wheeler, S. (2001) Genetic evidence and the origin of maize. *Latin American Antiquity* 12, 84–86.

Bennetzen, J.L., Ma, J. and Devos, K.M. (2005) Mechanisms of recent genome size variation in flowering plants. *Annals of Botany* 95, 127–132.

Ben-Ze'ev, N. and Zohary, D. (1973) Species relationships in the genus *Pisum. Israel Journal of Botany* 22, 73–91.

Bernstrom, P. (1952) Cytogenetic intraspecific studies in *Lamium.* I. *Hereditas* 38, 163–220.

Berry, M.A. (1985) The age of maize in the greater Southwest: a critical review. In: Ford, I. (ed.) *Prehistoric Food Production in North America.* Museum of Anthropology, University of Michigan, Ann Arbor, Michigan, pp. 279–308.

Besnard, G., Baradat, Ph., Chevalier, D., Tagmount, A. and Bervillé, A. (2001) Genetic differentiation in the olive complex (*Olea europaea*) revealed by RAPDs and RFLPs in the rRNA genes. *Genetic Resources and Crop Evolution* 48, 165–182.

Besnard, G., Khadari, B., Baradat, Ph. and Bervillé, A. (2002) Combination of chloroplast and mitochondrial DNA polymorphism to study cytoplasm genetic differentiation in the olive complex (*Olea europaea* L.). *Theoretical and Applied Genetics* 105, 139–144.

Besnard, G., Rubio de Casas, R. and Vagas, P. (2007) Plastid and nuclear DNA polymorphisms reveals historical processes of isolation and reticulation in the olive complex (*Olea europaea*). *Journal of Biogeography* 34, 736–752.

Besnard, G., Garcia-Verdugo, C., Rubio de Casas, R., Treier, U.A., Galland, N. and Vagas, P. (2008) Polyploidy in the olive complex (*Olea europaea*): evidence from flow cytometry and nuclear microsatellite analysis. *Annals of Botany* 101, 25–35.

Bettinger, R.L., Barton, L. and Morgan, C. (2010) The origins of food production in North China: a different kind of agricultural revolution. *Evolutionary Anthropology* 19, 9–21.

Bever, J.D. and Felber, F. (1992) The theoretical population genetics of autopolyploidy. *Oxford Surveys of Evolutionary Biology* 8, 185–217.

Bhattacharyya, M.K., Smith, A.M., Ellis, N., Hedley, C. and Martin, C. (1990) The wrinkled-seed character of pea described by Mendel is caused by a transposon-like insertion in a gene encoding starch branching enzyme. *Cell* 60, 115–122.

Billington, H.L., Mortimer, A.M. and McNeilly, T. (1988) Divergence and genetic structure in adjacent grass populations. I. Quantitative genetics. *Evolution* 42, 1267–1277.

Binford, L.F. (1968) Post-pleistocene adaptation. In: Binford, S.R. and Binford, L.R. (eds) *New Perspectives in Archeology.* Aldine, Chicago.

Bingham, E.T. (1980) Maximizing heterozygosity in autopolyploids. In: Lewis, W.D. (ed.) *Polyploidy: Biological Relevance.* Plenum Press, New York, pp. 471–490.

Birky, C.W. (1983) Relaxed cellular controls and organelle heredity. *Science* 222, 468–475.

Bisht, M.S. and Mukai, Y. (2001) Genomic *in situ* hybridization identifies genome donor of finger millet (*Eleusine coracana*). *Theoretical and Applied Genetics* 102, 825–832.

Blackman, B.K., Strasburg, J.L., Raduski, A.R., Michaels, S.D. and Rieseberg, L.H. (2010) The role of recently derived *FT* paralogs in sunflower domestication. *Current Biology* 20, 629–635.

Blake, N.K., Lehfeldt, B.R., Lavin, M. and Talbert, L.E. (1999) Phylogenetic reconstruction based on low copy DNA sequence data in an allopolyploid: the B genome of wheat. *Genome* 42, 351–360.

Blanc, G., Barakat, A., Guyot, R., Cooke, R. and Delseny, M. (2000) Extensive duplication and reshuffling in the Arabidopsis genome. *Plant Cell* 12, 1093–1101.

Bloom, W.L. (1976) Multivariate analysis of the introgressive replacement of *Clarkia nitans* by *Clarkia speciosa polyantha* (Onagraceae). *Evolution* 30, 412–424.

Böcher, T.W. (1947) Cytogenetic and biological studies in *Geranium robertianum* L. K. Danske Videsk. Selak. Skr. Biol. Medd. 20, 1–29.

Boffey, S.A. and Leech, R.M. (1982) Chloroplast DNA levels and the control of chloroplast division in light grown wheat leaves. *Plant Physiology* 69, 1387–1391.

Bogdanova, V.S., Galieva, E.R. and Kosterin, O.E. (2009) Genetic analysis of nuclear-cytoplasmic incompatatibility in pea associated with cytoplasm of an accession of wild subspecies *Pisum sativum* subsp. *elatius* (Bieb.) Schmahl. *Theoretical and Applied Genetics* 118, 801–809.

Bond, D.A. (1995) Faba bean: *Vicia faba* (Leguminosae – Papilionoideae). In: Smartt, J. and Simmonds, N.W. (eds) *Evolution of Crop Plants.* Longman Scientific & Technical, Harlow, UK, pp. 312–316.

Boster, J.S. (1985) Selection for perceptual distinctiveness: evidence from Aguaruna cultivars of *Manihot esculenta. Economic Botany* 39, 310–325.

Bousalem, M., Arnau, G., Hochu, I., Arnolin, R., Viader, V., Santoni, S. and David, J. (2006) Microsatellite segregation analysis and cytogenetic evidence for tetrasomic inheritance in the American yam

Dioscorea trifida and a new basic chromosome number in the *Dioscoreae. Theoretical and Applied Genetics* 113, 439–451.

Bousalem, M., Viader, V., Mariac, C., Gomez, R.-M., Hochu, I., Santoni, S. and David, J. (2010) Evidence of diploidy in the wild Amerindian yam, a putative progenitor of the endangered species *Dioscorea trifida* (Dioscoreaceae). *Genome* 53, 317–383.

Brace, C.L., Nelson, H. and Korn, N. (1979) *Atlas of Human Evolution.* Holt, Reinhart and Winston, New York.

Bradshaw, H.D., Otto, K.G., Frewen, B.G., McKay, J.K. and Schemske, D.W. (1998) Quantitative trait loci affecting differences in floral morphology between two species of monkeyflower (*Mimulus*). *Genetics* 149, 367–382.

Breese, E.L. (1989) *Regeneration and Multiplication of Germplasm Resources in Seed Banks: The Scientific Background.* International Board for Plant Genetic Resources, Rome.

Bres-Patry, C., Lorieux, M., Clément, G., Bangratz, M. and Ghesquière, A. (2001) Heredity and genetic mapping of domestication-related traits in a temperate *japonica* weedy race. *Theoretical and Applied Genetics* 102, 118–126.

Bretagnolle, F. and Thompson, J.D. (1995) Gametes with the somatic chromosomes number: mechanisms of their formation and role in the evolution of autopolyploid plants. *New Phytologist* 129, 1–22.

Breton, C., Terral, J.-F., Pinatel, C., Médail, F., Bonhomme, F. and Berville, A. (2009) The origins of the domestication of the olive tree. *C.R. Biologies* 332, 1059–1064.

Bretting, P.K. and Goodman, M.M. (1989) Karyotypic variation in Mesoamerican races of maize and its systematic significance. *Economic Botany* 43, 109–124.

Briggs, D. and Walters, S.M. (1984) *Plant Variation and Evolution.* Cambridge University Press, New York.

Briggs, F.N. and Knowles, P.F. (1967) *Introduction to Plant Breeding.* Reinhold Publishing Corporation, New York.

Bringhurst, R. (1990) Cytogenetics and evolution in American *Fragaria. HortScience* 25, 879–881.

Bringhurst, R.S. and Senanayake, Y.D.A. (1966) The evolutionary significance of natural *Fragaria chiloensis* x *F. vesca* hybrids resulting from unreduced gametes. *American Journal of Botany* 53, 1000–1006.

Bringhurst, R.S. and Voth, V. (1984) Breeding octoploid strawberries. *Iowa State Journal of Research* 58, 371–382.

Brooks, H.J. and Vest, G. (1988) Public programs on genetics and breeding of horticultural crops in the US. *HortScience* 20, 826–830.

Brown, A.H.D. (1988) Isozymes, plant population, genetic structure and genetic evolution. *Theoretical and Applied Genetics* 52, 145–157.

Brown, A.H.D. and Marshall, D.R. (1981) Evolutionary changes accompanying colonization in plants. In: Scudder, C.C.E. and Reveal, J.L. (eds) *Evolution Today.* Hunt Institute for Botanical Documentation, Carnegie-Mellon University, Pittsburgh, Pennsylvania, pp. 351–363.

Brown, A.H.D. and Marshall, D.R. (1995) A basic sampling strategy: theory and practice. In: Guarino, L., Rao, V.R. and Reid, R. (eds) *Collecting Plant Genetic Diversity.* CAB International, Wallingford, UK, pp. 75–92.

Brown, A.H.D., Doyle, J.L., Grace, J.P. and Doyle, J.J. (2002) Molecular phylogenetic relationships within and among diploid races of *Glycine tomentella* (Leguminosae). *Australian Systematics and Botany* 15, 37–47.

Brubaker, C.L. and Wendel, J.F. (1994) Reevaluating the origin of domesticated cotton (*Gossypium hirsutum* Malvaceae) using nuclear restriction fragment length polymorphisms (RFLPs). *American Journal of Botany* 81, 1309–1326.

Brunken, J., de Wet, J.M.J. and Harlan, J.R. (1977) The morphology and domestication of Pearl Millet. *Economic Botany* 31, 163–174.

Brush, S.B., Carney, H.J. and Human, Z. (1981) Dynamics of Andean potato agriculture. *Economic Botany* 35, 70–88.

Buckler, E.S. and Holtsford, T.P. (1996) *Zea* systematics, ribosomal ITS evidence. *Molecular Biology and Evolution* 13, 612–622.

Buerkle, C.A., Morris, R.J., Asmussen, M.A. and Rieseberg, L.H. (2000) The likelihood of homoploid hybrid speciation. *Heredity* 84, 441–451.

Buggs, R.J., Chamala, S., Wu, W., Gao, L., May, G.D., Schnable, P., Soltis, D.E., Soltis, P.S. and Barbazak, W.B. (2010) Characterization of duplicate gene evolution in the recent natural allopolyploid *Tragapogon miscellus* by next-generation sequencing and Sequenom iPLEX MassARRAY genotyping. *Molecular Ecology* 19, 132–146.

Burger, J.C., Chapman, M.A. and Burke, J.M. (2008) Molecular insights into the evolution of crop plants. *American Journal of Botany* 95, 113–122.

Burke, J.M. and Arnold, M.L. (2001) Genetics and the fitness of hybrids. *Annual Review of Genetics* 35, 31–52.

Burke, J.M., Tang, S., Knapp, S.J. and Rieseberg, L.H. (2002) Genetic analysis of sunflower domestication. *Genetics* 161, 1257–1267.

Burke, J.M., Lai, Z., Salmaso, M., Nakazato, T., Tang, S., Heesacker, A., Knapp, S. and Rieseberg, L.H. (2004) Comparative mapping and rapid karyotypic evolution in the genus *Helianthus. Genetics* 167, 449–457.

Burke, J.M., Tang, S., Knapp, S.J. and Rieseberg, L.H. (2005) Genetic consequences of selection during the evolution of cultivated sunflower. *Genetics* 171, 1933–1940.

Burnham, C. (1962) *Discussions in Cytogenetics.* Burgress, Minneapolis.

Burow, M.D., Simpson, C.E., Faries, M.W., Starr, J.L. and Paterson, A.H. (2009) Molecular biogeographic study of recently described B- and A- genome *Arachis* species, also providing new insights into the origins of cultivated peanut. *Genome* 52, 107–119.

Busbice, T.H. (1968) Effects of inbreeding on fertility in *Medicago sativa* L. *Crop Science* 8, 231–234.

Busbice, T.H. and Wilsie, C.P. (1966) Inbreeding depression and heterosis in autotetraploids with application to *Medicago sativa* L. *Euphytica* 15, 53–67.

Bush, G.L. (1975) Modes of animal speciation. *Annual Review of Ecology and Systematics* 6, 339–364.

Bush, M.B., Piperno, D.R. and Colinvaux, P.A. (1989) A 6,000 year history of Amazonian maize cultivation. *Nature* 340, 303–305.

Butterfass, T. (1979) *Patterns of Chloroplast Reproduction.* Springer-Verlag, New York.

Butzer, K.W. (1995) Biological transfer, agricultural change and environmental impacts of 1492. In: Duncan, R.R. (ed.) *International Germplasm Transfer: Past and Present.* Special Publication No. 23, CSSA, Madison, Wisconsin.

Byers, D.S. (ed.) (1967) *The Prehistory of the Tehuacan Valley.* Andover Foundation for Archaeological Research, University of Texas Press, Austin, Texas.

Cai, H.W. and Morishima, H. (2002) QTL clusters reflect character associations in wild and cultivated rice. *Theoretical and Applied Genetics* 104, 1217–1228.

Camargo, L.E.A. and Osborn, T.C. (1996) Mapping loci controlling flowering time in *Brassica oleracea. Theoretical and Applied Genetics* 92, 610–616.

Cambell, B.G. (1982) *Humankind Emerging,* 3rd edn. Little and Brown, Boston, Massachusetts.

Cambell, D.R. and Waser, N.M. (2001) Genotype-by-environment interaction and the fitness of plant hybrids in the wild. *Evolution* 55, 669–676.

Carlson, P.S. (1972) Locating genetic loci with aneuploids. *Molecular and General Genetics* 114, 273–280.

Carney, S.E. and Arnold, M.L. (1997) Differences in pollen-tube growth rate and reproductive isolation between Louisiana irises. *Journal of Heredity* 88, 545–549.

Carney, S.E., Cruzen, M.B. and Arnold, M.L. (1994) Reproductive interactions between hybridizing irises – analysis of pollen-tube growth and fertilization success. *American Journal of Botany* 81, 1169–1175.

Carr, D.E. and Dudash, M.R. (1996) Inbreeding depression in two species of *Mimulus* (Scrophulariaceae) with contrasting mating system. *American Journal of Botany* 83, 586–593.

Carr, G.D. and Kyhos, D.W. (1981) Adaptive radiation in the Hawaiian silversword alliance (Compositae-Madiinae). I. Cytogenetics of spontaneous hybrids. *Evolution* 35, 543–556.

Carson, H.L. (1971) Speciation and the founder principle. *University of Missouri Stadler Genetics Symposium* 3, 51–70.

Carson, H.L. and Templeton, A.R. (1984) Genetic revolutions in relation to speciation phenomena: the founding of new populations. *Annual Review of Ecology and Systematics* 15, 97–131.

Cavalli-Sforza, L.L. and Bodmer, W.F. (1971) *The Genetics of Human Populations.* W.H. Freeman, San Francisco, California.

Cavalli-Sforza, L.L. and Cavelli-Sforza, F. (1995) *The Great Human Diasporas.* Addison-Wesley, Reading, Massachusetts.

Challice, J.S. (1974) Rosaceae chemotaxonomy and the origins of the Pomoideae. *Botanical Journal of the Linnean Society* 69, 239–259.

Challice, J.S. and Kovanda, M. (1981) Chemotaxonomic studies in the family Rosaceae and the evolutionary origins of the subfamily Maloideae. *Preslia* 53, 289–304.

Chang, T.T. (1975) Exploration and survey in rice. In: Frankle, O.H. and Hawkes, J.G. (eds) *Crop Genetic Resources for Today and Tomorrow.* Cambridge University Press, Cambridge, pp. 159–165.

Chang, T.T. (1995) Rice, *Oryza sativa* and *Oryza glaberrima* (Gramineae – Oryzeae) In: Smartt, J. and Simmonds, N.W. (eds) *Evolution of Crop Plants.* Longman Scientific and Technical, Harlow, UK.

Chapman, M.A., Pashley, C.H., Wenzler, J., Hvala, J., Tang, S., Knapp, S.J. and Burke, J.M. (2008) A geneomic scan for selection reveals candidates for genes involved in the evolution of cultivated sunflower (*Helianthus annuus*). *The Plant Cell* 20, 2931–2945.

Charlesworth, D. and Charlesworth, B. (1987) Inbreeding depression and its evolutionary consequences. *Annual Review of Ecology and Systematics* 18, 237–268.

Chaudhary, B., Flagel, L., Stuper, R.M., Udall, J.A., Verma, N., Springer, N.M. and Wendel, J.F. (2009) Reciprocal silencing, transcriptional bias and functional divergence of homeologs in polyploidy cotton (*Gossypium*). *Genetics* 182, 503–517.

Chen, J., Isshiki, S., Tashiro, Y. and Miyazaki, S. (1997) Biochemical affinities between *Cucumis hystrix* Chakr. and two cultivated *Cucumis* species (*C. sativus* L. and *C. melo* L.). *Euphytica* 97, 139–141.

Chevreau, E., Lespinasse, Y. and Gallet, M. (1985) Inheritance of pollen enzymes and polyploid origin of apple (*Malus* x *domestica* Borkh.). *Theoretical and Applied Genetics* 71, 268–277.

Chikmawati, T., Skovmand, B. and Gustafson, J.P. (2005) Phylogenetic relationships among *Secale* species revealed by amplified fragment length polymorphisms. *Genome* 48, 792–801.

Childe, V.G. (1952) *New Light on the Most Ancient East.* Routledge and Kegan Paul, London.

Chomkos, S.A. and Crawford, G.N. (1978) Plant husbandry in prehistoric eastern North America: new evidence for its development. *American Antiquity* 43, 405–408.

Chooi, W.Y. (1971) Variation in nuclear DNA content in the genus *Vicia. Genetics* 68, 195–211.

Choumane, W. and Heizmann, P. (1988) Structure and variability of nuclear ribosomal genes in the genus *Helianthus. Theoretical and Applied Genetics* 76, 481–489.

Chyi, Y. and Weeden, N. (1984) Relative isozyme band intensities permit the identification of the 2n gamete parent for triploid apple cultivars. *HortScience* 19, 258–260.

Clark, G. (1967) *The Stone Hunters.* McGraw-Hill, New York.

Clausen, J. (1922) Studies on the collective species *Viola tricolor* L. II. *Botanisk Tidsskrift* 37, 363–416.

Clausen, J. (1926) Genetical and cytological investigations on *Viola tricolor* L. and *V. arvensis* Murr. *Hereditas* 8, 1–156.

Clausen, J. (1951) *Stages in the Evolution of Plant Species.* Cornell University Press, Ithaca, New York.

Clausen, J. and Heisey, W.M. (1958) *Experimental Studies on the Nature of Species. IV. Genetic Structure of Ecological Races.* Publication 615, Carnegie Institute of Washington, Washington, DC.

Clausen, J., Keck, D.D. and Heisey, W.M. (1940) *Experimental Studies on the Nature of Species. I. Effect of Varied Environments on Western North American Plants.* Publication No. 520, Carnegie Institute of Washington, Washington, DC.

Clegg, M.T. (1990) Molecular diversity in plant populations. In: Brown, A.H.D., Clegg, M.T., Kahler, A.L. and Weir, B.S. (eds) *Plant Population Genetics, Breeding, and Genetic Resources.* Sinauer Associates, Sunderland, Massachusetts, pp. 98–115.

Clegg, M.T. and Brown, A.H.D. (1983) The founding of plant populations. In: Schonewald-Cox, C.M., Chambers, B.M., MacBryde, B. and Thomas, W.L. (eds) *Conservation Genetics.* Benjamin Cummins, Menlo Park, California, pp. 216–228.

Clegg, M.T. and Durbin, M.L. (2000) Flower color variation: a model for the experimental study of evolution. *Proceedings of the National Academy of Sciences USA* 97, 7016–7023.

Clegg, M.T., Rawson, R.Y. and Thomas, K. (1984) Chloroplast DNA variation in pearl millet and related species. *Genetics* 106, 449–461.

Clegg, R.E., Allard, R. and Kahler, A. (1972) Is the gene the unit of selection? *Proceedings of the National Academy of Sciences USA* 69, 2474–2478.

Cleland, R.E. (1972) *Oenothera: Cytogenetics and Evolution.* Academic Press, New York.

Clutton-Brock, J. (1999) *A Natural History of Domesticated Animals.* Cambridge University Press, Cambridge.

Coates, D.J., Yen, D.E. and Gaffey, P.M. (1988) Chromosome variation in taro, *Colocasia esculenta*; implications for origin in the Pacific. *Cytologia* 53, 551–560.

Cock, J.H. (1982) Cassava, a basic energy source in the tropics. *Science* 218, 755–762.

Cohen, M.N. (1977) *The Food Crisis in Prehistory: Overpopulation and the Origins of Agriculture.* Yale University Press, New Haven, Connecticut.

Colwell, R. (1951) The use of radioactive isotopes in determining spore distribution patterns. *American Journal of Botany* 38, 511–523.

Colwell, R.E., Norse, E.A., Pimentel, D., Sharples, F.E. and Simberloff, D. (1985) Genetic engineering in agriculture. *Science* 229, 111–112.

Comai, L. (2000) Genetic and epigenetic interactions in allopolyploid plants. *Plant Molecular Biology* 43, 387–399.

Comai, L., Tyagi, A.P., Winter, K., Holmes-Davis, R., Reynolds, S.H., Stevens, Y. and Byers, B. (2000) Phenotypic instability and rapid gene silencing in newly formed *Arabidopsis* allotetraploids. *The Plant Cell* 12, 1551–1567.

Conner, A.J., Glare, T.R. and Nap, J.P. (2003) The release of genetically modified crops into the environment: Part II. Overview of ecological risk assessment. *Plant Journal* 33, 19–46.

Constable, G. (1973) *The Neanderthals.* Time-Life, New York.

Cook, L.M. and Soltis, P.M. (1999) Mating systems of diploid and allotetraploid populations of *Tragopogon* (Asteraceae). I. Natural populations. *Heredity* 82, 237–244.

Cook, L.M. and Soltis, P.S. (2000) Mating systems of diploid and allotetraploid populations of *Tragopogon* (Asteraceae). II. Artificial populations. *Heredity* 84, 410–415.

Coons, G.H. (1975) Interspecific hybrids between *Beta vulgaris* and the wild species of *Beta. Journal of the American Society of Sugar Beet Technology* 18, 281–386.

Coulibaly, S., Pasquet, R.S., Papa, R. and Gepts, P. (2002) AFLP analysis of the phenetic organization and genetic diversity of *Vigna unguiculata* L. Walp. reveals extensive gene flow between wild and domesticated types. *Theoretical and Applied Genetics* 104, 358–366.

Coursey, D.G. (1967) *Yams.* Longman, London.

Coyne, J.A., Barton, N.H. and Turelli, M. (2000) Is Wright's shifting balance process important in nature? *Evolution* 54, 306–317.

Crawford, D.J. (1990) *Plant Molecular Systematics: Macromolecular Approaches.* John Wiley & Sons, New York.

Crawford, G.W. (1992) Prehistoric plant domestication in East Asia. In: Cowan, C.W. and Watson, P.J. (eds) *The Origins of Agriculture.* Smithsonian Institution Press, Washington and London, pp. 7–38.

Crites, G.D. (1993) Domesticated sunflower in 5th millennium BP temporal context – New evidence from Middle Tennessee. *American Antiquity* 58, 146–148.

Cronn, R., Brothers, M., Klier, K., Bretting, P.K. and Wendel, J.F. (1997) Allozyme variation in domesticated annual sunflower and its wild relatives. *Theoretical and Applied Genetics* 95, 532–545.

Crow, J.F. (1945) A chart of the χ^2 and *t* distributions. *Journal of the American Statistical Association* 40, 375.

Crow, J.F. and Kimura, M. (1970) *An Introduction to Population Genetics Theory.* Harper and Row, New York.

Crowe, L.K. (1964) The evolution of outbreeding in plants. I. The angiosperms. *Heredity* 19, 435–457.

Crowe, L.K. (1971) The polygenic control of outbreeding in *Borago officinalis. Heredity* 27, 111–118.

Cubero, J.I. (1974) On the origin of *Vicia faba. Theoretical and Applied Genetics* 45, 45–47.

Cullis, C.A. (1987) The generation of somatic and heritable variation in response to stress. *American Naturalist* 130, 562–573.

Dale, P.J. (1992) Spread of engineered genes into wild relatives. *Plant Physiology* 100, 13–15.

D'Andrea, A.C., Klee, M. and Casey, J. (2001) Archaeobotanical evidence for pearl millet (*Pennisetum glaucum*) in sub-Saharan Africa. *Antiquity* 75, 341–348.

Daniels, J. and Roach, B.T. (1987) Taxonomy and evolution. In: Heinz, D.J. (ed.) *Sugarcane Improvement through Breeding.* Elsevier Press, Amsterdam, pp. 7–84.

Daniels, J., Roach, B.T., Daniels, C. and Paton, N.H. (1991) The taxonomic status of *Saccharum barberi* Jeswiet and *S. sinense* Roxb. *Sugarcane* 3, 11–16.

Dansi, A., Mignouna, H.D. and Zoundjihékpon, J. (1999) Morphological diversity, cultivar groups and possible descent in the cultivated yams (*Dioscorea cayenensis/D. rotundata*) complex in Benin Republic. *Genetic Resources and Crop Evolution* 46, 371–388.

Dansi, A., Mignouna, H.D., Pillay, M. and Zok, S. (2001) Ploidy variation in the cultivated yams (*Dioscorea cayenensis-Dioscorea rotundata* complex) from Cameroon as determined by flow cytometry. *Euphytica* 119, 301–307.

Dark, S.O.S. (1971) Experiments in the cross pollination of sugar beet in the field. *Journal of the National Institute of Agricultural Biology* 12, 242–266.

Darlington, C.D. and Mather, K. (1949) *The Elements of Genetics.* G. Allen, London.

Darrow, G.M. (1966) *The Strawberry. History, Breeding and Physiology.* Holt, Rinehart and Winston, New York.

Da Silva, J.A.G., Sorrels, M.E., Burnquist, W.L. and Tanksley, S.D. (1993) RFLP linkage map and genome analysis of *Saccharum spontaneum. Genome* 36, 782–791.

Davies, D.R. (1995) Peas: *Pisum sativum* (Leguminosae – Papilionoideae). In: Smartt, J. and Simmonds, N.W. (eds) *Evolution of Crop Plants*. Longman Scientific and Technical, Harlow, UK, pp. 294–296.

Davies, M.S. and Hillman, G.C. (1992) Domestication of cereals. In: Chapman, G.P. (ed.) *Grass Evolution and Domestication*. Cambridge University Press, Cambridge, pp. 199–244.

Dean, C. and Leech, R.M. (1982) Genome expression during normal leaf development. *Plant Physiology* 70, 1605–1608.

Debouck, D.G. (2000) Genetic resources of *Phaseolus* beans: patterns in time, space and people. In: Fueyo, M.A., Gonzalez, A.J., Ferreira, J.J. and Giraldez, R. (eds) *II. Seminario de Judía de la Península Ibérica*. Actas de la Asociación Espanola de Leguminosas. Villaviciosa, Asturias, pp. 17–39.

Debouck, D.G. and Smartt, J. (1995) Beans *Phaseolus* spp. (Leguminosae-Papilionoideae). In: Smartt, J. and Simmonds, N.W. (eds) *Evolution of Crop Plants*. Longman Scientific and Technical, Harlow, UK, pp. 287–294.

DeCandolle, A. (1886 – reprint 1959) *Origin of Cultivated Plants*. Hafner Publishing Company, New York, 468 pp.

Decker, D.S. (1988) Origin(s), evolution and systematics of *Cucurbita pepo* (Cucurbitaceae). *Economic Botany* 42, 4–15.

Decker-Walters, D.S. (1990) Evidence for multiple domestications of *Curcurbita pepo*. In: Bates, D.M. (ed.) *Biology and Utilization of the Cucurbitaceae*. Cornell University Press, Ithaca, New York, pp. 46–101.

Decker-Walters, D., Staub, J., López-Sesé, A. and Nakata, E. (2001) Diversity in landraces and cultivars of bottle gourd (*Lagenaria siceraria*; Cucurbitaceae) as assessed by random amplified polymorphic DNA. *Genetic Resources and Crop Evolution* 48, 369–380.

Deerr, N. (1949) *The History of Sugar*. Wiley, New York.

Delgado-Salinas, A., Bibler, R. and Lavin, M. (2006) Phylogeny of the genus *Phaseolus* (Leguminaceae): a recent diversification in an ancient landscape. *Systematic Botany* 31, 779–791.

De Moraes, A.P., Filho, W. and Guerra, M. (2007) Karyotype diversity and the origin of grapefruit. *Chromosome Research* 15, 115–121.

De Nettancourt, D. (1977) *Incompatibility in Angiosperms*. Springer-Verlag, Berlin.

De Nettancourt, D. (2001) *Incompatibility and Incongruity in Wild and Culivated Plants*. Springer-Verlag, Berlin.

De Pamphilis, C.W. and Palmer, J.D. (1990) Loss of photosynthetic and chlororespiratory genes from the plastid genome of a parasitic flowering plant. *Nature* 348, 337–339.

Denham, T.P., Haberle, G., Lentfer, C., Fullagar, R., Field, J., Therin, M., Porch, M. and Winsborough, B. (2003) Origins of agriculture at Kuk swamp in the Highlands of New Guinea. *Science* 301, 189–193.

Denham, T.P., Haberle, G. and Lentfer, C. (2004) New evidence and revised interpretations of early agriculture in New Guinea. *Antiquity* 78, 839–857.

Den Nijs, A.P.M. and Visser, D.L. (1985) Relationships between African species of the genus *Cucumis* L. estimated by the production, vigor and fertility of F_1 hybrids. *Euphytica* 34, 279–290.

Deu, M., DeLeon, D.G., Glazmann, J.C., DeGremont, I., Chantereau, J., Lanaud, C. and Hamon, P. (1994) RFLP diversity in cultivated *Sorghum* in relation to racial differentiation. *Theoretical and Applied Genetics* 88, 838–844.

Devlin, B. and Ellstrand, N.C. (1990) The development and application of a refined method for estimating gene flow from an angiosperm paternity analysis. *Evolution* 44, 248–258.

Devos, K.M. and Gale, M.D. (2000) Genome relationships: the grass model in current research. *Plant Cell* 12, 637–646.

Devos, K.M., Chinoy, M.D., Liu, C.J. and Gale, M.D. (1992a) RFLP-based genetic map of the homologous group 3 chromosomes of wheat and rye. *Theoretical and Applied Genetics* 83, 931–939.

Devos, K.M., Atkinson, M.D., Chinoy, M.D., Liu, C.J. and Gale, M.D. (1992b) Chromosomal rearrangements in rye genome relative to that of wheat. *Theoretical and Applied Genetics* 85, 673–680.

deWet, J.M.J. (1968) Diploid-tetraploid-haploid cycles and the origin of variability in *Dichanthium* agamospecies. *Evolution* 22, 394–397.

deWet, J.M.J. (1978) Systematics and evolution of *Sorghum* section *Sorghum* (Gramineae). *American Journal of Botany* 65, 477–484.

deWet, J.M.J. and Huckabay, J.P. (1967) The origin of *Sorghum bicolor*. II. Distribution and domestication. *Evolution* 21, 787–802.

deWet, J.M.J., Oestry-Stidd, L.L. and Cubero, J.I. (1979) Origins and evolution of the foxtail millet *Setaria italica*. *Journal of Agriculture and Traditional Botanical Applications* 26, 53–64.

deWet, J.M.J., Newell, C.A. and Brink, D.E. (1984a) Counterfeit hybrids between *Tripsium* and *Zea* (Graminineae). *American Journal of Botany* 71, 245–251.

deWet, J.M.J., Rao, K.E.P., Brink, D.E. and Mengesha, M.H. (1984b) Systematics and evolution of *Eleusine coracana* (Gramineae). *American Journal of Botany* 71, 550–557.

Dewey, D. (1980) Some applications and misapplications of induced polyploidy in plant breeding. In: Lewis, W.D. (ed.) *Polyploidy: Biological Relevance.* Plenum Press, New York, pp. 445–476.

Dewey, D.R. (1966) Inbreeding depression in diploid, tetraploid and hexaploid crested wheat. *Crop Science* 6, 144–147.

Diamond, J. (1998) *Guns, Germs and Steel: the Fates of Human Societies.* W.W. Norton, New York.

Diaz, J., Schmiediche, P. and Austin, D.F. (1966) Polygon of crossibility between eleven species of *Ipomoea*: Section *Batatas* (Convolvulaceae). *Euphytica* 88, 189–200.

Dice, L.R. (1945) Measures of the amount of ecologic association between species. *Ecology* 26, 297–302.

Dickau, R., Ranere, A.J. and Cooke, R.G. (2007) Starch grain evidence for the preceramic dispersals of maize and root crops into tropical dry and humid forests of Panama. *Proceedings of the National Academy of Sciences USA* 104, 3651–3656.

Dickson, E.E., Kresovich, S. and Weeden, N.F. (1991) Isozymes in North American *Malus* (Rosaceae): hybridization and species differentiation. *Systematic Botany* 16, 363–375.

Dilcher, D. (2000) Toward a new synthesis: major evolutionary trends in the angiosperm fossil record. *Proceedings of the National Academy of Sciences USA* 97, 7030–7036.

Dillehay, T.D. (2000) *The Settlement of the Americas.* Basic Books, New York.

Dillehay, T.D., Rossen, J., Andres, T.C. and Williams, D.E. (2007) Preceramic adoption of peanut, squash and cotton in Northern Peru. *Science* 316, 1890–1893.

Djé, Y., Heuertz, M., Lefebvre, C. and Vekemans, X. (2000) Assessment of genetic diversity within and among germplasm accessions in cultivated sorghum using microsatellite markers. *Theoretical and Applied Genetics* 100, 918–925.

Dobzhansky, T. (1970) *Genetics of the Evolutionary Process.* Columbia University Press, New York.

Dobzhansky, T. and Pavlovsky, O. (1953) Indeterminate outcome of certain experiments on *Drosophila* populations. *Evolution* 7, 198–210.

Dobzhansky, T., Ayala, F.J., Stebbins, G.L. and Valentine, J.W. (1977) *Evolution.* W.H. Freeman, San Francisco, California.

Doebley, J. (1989) Isozymic evidence and the evolution of crop plants. In: Soltis, D.E and Soltis, P.S. (eds) *Isozymes in Plant Biology.* Advances in Plant Sciences Series Vol. 4. Dioscorides Press, Portland, Oregon.

Doebley, J. (1990) Molecular evidence for gene flow among *Zea* species. *Bioscience* 40, 443–448.

Doebley, J. and Iltis, H.H. (1980) Taxonomy of *Zea* (Gramininae). I. A subgeneric classification with key to taxa. *American Journal of Botany* 67, 982–993.

Doebley, J. and Stec, A. (1993) Inheritance of the morphological differences between maize and teosinte: comparison of results for two F_2 populations. *Genetics* 134, 559–570.

Doebley, J., Goodman, M.M. and Stuber, C.W. (1984) Isozyme variation from maize from the southwestern United States: taxonomy and anthropological implications. *Maydica* 28, 97–120.

Doebley, J., Goodman, M.M. and Stuber, C.W. (1985) Isozyme variation in the races of maize from Mexico. *American Journal of Botany* 72, 629–639.

Doebley, J., Goodman, M.M. and Stuber, C.W. (1987) Patterns of isozyme variation between maize and Mexican annual teosinte. *Economic Botany* 41, 234–246.

Doebley, J., Stec, A., Wendel, J. and Edwards, M. (1990) Genetic and morphological analysis of a maize-teosinte F_2 population: implications for the origin of maize. *Proceedings of the National Academy of Sciences USA* 87, 9888–9892.

Doebley, J., Stec, A. and Gustus, C. (1995) *Teosinte branched 1* and the origin of maize: evidence for epistasis and the evolution of dominance. *Genetics* 141, 333–346.

Doebley, J.F., Gaut, B.S. and Smith, B.D. (2006) The molecular genetics of crop domestication. *Cell* 127, 1309–1321.

Doganlar, S., Frary, A., Daunay, M.-C., Lester, R.N. and Tanksley, S.D. (2002) Conservation of gene function in the *Solanaceae* as revealed by comparative mapping of domestication traits in eggplant. *Genetics* 161, 1713–1726.

Doggett, H. (1988) *Sorghum.* Longman, London.

Doggett, H. and Rao, K.E.P. (1995) Sorghum, *Sorghum bicolor* (Gramineae-Androponeae). In: Smartt, J. and Simmonds, N.W. (eds) *Evolution of Crop Plants.* Longman Scientific and Technical, Harlow, UK, pp. 173–180.

Dong, S. and Adams, K.L. (2011) Differential contributions to the transcriptone of duplicated genes in response to abiotic stresses in natural and synthetic polyploids. *New Phytologist* 190, 1045–1057.

Donovan, L.A., Rosenthal, D.R., Sanchez-Velenosi, M., Rieseberg, L.H. and Ludwig, F. (2010) Are hybrid species more fit than ancestral parent species in the current hybrid species habitats? *Journal of Evolutionary Biology*. Doi:10.111/j.1420-9101.2010.01950.x

Doolittle, W.F. and Sapienza, C. (1980) Selfish genes, the phenotype paradigm and genome evolution. *Nature* 284, 601–603.

Dorweiler, J.E. and Doebley, J. (1997) Developmental analysis of *teosinte glume architecture 1*: a key locus in the evolution of maize. *American Journal of Botany* 84, 1313–1322.

Douches, D.S. and Quiros, C.F. (1988) Genetic strategies to determine the mode of 2n gamete production in diploid potatoes. *Euphytica* 38, 247–260.

Dowrick, V.P.J. (1956) Heterostyly and homostyly in *Primula obconica*. *Heredity* 10, 219–236.

Doyle, J. (1985) *Altered Harvest: Agriculture, Genetics and the Fate of the World's Food Supply*. Viking Press, New York.

Doyle, J.J. and Beachy, R.N. (1990) Ribosomal gene variation in soybean (*Glycine*) and its relatives. *Theoretical and Applied Genetics* 70, 369–376.

Doyle, J.J., Doyle, J.L. and Brown, A.H.D. (1990) A chloroplast-DNA phylogeny of the wild perennial relatives of soybean (*Glycine* subgenus *Glycine*): congruence with morphological and crossing groups. *Evolution* 44, 371–389.

Doyle, J.J., Doyle, J.L. and Brown, A.H.D. (1999) Origins, colonization, and linkage recombination in a widespread perennial soybean polyploid complex. *Proceedings of the National Academy of Sciences USA* 96, 10741–10745.

Doyle, J.J., Doyle, J.L. and Brown, A.H.D. (2002) Genomes, multiple origins, and linkage recombination in the *Glycine tomentella* (Leguminosae) polyploidy complex: histone H3-D sequences. *Evolution* 56, 1388–1402.

Doyle, J.J., Doyle, J.L., Rauscher, J.T. and Brown, A.H.D. (2004) Evolution of the perennial soybean polyploidy complex (*Glycine* subgenus *Glycine*): a study of contrasts. *Biological Journal of the Linnean Society* 82, 583–597.

Drossou, A., Katsiotis, A., Leggett, J.M., Loukas, M. and Tsakas, S. (2004) Genome and species relationships in *Avena* based on RAPD and AFLP molecular markers. *Theoretical and Applied Genetics* 109, 48–54.

Duc, G., Bao, S., Baum, M., Redden, B., Sadiki, M., Suso, M.J., Vishniakova, M. and Zong, X. (2010) Diversity, maintenance and use of *Vicia faba* L. genetic resources. *Field Crop Research* 115, 270–278.

Dudash, M.R., Carr, D.E. and Fenster, C.B. (1997) Five generations of enforced selfing and outcrossing in *Mimulus guttatus*: inbreeding variation at the population and family level. *Evolution* 51, 54–65.

Dumont, R., Dansi, A., Vernier, P. and Zoundjihekpon, J. (2005) Biodiversité et domestication des ignames en Afrique de L'Ouest: pratiques traditionnelles conduisant à *Dioscorea rotundata*. Repères, CIRAD, 120 pp.

Dunbier, M.W. and Bingham, E.T. (1975) Maximum heterozygosity in alfalfa: results using haploid-derived autotetraploids. *Crop Science* 15, 527–531.

Durbin, M.L., Learn, G.H., Hutty, G.A. and Clegg, M.L. (1995) Evolution of the chalcone synthase gene family in the genus *Ipomoea*. *Proceedings of the National Academy of Sciences USA* 92, 3338–3342.

Durbin, M.L., McCaig, B. and Clegg, M.T. (2000) Molecular evolution of chalcone synthase multigene family in the morning glory family. *Plant Molecular Biology* 42, 79–92.

Durbin, M.L., Denton, A.L. and Clegg, M.T. (2001) Dynamics of mobile element activity in chalcone synthase loci in the common morning glory (*Ipomoea purpurea*). *Proceedings of the National Academy of Sciences USA* 98, 5084–5089.

Dvořák, J., Di Terlizzi, P., Zhang, H.-B. and Resta, P. (1993) The evolution of polyploid wheats: identification of the A genome donor species. *Genome* 36, 21–31.

Dvořák, J., Luo, M.-C., Yang, Z.-L. and Zhang, H.-B. (1998) The structure of the *Aegilops tauschii* genepool and the evolution of hexaploid wheat. *Theoretical and Applied Genetics* 97, 657–670.

East, E.M. (1916) Studies on size inheritance in *Nicotiana*. *Genetics* 1, 164–176.

Edwards, A.W.F. and Cavalli-Sforza, L.L. (1964) Reconstruction of evolutionary trees. In: Heywood, V.E. and McNeill, J. (eds) *Phenetic and Phylogenetic Classification*. Systematics Association, London, pp. 67–76.

Ehrendorfer, F. (1979) Polyploidy and distribution. In: Heywood, V.E. and McNeill, J. (eds) *Phenetic and Phylogenetic Classification*. Systematics Association, London.

Ehrlich, P.R. and Raven, P.H. (1969) Differentiation of populations. *Science* 165, 1228–1232.

Eldridge, N. and Gould, S.J. (1972) Punctuated equilibria: an alternative to phyletic gradualism. In: Schopf, T.J.M. (ed.) *Models in Paleobiology*. Freeman Cooper, San Francisco, California, pp. 82–115.

Ellstrand, N.C. (2001) When transgenes wander, should we worry? *Plant Physiology* 125, 1543–1545.

Ellstrand, N.C. (2003) *Dangerous Liaisons? When cultivated plants mate with their wild relatives*. Johns Hopkins University Press, Baltimore, Maryland.

Ellstrand, N.C. and Hoffman, C.A. (1990) Hybridization as an avenue of escape for engineered genes. *BioScience* 40, 438–442.

Ellstrand, N.C. and Marshall, P.L. (1985) Interpopulational gene flow by pollen in wild radish, *Raphanus sativus. American Naturalist* 126, 606–616.

Ellstrand, N.C. and Schierenbeck, K.A. (2000) Hybridization as a stimulus for the evolution of invasiveness in plants? *Proceedings of the National Academy of Sciences USA* 97, 7043–7050.

Ellstrand, N.C., Whitkus, R. and Rieseberg, L.H. (1996) Distribution of spontaneous plant hybrids. *Proceedings of the National Academy of Sciences USA* 93, 5090–5093.

Ellstrand, N.C., Prentice, H.C. and Hancock, J.F. (1999) Gene flow and introgression from domesticated plants into their wild relatives. *Annual Review of Ecology and Systematics* 30, 539–563.

Emms, S.K. and Arnold, M.L. (1997) The effect of habitat on parental and hybrid fitness: transplant experiments with Louisiana irises. *Evolution* 51, 1112–1119.

Emms, S.K. and Arnold, M.L. (2000) Site-to-site differences in pollinator visitation patterns in a Louisiana iris hybrid zone. *Oikos* 91, 568–578.

Emory, K.P. and Sinoto, Y.H. (1964) Préhistoire de la polynésie. *Journal de la Société des Océanistes* 20, 39–41.

Endrizzi, J.E., Turcotte, E.L. and Kohel, R.J. (1985) Genetics, cytology and evolution of *Gossypium. Advances in Genetics* 23, 271–375.

Engles, J.M.M., Ramanatha, V., Brown, A.H.D. and Jackson, M.T. (eds) (2002) *Managing Plant Genetic Diversity*. CAB International, Wallingford, UK.

Eshbaugh, W.H. (1975) Genetic and biochemical systematic studies of chili peppers (*Capsicum*-Solanaceae). *Bulletin of the Torrey Botanical Club* 102, 396–403.

Eshbaugh, W.H. (1980) The taxonomy of the genus *Capsicum* (Solanaceae). *Phytologia* 47, 153–166.

Eshbaugh, W.H. (1993) Peppers: history and exploitation of a serendipitous new crop discovery. In: Janick, J. and Simon, J.E. (eds) *New Crops*. Wiley, New York, pp. 132–139.

Eshbaugh, W.H., Guttman, S.I. and McLeod, M.J. (1983) The origin and evolution of domesticated *Capsicum* species. *Journal of Ethnobiology* 3, 49–54.

Eubanks, M.W. (2001a) The mysterious origin of maize. *Economic Botany* 55, 492–514.

Eubanks, M.W. (2001b) An interdisciplinary perspective on the origin of maize. *Latin American Antiquity* 12, 91–98.

Eyre-Walker, A., Gaut, R.G., Hilton, H., Feldman, D.L. and Gaut, B.S. (1998) Investigation of the bottleneck leading to the domestication of maize. *Proceedings of the National Academy of Sciences USA* 95, 4441–4446.

Faegri, K.L. and van der Pijl, V. (1979) *The Principles of Pollination Ecology*. Pergamon Press, Oxford.

Falconer, D.S. and Mackay, T.F.C. (1996) *Introduction to Quantitative Genetics*. Burgess Press, New York.

Fedoroff, N. (2000) Transposons and genome evolution in plants. *Proceedings of the National Academy of Sciences USA* 97, 7002–7007.

Feldman, M. (1963) Evolutionary studies in the *Aegilops-Triticum* group with special emphasis on causes of variability in the polyploid species of section Pleinoathera. PhD thesis, Hebrew University, Jerusalem.

Feldman, M., Galili, G. and Levy, A.A. (1986) Genetic and evolutionary aspects of allopolyploidy in wheat. In: Barigozzi, C. (ed.) *The Origin and Domestication of Cultivated Crops*. Elsevier, Amsterdam, pp. 83–101.

Feldman, M., Liu, B., Segal, G., Addo, S., Levy, A.A. and Vega, J.M. (1997) Rapid elimination of low-copy DNA sequences in polyploidy wheat: a possible mechanism for differentiation of homologous chromosomes. *Genetics* 147, 1381–1387.

Fisher, R.A. (1930) *The Genetical Theory of Natural Selection*. Clarendon, Oxford, UK.

Flagel, L.E. and Wendel, J.F. (2009) Gene duplication and evolutionary novelty in plants. *New Phytologist* 183, 557–564.

Flagel, L.E. and Wendel, J.F. (2010) Evolutionary rate variation, genomic dominance and duplicate gene expression evolution during allopolyploid cotton speciation. *New Phytologist* 186, 184–193.

Flagel, L., Udall, J., Nettleton, D. and Wendel, J. (2008) Duplicate gene expression in allopolyploid *Gossypium* reveals two temporally distinct phases of expression evolution. *BMC Biology* 6, 17.

Flannery, K.V. (1968) Archeological systems theory and early Mesoamerica. In: Meggers, B.J. (ed.) *Anthropological Archeology in the Americas*. Anthropological Society of Washington, Washington, DC, pp. 67–87.

Flor, H.H. (1954) *Identification of Races of Flax Rust by Lines with Single Rust-conditioning Genes*. Technical Bulletin No. 1087, US Department of Agriculture, Washington, DC.

Folkertsma, R.T., Frederick, H., Rattunde, W., Chandra, S., Raju, G.S. and Hash, C.T. (2005) The pattern of genetic diversity of Guinea-race *Sorghum bicolor* (L.) Moench land races as revealed with SSR markers. *Theoretical and Applied Genetics* 111, 399–409.

Ford, E.B. (1975) *Ecological Genetics: The Evolution of Super-genes*. Chapman & Hall, London.

Ford, R.I. (1985) *Prehistoric Food Production in North America*. Museum of Anthropology, University of Michigan, Ann Arbor, Michigan.

Ford-Lloyd, B.V. and Williams, J.T. (1975) A revision of *Beta* section *Vulgaris* (Chenopodiaceae) with new light on the origin of cultivated beets. *Botanical Journal of the Linnean Society* 71, 89–102.

Forsline, P., Dickson, E. and Djangalieu, A. (1994) Collection of wild *Malus, Vitus* and other fruit species genetic resources in Kasakstan and neighboring republics. *HortScience* 29, 433.

Forsline, P.L., Towill, L.E., Waddell, J.W., Sushnoff, C., Lamboy, W.F. and McFerson, J.R. (1998) Recovery and longevity of cryopreserved dormant apple buds. *Journal of the American Society for Horticultural Science* 123, 365–370.

Fowler, N.L. and Levin, D.A. (1984) Ecological constraints on the establishment of a novel polyploid in competition with its diploid progenitor. *American Naturalist* 124, 703–711.

Fowler, S. and Thomashow, M.F. (2002) Arabidopsis transcriptome profiling indicates that multiple regulatory pathways are activated during cold acclimation in addition to the CBF cold response pathway. *Plant Cell* 14, 1675–1690.

Frankel, O. and Munday, A. (1962) The evolution of wheat. In: *The Evolution of Living Organisms*. Symposium of the Royal Society, Victoria, Australia.

Franklin-Tong, V.E. (2008) *Self-incompatibility in Flowering Plants*. Springer, Berlin.

Frederikson, S. and Peterson, G. (1998) A taxonomic revision of *Secale*. *Nordic Journal of Botany* 18, 399–420.

Freeling, M. (1973) Simultaneous induction by anaerobiosis or 2,4-D of multiple enzymes specified by two unlinked genes: differential *Adh-1-Adh-2* expression in maize. *Molecular and General Genetics* 127, 215–227.

Fregene, M.A., Vargas, J., Ikae, J., Angel, F., Tohme, J., Asiedu, R.A., Akoroda, M.O. and Roca, W.M. (1994) Variability of chloroplast DNA and nuclear ribosomal DNA in cassava (*Manihot esculenta* Crantz) and its wild relatives. *Theoretical and Applied Genetics* 89, 719–727.

Fregene, M., Angel, F., Gomez, R., Rodriquez, F., Chavarriaga, P., Roca, W., Tohme, J. and Bonierbale, M. (1997) A molecular genetic map of cassava (*Manihot esculenta* Crantz). *Theoretical and Applied Genetics* 95, 431–441.

Friesen, L.F., Jones, T.L., Acker, R.C.V. and Morrison, I.N. (2000) Identification of *Avena fatua* populations resistant to imazamethabenz, flamprop and fenoxaprop-P. *Weed Science* 48, 532–540.

Fryxell, P.A. (1957) Mode of reproduction of higher plants. *Botanical Review* 23, 135–233.

Fryxell, P.A., Craven, L.A. and Stewert, J.M. (1992) A revision of *Gossypium* section *Grandicalyx* (Malvaceae) including the description of six new species. *Systematic Botany* 17, 91–114.

Fu, Y.-B. and Williams, D.J. (2008) AFLP variation in 25 *Avena* species. *Theoretical and Applied Genetics* 117, 333–342.

Fukunaga, K. and Kato, K. (2003) Mitochondrial DNA variation in foxtail millet, *Setaria italica* (L.) P. Beauv. *Euphytica* 129, 7–13.

Fukunaga, K., Ichitani, K. and Kawase, M. (2006) Phylogenetic analysis of the rDNA intergenetic spacer subrepeats and its implication for the domestication history of foxtail millet, *Setaria italica*. *Theoretical and Applied Genetics* 113, 261–269.

Fullagar, R., Field, J., Denham, T. and Lentfer, C. (2006) Early and mid Holocene tool-use and processing of taro (*Colocasia esculenta*), yam (*Dioscorea* sp.) and other plants at Kuk Swamp in the highlands of Papua New Guinea. *Journal of Archaeological Science* 33, 595–614.

Fuller, D. (2007) Contrasting patterns of crop domestication and domestication rates: recent archaeobotanical insights from the Old World. *Annals of Botany* 100, 903–924.

Fuller, D., Korisettar, R., Vankatasubbaiah, P.C. and Jones, M.K. (2004) Early plant domestications in southern India: Some preliminary archaeobotanical results. *Vegetation History and Archaeobotany* 13, 115–129.

Fuller, D.Q., Harvey, E. and Qin, L. (2007) Presumed domestication? Evidence for wild rice cultivation and domestication in the fifth millennium BC of the lower Yangtse region. *Antiquity* 81, 316–331.

Fuller, D.Q., Qin, L., Zheng, Y., Zhao, Z., Chen, X., Hosoya, L.A. and Sun, G.-P. (2009) The domestication process and domestication rate in rice: spikelet bases from the lower Yangtze. *Science* 323, 1607–1610.

Futuyma, D.J. (1979) *Evolutionary Biology.* Sinauer Associates, Sunderland, Massachusetts.

Gaeta, R.T., Pires, J.C., Iniguez-Luy, F., Leon, E. and Osborn, T.C. (2007) Genomic changes in resynthesized *Brassica napus* and their effect on gene expression and phenotype. *The Plant Cell* 19, 3403–3417.

Galinat, W.C. (1973) Intergenomic mapping of maize, teosinte and *Tripsicum. Evolution* 27, 644–655.

Ganal, M.W., Altmann, T. and Röder, M.S. (2009) SNP identification in crop plants. *Current Opinions in Plant Biology* 12, 211–217.

Gaut, B.S. and Doebley, J.F. (1997) DNA sequence evidence for the segmental allopolyploid origin of maize. *Proceedings of the National Academy of Sciences USA* 94, 6809–6814.

Ge, S., Sang, T., Lu, B.-R. and Hong, D.-Y. (1999) Phylogeny of rice genomes with emphasis on origins of allopolyploid species. *Proceedings of the National Academy of Sciences USA* 96, 14400–14405.

Ge, S., Sang, T., Lu, B.-R. and Hong, D.-Y. (2001) Rapid and reliable identification of rice genomes by RFLP analysis of PCR-amplified Adh genes. *Genome* 44, 1136–1142.

Gepts, P. (1988) A middle American and an Andean common gene pool. In: Gepts, P. (ed.) *Genetic Resources of Phaseolus Beans.* Kluwer Academic, Dordrecht, the Netherlands, pp. 375–390.

Gepts, P. (1998) Origin and evolution of common bean: past events and recent trends. *HortScience* 33, 1124–1130.

Gepts, P. (2004) Introduction of transgenic crops in centers of origin and domestication. In: Kleinman, D.L., Kinchy, A.J. and Handelsman, J. (eds) *Controversies in Science and Technology: From Maize to Menopause.* University of Wisconsin Press, Madison, Wisconsin, pp. 119–134.

Gepts, P. and Bliss, F.A. (1985) F_1 weakness in the common bean. *Journal of Heredity* 76, 447–450.

Gepts, P. and Papa, R. (2003) Possible effects of (trans)gene flow from crops on the genetic diversity from landraces and wild relatives. *Environmental Biosafety Research* 2, 89–103.

Gibbons, A. (2001a) The riddle of coexistence. *Science* 291, 1725–1729.

Gibbons, A. (2001b) The peopling of the Pacific. *Science* 291, 1735–1737.

Gibson, G. and Dworkin, I. (2004) Uncovering cryptic genetic variation. *Nature Reviews Genetics* 5, 681–690.

Gilbert, N.E. (1967) Additive combining abilities fitted to plant breeding data. *Biometrics* 23, 45–49.

Gilmour, S.J., Sebolt, A.M., Salazar, M.P., Everard, J.D. and Thomashow, M.F. (2000) Overexpression of the Arabidopsis CBF3 transcriptional activator mimics multiple biochemical changes associated with cold acclimation. *Plant Physiology* 124, 1854–1865.

Golenberg, E.M. (1989) Estimation of gene flow and genetic neighborhood size by indirect methods in a selfing annual *Triticum dioccoides. Evolution* 41, 1326–1334.

Gomes, J., Jadric, S., Winterhatter, M. and Brkic, S. (1982) Alcohol dehydrogenase isoenzymes in chickpea cotyledons. *Phytochemistry* 21, 1219–1224.

Goodenough, C.J. and Wade, M.J. (2000) The ongoing hypothesis: a reply to Coyne, Barton and Turelli. *Evolution,* 317–324.

Goodman, M.M. (1990) Genetic and germ plasm stocks worth saving. *Journal of Heredity* 81, 11–16.

Goodrich, C.E. (1863) The potato. Its diseases – with incidental remarks on its sorts and culture. *Transactions of the New York State Agricultural Society* 23, 103–134.

Goodwin, T.W. and Mercer, E.I. (1972) *Introduction to Plant Biochemistry.* Pergamon Press, London.

Gordon, H. and Gordon, M. (1957) Maintenance of polymorphism by potentially injurious genes in eight natural populations of the platyfish, *Xiphophorus maculates. Journal of Genetics* 5, 44–51.

Gornall, R.J. (1983) Recombination systems and plant domestication. *Biological Journal of the Linnean Society* 20, 375–383.

Gottlieb, L.B. (1974) Genetic stability in a peripheral isolate of *Stephanomeria exigua* ssp. *coronaria* that fluctuates in population size. *Genetics* 76, 551–556.

Gottlieb, L.B. (1977a) Phenotypic variation in *Stephanomeria exigua* ssp. *coronaria* (Compositae) and its recent derivative species 'malheurensis'. *American Journal of Botany* 64, 873–880.

Gottlieb, L.B. (1977b) Evidence for duplication and divergence of the structural gene for PGI in diploid species of *Clarkia. Genetics* 86, 289–307.

Gottlieb, L.D. (1981) Electrophoretic evidence and plant populations. In: Reinhold, L., Harborne, J.B. and Swain, T. (eds) *Progress in Phytochemistry,* Vol. 7. Pergamon Press, Oxford, pp. 1–46.

Gottleib, L.D. (1982) Conservation and duplication of isozymes in plants. *Science* 216, 373–380.

Gottleib, L.D. (1984) Genetics and morphological evolution in plants. *American Naturalist* 123, 681–709.

Gottleib, L.D. and Ford, V.S. (1997) A recently silenced, duplicate *PgiC* locus in *Clarkia*. *Molecular Biology and Evolution* 14, 125–132.

Gottleib, L.D. and Greve, L.C. (1981) Biochemical properties of duplicated isoenzymes of phosphoglucose isomerase in the plant *Clarkia xantiana*. *Biochemical Genetics* 19, 155–172.

Gottleib, L.D. and Higgins, R.C. (1984) Phosphoglucose isomerase expression in species of *Clarkia* with and without a duplication in the coding gene. *Genetics* 107, 131–140.

Gould, S.J. and Eldridge, N. (1977) Punctuated equilibria: the tempo and mode of evolution reconsidered. *Paleobiology* 3, 115–151.

Grandillo, S. and Tanksley, S.D. (1996) QTL analysis of horticultural traits differentiating the cultivated tomato from the closely related species *Lycopersicon pimpinellifolium*. *Theoretical and Applied Genetics* 92, 935–951.

Grandillo, S. and Tanksley, S.D. (1999) Identifying the loci responsible for natural variation in fruit size and shape in tomato. *Theoretical and Applied Genetics* 99, 978–987.

Grant, V. (1952) Genetic and taxonomic studies in *Gilia*. II. *Gilia capitates abrotanifolia*. *Aliso* 2, 363–373.

Grant, V. (1958) The regulation of recombination in plants. *Cold Spring Harbor Symposium of Quantitative Biology* 23, 337–363.

Grant, V. (1963) *The Origin of Adaptations*. Columbia University Press, New York.

Grant, V. (1966) Linkage between viability and fertility in a species cross in *Gilia*. *Genetics* 54, 867–880.

Grant, V. (1975) *The Genetics of Flowering Plants*. Columbia University Press, New York.

Grant, V. (1977) *Organismic Evolution*. W.H. Freeman, San Francisco, California.

Grant, V. (1981) *Plant Speciation*, 2nd edn. Columbia University Press, New York.

Grant, V. (1985) *The Evolutionary Process: a Critical Review of Evolutionary History*. Columbia University Press, New York.

Grant, V. and Grant, A. (1954) Genetic and taxonomic studies in *Gilia*. VII. The woodland *Gilias*. *Aliso* 3, 59–91.

Grant, V. and Grant, A. (1960) Genetic and taxonomic studies in *Gilia*. XI. Fertility relationships of the cobwebby *Gilias*. *Aliso* 3, 203–287.

Grant, V. and Grant, A. (1964) Genetic and taxonomic studies in *Gilia*. XI. Fertility relationships of the diploid cobwebby *Gilias*. *Evolution* 18, 196–212.

Grassi, F., Labra, M., Imazio, S., Spada, A., Sgorbati, S., Scienza, A. and Sala, F. (2003) Evidence of a secondary grapevine domestication centre detected by SSR analysis. *Theoretical and Applied Genetics* 107, 1315–1320.

Gray, A.R. (1982) Taxonomy and evolution of broccoli (*Brassica oleraceae* var. *italica*). *Economic Botany* 36, 397–410.

Green, P.S. (2002) A revision of *Olea* L. (Oleaceae). *Kew Bulletin* 57, 91–140.

Griffing, B. (1956) A generalized treatment of the use of diallele crosses in quantitative inheritance. *Heredity* 10, 31–50.

Grillo, M.A., Li, C., Fowlkes, A.M., Briggeman, T.M., Zhou, A., Schemske, D.W. and Sang, T. (2009) Genetic architecture for the adaptive origin of annual wild rice, *Oryza nivara*. *Evolution* 63, 870–883.

Grime, J.P. and Hunt, R. (1975) Relative growth rate: its range and adaptive significance in a local flora. *Journal of Ecology* 63, 393–422.

Gross, B.L. and Olsen, K.M. (2010) Genetic perspectives on crop domestication. *Trends in Plant Science* 15, 529–537.

Gross, B.L. and Rieseberg, L.H. (2005) The ecological genetics of homoploid hybrid speciation. *Journal of Heredity* 96, 241–252.

Gross, B.L. and Rieseberg, L.H. (2007) Selective sweeps in the homoploid hybrid species *Helianthus deserticola*: evolution in concert across species and origins. *Molecular Ecology* 16, 5246–5258.

Gross, B.L., Schwarzbach, A.E. and Rieseberg, L.H. (2003) Origin(s) of the diploid hybrid species *Helianthus deserticola* (Asteraceae). *American Journal of Botany* 90, 1708–1719.

Gross, B.L., Kane, N.C., Lexer, C., Ludwig, F., Rosenthal, D.M., Donovan, L.A. and Rieseberg, L.H. (2004) Reconstructing the origin of *Helianthus deserticola*: survival and selection on the desert floor. *American Naturalist* 164, 145–156.

Grun, P. (1990) The evolution of the cultivated potato. *Economic Botany* 44, 39–55.

Guarino, L., Rao, V.R. and Reid, R. (1995) *Collecting Plant Genetic Diversity*. CAB International, Wallingford, UK.

Guo, J., Wang, Y., Song, C., Zhou, J., Qiu, L., Huang, H. and Wang, Y. (2010) A single origin and moderate bottleneck during domestication of soybean (*Glycine max*): implications from microsatellites and nucleotide sequences. *Annals of Botany* 106, 505–514.

Gustafsson, O. (1948) Polyploidy, life form and vegetative reproduction. *Hereditas* 34, 1–22.

Haghighi, K.R. and Asher, P.D. (1988) Fertile, intermediate hybrids between *Phaseolis vulgaris* and *P. acutifolius* from congruity backcrossing. *Sexual Plant Reproduction* 1, 51–58.

Hahn, S., Bai, K.V. and Asiedu, R.A. (1990) Tetraploids, triploids and 2n pollen from diploid interspecies crosses with cassava. *Theoretical and Applied Genetics* 79, 433–439.

Hahn, S.K. (1909) *Die Entstehurg den Pfuzkultun.* C. Winter, Heidelberg.

Hails, R.S. (2000) Genetically modified plants: the debate continues. *Trends in Ecology and Evolution* 15, 14–18.

Haldane, J.B.S. (1957) The cost of natural selection. *Journal of Genetics* 55, 511–524.

Haldane, J.B.S. (1960) More precise expressions for the cost of natural selection. *Journal of Genetics* 57, 351–360.

Hamblin, M.T., Warburton, M.L. and Buckler, E.S. (2007) Empirical comparison of simple sequence repeats and single nucleotide polymorphisms in assessment of maize diversity and relatedness. *PLoS ONE* 2(12), e1367. doi:10.1371/journal.pone.0001367

Hammer, K., Fitatenko, A.A. and Korzun, V. (2000) Microsatellite markers – a new tool for distinguishing diploid wheat species. *Genetic Resources and Crop Evolution* 47, 497–505.

Hamon, P. and Touré, B. (1990) Characterization of traditional yam varieties belonging to the *Dioscorea cayenensis-rotundata* complex by their isozyme patterns. *Euphytica* 46, 101–107.

Hamrick, J.L. and Godt, M.J.W. (1997) Allozyme diversity in cultivated crops. *Crop Science* 37, 26–30.

Hamrick, J.R. and Allard, R.W. (1972) Microgeographic variation in allozyme frequencies in *Avena sativa*. *Proceedings of the National Academy of Sciences USA* 69, 2100–2104.

Hanada, K., Vallejo, V., Nobuta, K., Slotkin, R.K., Lisch, D., Meyers, B.C. and Jiang, N. (2009) The functional role of Pack-MULESs in rice inferred from purifying selection and expression profile. *The Plant Cell* 21, 25–38.

Hancock, J.F. (1999) *Strawberries.* CAB International, Wallingford, UK.

Hancock, J.F. (2003) A framework for assessing the risk of transgenic crops. *BioScience* 53, 512–519.

Hancock, J.F. and Bringhurst, R.S. (1981) Evolution in California populations of diploid and octoploid *Fragaria* (Rosaceae): a comparison. *American Journal of Botany* 68, 1–5.

Handel, S.N. (1983a) Contrasting gene flow patterns and genetic subdivision in adjacent populations of *Cucumis sativus* (Cucurbitae). *Evolution* 37, 760–771.

Handel, S.N. (1983b) Pollination ecology, plant population structure and gene flow. In: Real, L. (ed.) *Pollination Biology.* Academic Press, New York, pp. 163–211.

Hanelt, P. (1972) Zur Geschriche des Anbaues von *Vicia faba* und ihre Gliederung. *Kulturpflanzer* 20, 209–223.

Hansche, P.E., Bringhurst, R.S. and Voth, V. (1968) Estimates of genetic and environmental parameters in the strawberry. *Proceedings of the American Society for Horticultural Science* 92, 338–345.

Harborne, J.B. (1982) *Introduction to Ecological Biochemistry.* Academic Press, London and New York.

Hardy, G.H. (1908) Mendelian proportions in a mixed population. *Science* 28, 49–50.

Harlan, J.R. (1965) The possible role of weed races in the evolution of cultivated plants. *Euphytica* 14, 173–176.

Harlan, J.R. (1967) Agricultural origins: centers and non-centers. *Science* 174, 468–474.

Harlan, J.R. (1976) Plant and animal distribution in relation to domestication. *Philosophical Transactions of the Royal Botanical Society, London* 275, 13–25.

Harlan, J.R. (1992a) *Crops and Man.* American Society of Agronomy, Madison, Wisconsin.

Harlan, J.R. (1992b) Indigenous African agriculture. In: Cowen, C.W. and Watson, P.J. (eds) *The Origins of Agriculture.* Smithsonian Institution Press, Washington, DC, pp. 59–70.

Harlan, J.R. and deWet, J.M.J. (1971) Toward a rational classification of cultivated plants. *Taxonomy* 20, 509–517.

Harlan, J.R. and deWet, J.M.J. (1972) A simplified classification of cultivated sorghum. *Crop Science* 12, 172–176.

Harlan, J.R. and deWet, M.J. (1975) On Ö. Winge and a prayer: the origins of polyploidy. *Botanical Review* 41, 361–390.

Harlan, J.R. and Zohary, D. (1966) Distribution of wild wheats and barley. *Science* 153, 1074–1080.

Harlan, J.R., deWet, M.J. and Price, E.G. (1973) Comparative evolution of cereals. *Evolution* 27, 311–325.

Harris, S.A., Robinson, J.P. and Juniper, B.E. (2002) Genetic clues to the origin of the apple. *Trends in Genetics* 18, 426–430.

Harrison, R.E., Luby, J.J. and Furnier, G.R. (1997a) Chloroplast DNA restriction fragment variation among strawberry (*Fragaria* spp.) taxa. *Journal of the American Society for Horticultural Science* 122, 63–68.

Harrison, R.E., Luby, J.J., Furnier, G.R. and Hancock, J.F. (1997b) Morphological and molecular variation among populations of octoploid *Fragaria virginiana* and *F. chiloensis* (Rosaceae) from North America. *American Journal of Botany* 84, 612–620.

Hart, G.E. (1988) Genetics and evolution of multilocus isozymes in hexaploid wheat. In: Rattazzi, M.C., Scandalios, J.G. and Whitt, G.S. (eds) *Isozymes*, Vol. 10: *Genetics and Evolution*. A.R. Liss, New York.

Harter, A.V., Gardner, K.A., Falush, D., Lentz, D.L., Bye, R.A. and Rieseberg, L.H. (2004) Origin of extant domesticated sunflowers in eastern North America. *Nature* 430, 201–205.

Hartl, D.L. (1980) *Principles of Population Genetics*. Sinauer Associates, Sunderland, Massachusetts.

Hatchett, J.H. and Gallum, R. (1970) Genetics of the ability of the Hessian fly, *Mayetiola destructor*, to survive on wheats having different genes for resistance. *Annual Report of the Entomological Society of America* 63, 1400–1407.

Hauber, D.P. (1986) Autotetraploidy in *Haplopappus spinulosus* hybrids: evidence from natural and synthetic tetraploids. *American Journal of Botany* 73, 1595–1606.

Havey, M.J. and Muehlbauer, F.J. (1989) Variability for restriction fragment lengths and phylogenies in lentil. *Theoretical and Applied Genetics* 77, 839–843.

Haviland, W.A. (2002) *Human Evolution and Prehistory*. Holt, Rinehard and Winston, New York.

Hawkes, J.G. (1967) History of the potato. *Journal of the Royal Horticulture Society* 92, 207–224.

Hawkes, J.G. (1990) *The Potato: Evolution, Biodiversity and Genetic Resources*. Belhaven Press, Washington, DC.

Hawtin, G.C., Singh, K.B. and Sexena, M.C. (1980) Some recent developments in the understanding and improvement of *Cicer* and *Lens*. In: Summerfield, R.J. and Bunting, A.H. (eds) *Advances in Legume Science*. Royal Botanical Gardens, Kew.

Hedrick, P.W. (1980) Hitchhiking: a comparison of linkage and partial selfing. *Genetics* 94, 791–808.

Hedrick, P.W. and Holden, L. (1979) Hitch-hiking: an alternative to coadaptation for the barley and slender oat examples. *Heredity* 43, 79–86.

Hedrick, U.P. (1917) *The Peaches of New York*. New York Experiment Station, Ithaca, New York.

Hedrick, U.P. (1983) *Genetics of Populations*. Jones and Bartlett Publishers, Boston, Massachusetts.

Hegarty, M.J., Jones, J.M., Wilson, I.D., Barker, G.L., Coghill, J.A., Sanchez-Baracaldo, P., Liu, G., Buggs, R.J.A., Abbott, R.J., Edwards, K.J. and Hiscock, S.J. (2005) Development of anonymous cDNA microarrays to study changes to the *Senecio* floral transcriptome during hybrid speciation. *Molecular Ecology* 14, 2493–2510.

Hegarty, M.J., Barker, G.L., Wilson, I.D., Abbott, K.S., Edwards, K.J. and Hiscock, S.J. (2006) Transcriptome shock after interspecific hybridization in *Senecio* is ameliorated by genome duplication. *Current Biology* 16, 1652–1659.

Hegarty, M.J., Barker, G.L., Brennan, A.C., Edwards, K.J., Abbott, K.S. and Hiscock, S.J. (2008) Changes to gene expression associated with hybrid speciation in plants: further insights from transcriptome studies in *Senecio*. *Philosophical Transactions Series B* 362, 3055–3069.

Hegarty, M.J., Barker, G.L., Brennan, A.C., Edwards, K.J, Abbott, K.S. and Hiscock, S.J. (2009) Extreme changes to gene expression associated with homoploid hybrid speciation. *Molecular Ecology* 18, 877–889.

Hegde, S.G., Nason, J.D., Clegg, J.M. and Ellstrand, N.C. (2006) The evolution of California's wild radish has resulted in the extinction of its progenitors. *Evolution* 60, 1187–1197.

Heiser, C.B. (1947) Hybridization between the sunflower species *Helianthus annuus* and *H. petiolaris*. *Evolution* 1, 249–262.

Heiser, C.B. (1949) Study in the evolution of the sunflower species *Helianthus annuus* and *H. bolanderi*. *University of California Publications in Botany* 23, 157–208.

Heiser, C.B. (1951) Hybridization in the annual sunflowers: *Helianthus annuus* x *H. debilis* var. *cucumerifolius*. *Evolution* 5, 42–51.

Heiser, C.B. (1958) Three new annual sunflowers (*Helianthus*) from the southwestern United States. *Rhodora* 60, 272–283.

Heiser, C.B. (1973) Introgression revisited. *Botanical Review* 39, 347–366.

Heiser, C.B. (1976) *The Sunflower*. University of Oklahoma Press, Norman, Oklahoma.

Heiser, C.B. (1990) *Seed to Civilization: The Story of Food*. Harvard University Press, Cambridge, Massachuetts.

Heiser, C.B. (1995a) Sunflowers: *Helianthus* (Compositae). In: Smartt, J. and Simmonds, N.W. (eds) *Evolution of Crop Plants*. Longman Scientific & Technical, Harlow, UK, pp. 51–53.

Heiser, C.B. (1995b) Peppers: *Capsicum* (Solanaceae) In: Smartt, J. and Simmonds, N.W. (eds) *Evolution of Crop Plants*. Longman Scientific & Technical, Harlow, UK, pp. 449–451.

Heiser, C.B. (2008) The sunflower (*Helianthus annuus*) in Mexico: further evidence for a North American domestication. *Genetic Resources and Crop Evolution* 55, 9–13.

Helbaek, H. (1959) Domestication of food plants in the Old World. *Science* 130, 365–372.

Helentjaris, T.D., Weber, D.F. and Wright, S. (1988) Identification of the genomic locations of duplicated nucleotide sequences in maize by analysis of restriction fragment length polymorphisms. *Genetics* 118, 353–363.

Hemingway, J.S. (1995) Mustards: *Brassica* ssp. and *Sinus alpa*. In: Smartt, J. and Simmonds, N. (eds) *Evolution of Crop Plants*. Longman Scientific & Technical, Harlow, UK, pp. 82–86.

Hermsen, J.G.T. (1984) Nature, evolution and breeding of polyploids. *Iowa State Journal of Research* 58, 421–436.

Heslop-Harrison, J.S. (2000) Comparative genome organization in plants: from sequence and markers to chromatin and chromosomes. *The Plant Cell* 12, 617–635.

Heslop-Harrison, J.S. and Schwarzacher, T. (2007) Domestication, genomics and the future for banana. *Annals of Botany* 100, 1073–1084.

Hesse, C.O. (1975) Peaches. In: Janick, J. and Moore, J.N. (eds) *Advances in Fruit Breeding*. Purdue University Press, West Lafayette, Indiana, pp. 285–335.

Heun, M., Schäfer-Pregl, R., Klawan, D., Castagna, R., Accerbi, M., Borghi, B. and Salamini, F. (1997) Site of Einkorn wheat domestication identified by DNA fingerprinting. *Science* 278, 1312–1314.

Hillman, G. (1975) The plant remains from Tell Abu Hureyra: a preliminary report. *Proceedings of the Prehistoric Society* 41, 70–73.

Hillman, G. (2000) Plant food economy of Abu Hureyra. In: Moore, A., Hillman, G. and Legge, T. (eds) *Village on the Euphrates, from Foraging to Farming at Abu Hureyra*. Oxford University Press, New York, pp. 372–392.

Hillman, G.C. and Davis, M.S. (1990) Measured domestication rates in wild wheats and barley under primitive cultivation, and their archeological implications. *Journal of World Prehistory* 4, 157–222.

Hilu, K.W. (1983) The role of single-gene mutations in the evolution of flowering plants. *Evolutionary Biology* 16, 97–128.

Hilu, K.W. (1994) Evidence from RAPD markers on the evolution of *Echinochloa* millets (Poaceae). *Plant Systematics and Evolution* 189, 247–257.

Hilu, K.W. and Johnson, J.L. (1992) Ribosomal DNA variation in finger millet and wild species of *Eleusine* (Poaceae). *Theoretical and Applied Genetics* 83, 895–902.

Hilu, K.W., De Wet, J.M.J. and Harlan, J.R. (1979) Archaeobotanical studies of *Eleusine coracana* ssp. *coracana* (Finger millet). *American Journal of Botany* 66, 330–333.

Ho, P.T. (1969) The origin of Chinese agriculture. In: Reed, C.A. (ed.) *Origins of Agriculture*. Mouton, The Hague, pp. 418–434.

Hoang-Tang and Liang, G.H. (1988) The genomic relationship between cultivated sorghum (*Sorghum bicolor* (L.) Moench) and Johnsongrass (*S. halepense* (L.) Pers.): a reevaluation. *Theoretical and Applied Genetics* 76, 277–284.

Hoang-Tang, Dube, S.K., Liang, G.H. and Kung, S.D. (1991) Possible repetitive DNA markers for Eusorghum and Parasorghum and their potential use in examining phylogenetic hypothesis about the origin of *Sorghum* species. *Genome* 34, 241–250.

Hochu, I., Santoni, S. and Bousalem, M. (2006) Isolation, characterization and cross-species amplification of microsatellite DNA loci in the tropical American yam *Dioscorea trifida*. *Molecular Ecology Notes* 6, 137–140.

Hodgson, R.W. (1961) Taxonomy and nomenclature in Citrus. In: Price, W.C. (ed.) *Proceedings of the Second Conference of the International Organization of Citrus*. University of Florida Press, Gainesville, Florida, pp. 1–7.

Hogenboom, N.G. (1973) A model for incongruity in intimate partner relationships. *Euphytica* 22, 219–233.

Hogenboom, N.G. (1975) Incompatibility and incongruity: two different mechanisms for the non-functioning of intimate partner relationships. *Proceedings of the Royal Society of London Series B* 188, 361–375.

Hoisington, A.M.R. and Hancock, J.F. (1981) Effect of allopolyploidy on the activity of selected enzymes in *Hibiscus. Plant Systematics and Evolution* 138, 189–198.

Hokanson, K. and Hancock, J. (2000) Early-acting inbreeding depression in three species of *Vaccinium. Sexual Plant Reproduction* 13, 145–150.

Hokanson, S., McFerson, J., Forsline, P., Lamboy, W., Luby, J., Djangaliev, A. and Aldwinckle, H. (1997a) Collecting and managing wild *Malus* germplasm in its center of diversity. *HortScience* 32, 173–176.

Hokanson, S.C., Grumet, R. and Hancock, J.F. (1997b) Effect of border rows and trap/donor ratios on pollen-mediated gene movement. *Ecological Applications* 7(3), 1075–1081.

Holliday, R.J. and Putwain, P.D. (1980) Evolution of herbicide resistance in *Senecio vulgaris*: variation in susceptibility to simazine between and within populations. *Journal of Applied Ecology* 17, 779–792.

Hopf, M. (1969) Plant remains and early farming at Jericho. In: Ucho, P.J. and Dimbleby, G.M. (eds) *Domestication and Exploitation of Plants and Animals*. Aldine, Chicago, Illinois, pp. 355–399.

Hopf, M. (1983) Jerico plant remains. In: Kenyon, K.M. and Holland, T.A. (eds) *Excavations at Jericho*, Vol. 5. British School of Archaeology in Jerusalem, London, pp. 576–621.

Hopf, M. (1986) Archaeological evidence of the spread and use of some members of the Leguminosae family. In: Barigozzi, C. (ed.) *The Origin and Domestication of Cultivated Crops*. Elsevier Science Publishers, New York.

Hosaka, K. (1993) Similar introduction and incorporation of potato chloroplast DNA into Japan and Europe. *Japanese Journal of Genetics* 68, 55–61.

Hosaka, K. (1995) Successive domestication and evolution of the Andean potatoes as revealed by chloroplast restriction endonuclease analysis. *Theoretical and Applied Genetics* 90, 356–363.

Hovav, R., Udall, J.A., Chaudhary, B., Flagel, L., Rapp, R. and Wendel, J.F. (2008) Partitioned expression of duplicated genes during development and evolution of a single cell in a polyploid plant. *Proceedings of the National Academy of Sciences USA* 105, 6191–6195.

Hu, G., Hawkins, J.S., Grover, C.E. and Wendel, J.F. (2010) The history and disposition of transposable elements in polyploid *Gossypium. Genome* 53, 599–607.

Huang, J.C. and Sun, M. (2000) Genetic diversity and relationships of sweetpotato and its wild relatives in *Ipomoea* series *Batatas* (Convolvulaceae) as revealed by inter-simple sequence repeat (ISSR) and restriction analysis of chloroplast DNA. *Theoretical and Applied Genetics* 100, 1050–1060.

Huang, S., Sirikhachornkit, A., Sun, X., Faris, J., Gill, B., Haselkorn, R. and Gornichi, P. (2002) Genes encoding plastid acetyl-CoA carboxylase and 3-phosphoglycerate kinase of the *Triticum/Aegilops* polyploid wheat. *Proceedings of the National Academy of Sciences USA* 99, 8133–8138.

Huang, S., Li, R., Zhang, Z. *et al.* (93 others) (2009) The genome of the cucumber, *Cucumis sativus* L. *Nature Genetics* 41, 1275–1283.

Hublin, J.J. (2009) The origin of Neanderthals. *The Proceedings of the National Academy of Sciences USA* 106, 16022–16027.

Hucl, P. and Scoles, G.J. (1985) Interspecific hybridization in the common bean: a review. *HortScience* 20, 352–357.

Huether, C.A. (1968) Exposure of natural variability underlying the pentamerous corolla constancy in *Linanthus androsaceus* ssp. *androsaceus. Genetics* 60, 123–146.

Huether, C.A. (1969) Constancy of the pentamerous corolla phenotype in natural populations of *Linanthus. Evolution* 23, 572–588.

Hummer, K.E., Nathewet, P. and Yanagi, T. (2009) Decaploidy in *Fragaria iturupensis* (Rosaceae). *American Journal of Botany* 96, 713–716.

Hunt, H.V., Vander Linden, M., Liu, X., Motuzaite-Matuzeviciute, G., Colledge, S. and Jones, M.K. (2008) Millets across Eurasia: chronology and context of early records of the genera *Panicum* and *Setaria* from archaeological sites in the Old World. *Vegetation History and Archaeobotany* 17 (Suppl. 1), S5–S18.

Husband, B.C. and Schemske, D.W. (1996) Evolution of the magnitude and timing of inbreeding depression in plants. *Evolution* 50, 54–70.

Husband, B.C. and Schemske, D.W. (1997) The effect of inbreeding in diploid and tetraploid populations of *Epilobium angustifolium* (Onagraceae): implications for the genetic basis of inbreeding depression. *Evolution* 51, 737–746.

Husband, B.C. and Schemske, D.W. (1998) Cytotype distribution at a diploid-tetraploid contact zone in *Chamerion (Epilobium) angustifolium* (Onagraceae). *American Journal of Botany* 85, 1688–1694.

Husband, B.C., Ozimec, B., Martin, S.L. and Pollock, L. (2008) Mating consequences of polyploidy evolution: current trends and insights from synthetic polyploids. *International Journal of Plant Science* 169, 195–206.

Husted, L. (1936) Cytological studies of the peanut *Arachis* II. Chromosome number, morphology and behavior and their application to the origin of the cultivated forms. *Cytologia* 7, 396–423.

Hutchinson, E.S., Price, S.C., Kahler, A.L., Morris, M.I. and Allard, R.W. (1983) An experimental verification of segregation theory in a diploidized tetraploid: Esterase loci in *Avena barbata. Journal of Heredity* 74, 381–383.

Huttley, G.A., Durbin, M.L., Glover, D.E. and Clegg, M.T. (1997) Nucleotide polymorphism in the chalcone synthase-A locus and evolution of the chalcone synthase multigene family of common morning glory *Ipomoea purpurea. Molecular Evolution* 6, 549–558.

Hymowitz, T. (1995) Soybean: *Glycine max* (Leguminosae – Papilionoidae) In: Smartt, J. and Simmonds, N.W. (eds) *Evolution of Crop Plants.* Longman Scientific and Technical, Harlow, UK, pp. 261–266.

Hymowitz, T. and Harlan, J.R. (1983) Introduction of soybeans to North America by Samuel Bowen in 1765. *Economic Botany* 37, 371–379.

Hymowitz, T., Singh, R.J. and Kollipara, K.P. (1998) The genomes of *Glycine. Plant Breeding Reviews* 16, 289–317.

Iltis, H.H. (1983) From teosinte to maize: The catastrophic sexual transmutation. *Science* 222, 886–894.

Iltis, H.H. (2000) Homeotic sexual translocations and the origin of maize (*Zea mays*, Poaceae): a new look at an old problem. *Economic Botany* 54, 7–42.

Iltis, H.H. and Doebley, J.F. (1980) Taxonomy of *Zea* (Gramineae). II. Subspecific categories in the *Zea mays* complex and a genetic synopsis. *American Journal of Botany* 67, 994–1004.

Irula, M., Rubio, J., Cubero, J.I., Gil, J. and Millán, T. (2002) Phylogenetic analysis in the genus *Cicer* and cultivated chickpea using RAPD and ISSR markers. *Theoretical and Applied Genetics* 104, 643–651.

Iwanaga, M. and Peloquin, S.J. (1982) Origin and evolution of cultivated tetraploid potatoes via 2n gametes. *Theoretical and Applied Genetics* 61, 161–169.

Izawa, T., Konishi, S., Shomura, A. and Yano, M. (2009) DNA changes tell us about rice domestication. *Current Opinion in Plant Biology* 12, 185–192.

Jaaska, V. (1983) *Secale* and *Triticale*. In: Tanksley, S.D. and Orton, T.J. (eds) *Isozymes in Plant Breeding and Evolution.* Elsevier, Amsterdam, pp. 79–101.

Jaccard, P. (1908) Nouvelles recherches sur la distribution florale. *Bulletin de la Société Vaudoise des Sciences* 44, 223–270.

Jackson, R.C. and Murry, B.G. (1983) Colchicine induced quadrivalent formation in *Helianthus*: evidence of ancient polyploidy. *Theoretical and Applied Genetics* 64, 219–222.

Jain, S.K. and Bradshaw, A.D. (1966) Evolutionary divergence among adjacent plant populations. I. The evidence and its theoretical basis. *Heredity* 21, 407–441.

James, J. (1979) New maize-x-*Tripsicum* hybrids. *Euphytica* 28, 239–247.

James, S.H. (1965) Complex hybridity of *Isotoma petraea*. I. The occurance of interchange heterozygosity, autogamy and a balanced lethal system. *Heredity* 20, 341–353.

Janick, J., Cummins, J., Brown, S. and Hemmat, M. (1996) Apples. In: Janick, J. and Moore, J. (eds) *Tree and Nut Fruits*, Vol 1. John Wiley & Sons, New York, pp. 1–79.

Jannoo, N., Grivet, L., Sequin, M., Paulet, F., Domaingue, R., Rao, P.R., Dookun, A., D'Hont, A. and Glaszmann, J.C. (1999) Molecular investigation of the genetic base of sugarcane cultivars. *Theoretical and Applied Genetics* 99, 171–184.

Janushevich, Z.V. (1984) The specific composition of wheat finds from ancient agricultural centers in the USSR. In: Zeist, W.V. and Casparie, W.A. (eds) *Plants and Ancient Man.* Balkema, Rotterdam, pp. 267–276.

Janzen, D.H. (1966) Coevolution of mutalism between ants and acacias in Central America. *Evolution* 20, 249–275.

Jarvis, D.I. and Hodgkin, T. (1999) Wild relatives and crop cultivars: detecting natural introgression and farmer selection of new genetic combinations in agrecosystems. *Molecular Ecology* 8, S159–S173.

Jasieniuk, M. and Maxwell, B.D. (1994) Population genetics and the evolution of herbicide resistance in weeds. *Phytoprotection* 75, 25–35.

Jenkins, J.A. (1948) The origin of the cultivated potato. *Economic Botany* 2, 379–392.

Jennings, D.L. (1995) Cassava: *Manihot esculenta* (Euphorbiaceae). In: Smartt, J. and Simmonds, N.W. (eds) *Evolution of Crop Plants*. Longman Scientific & Technical, Harlow, UK, pp. 129–132.

Jiang, L. and Liu, L. (2006) New evidence for the origins of sedentism and rice domestication in the lower Yangzi River, China. *Antiquity* 80, 355–361.

Jiang, N., Bao, Z., Zhang, X., Eddy, S.R. and Wessler, S.R. (2004) Pack-MULE transposable elements mediate gene evolution in plants. *Nature* 431, 569–573.

Jing, R., Vershinin, A., Grzebyta, J., Shaw, P., Smýkal, P., Marshall, D., Ambrose, M.J., Ellis, T.H.N. and Flavell, A.J. (2010) The genetic diversity and evolution of field pea (*Pisum*) studied by high throughput retroposon based insertion polymorphism (RIP) marker analysis. *BMC Evolutionary Biology* 10, 44.

Jobst, J., King, K. and Hemleben, V. (1998) Molecular evolution of the internal transcribed spacers (ITS1 and ITS2) and phylogenetic relationships among species of the family Cucurbitaceae. *Molecular Phylogenetics and Evolution* 9, 204–219.

Johannsen, W. (1903) *Uber Enblichkeit in Populationen und in reinen Linien*. Gustav Fisher, Jena, Germany.

Johns, T.A. and Keen, S.L. (1986) Ongoing evolution of the potato on the Altiplano of western Bolivia. *Economic Botany* 40, 409–424.

Johnson, A.W. and Packer, J.G. (1965) Polyploidy and environment in arctic Alaska. *Science* 148, 237–239.

Johnston, J.A., Wesselingh, R.A., Bouck, A.C., Donovan, L.A. and Arnold, M.L. (2001) Intimately linked or hardly speaking? The relationship between genotype and environmental gradients in a Lousiana iris population. *Molecular Ecology* 10, 673–681.

Johnston, M.O. and Schoen, D.J. (1996) Correlated evolution of self-fertilization and inbreeding depression: an experimental study of nine populations of *Amsinckia* (Boraginaceae). *Evolution* 50, 1478–1491.

Jones, A. (1967) Should Nishiyama's K123 (*Ipomoea trifida*) be designated *I. batatas*? *Economic Botany* 21, 163–166.

Jones, A. (1990) Unreduced pollen from a wild tetraploid relative of sweet potato. *Journal of the American Society of Horticulture* 115, 512–516.

Jones, H., Leigh, F.J., Mackay, I., Bower, M.A., Smith, L.M.J., Charles, M.P., Jones, G., Jones, M.K., Brown, T.A. and Powell, W. (2008) Population-based resequencing reveals that the flowering time adaptation of cultivated barley originated east of the Fertile Crescent. *Molecular Biology and Evolution* 25, 2211–2219.

Jones, M.K. and Liu, X. (2009) Origins of agriculture in East Asia. *Science* 324, 730–731.

Josefsson, E. (1967) Distribution of thioglucosides in different parts of *Brassica* plants. *Phytochemistry* 32, 151–159.

Joseph, M.C., Randall, D.D. and Nelson, C.J. (1981) Photosynthesis in polyploid tall fescue. II. Photosynthesis and ribulose-1,5 bisphosphate carboxylase of polyploidy tall fescue. *Plant Physiology* 68, 894–898.

Juniper, B.E., Watkins, R. and Harris, S.A. (1999) The origin of the apple. *Acta Horticulturae* 484, 27–33.

Kahler, A.L., Allard, R.W. and Miller, R.D. (1984) Mutation rates for enzyme and morphological loci in barley (*Hordeum vulgare* L.). *Genetics* 106, 729–734.

Kalton, R.R., Smit, A.G. and Leffel, R.C. (1952) Parent-inbred progeny relationships of selected orchardgrass clones. *Agronomy Journal* 44, 481–486.

Kaplan, L. and Lynch, T.F. (1999) *Phaseolus* (Fabaceae) in archeology: AMS radioactive carbon dates and their significance in pre-Columbian agriculture. *Economic Botany* 53, 261–272.

Kayser, M., Brauer, S., Weiss, G., Schiefenhovel, W., Underhill, P.A. and Stoneking, M. (2001) Independent histories of human Y chromosomes from Melanesia and Australia. *American Journal of Human Genetics* 68, 173–190.

Kennedy-O'Byrne, J. (1957) Notes on African grasses XXIX: a new species of *Eleusine* from tropical South Africa. *Kew Bulletin* 1, 65–72.

Kerem, Z., Lev-Yadun, S., Gopher, A., Weinberg, P. and Abbo, S. (2007) Chickpea cultivation in the Neolithic Levant through the nutritional perspective. *Journal of Archeological Science* 34, 1289–1293.

Kerster, H. and Levin, D. (1968) Neighborhood size in *Lithospermum caroliniense*. *Genetics* 60, 577–587.

Khairallah, M., Adams, M.W. and Sears, B.B. (1990) Mitochondrial DNA polymorphisms of Malawian bean lines: further evidence for two gene pools. *Theoretical and Applied Genetics* 80, 753–761.

Khush, G.S. (1962) Cytogenetics and evolutionary studies in *Secale*. II. Interrelationships of the wild species. *Evolution* 16, 484–496.

Khush, G.S. and Stebbins, G.L. (1961) Cytogenetic and evolutionary studies in *Secale*. I. Some new data on the ancestry of *S. cereale*. *American Journal of Botany* 48, 721–730.

Kiang, Y.T., Antonovics, J. and Wu, L. (1979) The extinction of wild rice (*Oryza perennis formosana*) in Taiwan. *Journal of Asian Ecology* 1, 1–9.

Kihara, H. (1954) Considerations on the evolution and distribution of *Aegilops* species based on the analyzer method. *Cytologia* 19, 336–357.

Kihara, H., Yamashita, H. and Tanaka, M. (1959) Genomes of 6 species of *Aegilops*. *Wheat Information Service* 8, 3–5.

Kilian, B., Özkan, H., Kohl, J., von Haeseler, A., Barale, F., Deusch, O., Brandolini, A., Yucel, C., Martin, W. and Salamini, F. (2006) Haplotype structure at seven barley genes: relevance to gene pool bottlenecks, phylogeny of ear type and site of barley domestication. *Molecular Genetics and Genomics* 276, 230–241.

Kilian, B., Özkan, H., Walthers, A., Kohl, J., Dagan, T., Salamini, F. and Martin, W. (2007) Molecular diversity at 18 loci in 321 wild and 92 domesticate lines reveal no reduction of nucleotide diversity during *Triticum monococcum* (Einkorn) domestication: implications for the origin of agriculture. *Molecular Biology and Evolution* 24, 2657–2688.

Kishima, Y., Mikami, T., Sugiura, M. and Kinoshita, T. (1987) *Beta* chloroplast genomes: analysis of fraction 1 protein and chloroplast DNA variation. *Theoretical and Applied Genetics* 73, 330–336.

Kislev, M.E., Weiss, E. and Hartmann, A. (2004) Impetus for sowing and the beginning of agriculture: ground collecting of wild cereals. *Proceedings of the National Academy of Sciences USA* 101, 2692–2695.

Klee, M., Zach, B. and Stika, H.-P. (2004) Four thousand years of plant exploration in the Lake Chad basin (Nigeria), part III: plant impressions in potherds from the final stone age Gajiganna culture. *Vegetation History and Archaeobotany* 13, 131–142.

Klinger, T. and Ellstrand, N.C. (1994) Engineered genes in wild populations: fitness of weed-crop hybrids of *Raphanus sativus*. *Ecological Applications* 4, 117–120.

Kloppenburg, J.R. (1988) *First the Seed*. Cambridge University Press, Cambridge.

Knight, J. (2003) A dying breed. *Nature* 421, 568–570.

Knight, R.L. (1948) The role of major genes in the evolution of economic characters. *Journal of Genetics* 48, 370–387.

Kobayashi, K. (1984) Proposed polyploid complex of *Ipomoea trifida*. In: *Proceedings of the Sixth Symposium of the International Society for Tropical Fruit Crops*. International Potato Center, Lima, Peru, pp. 561–568.

Kochert, G., Thomas Stalker, H., Gimenes, M., Galgaro, L., Romero Lopes, C. and Moore, K. (1996) RFLP and cytogenetic evidence on the origin and evolution of allotetraploid domesticated peanut, *Arachis hypogaea* (Leguminosae). *American Journal of Botany* 83, 1282–1291.

Koinange, E.M.K., Singh, S.P. and Gepts, P. (1996) Genetic control of the domestication syndrome in common bean. *Crop Science* 36, 1037–1045.

Kollipara, K.P., Sing, R.J. and Hymowitz, T. (1997) Phylogenetic and genomic relationships in common bean. *Crop Science* 36, 1037–1045.

Komatsuda, T., Pourkheirandish, M., He, C., Azhaguvel, P., Kanamori, H., Perovic, D., Stein, N., Graner, A., Wicker, T., Tagiri, A., Lundqvist, U., Fujimura, T., Matsuoka, M., Matsumoto, T. and Yano, M. (2007) Six-rowed barley originated from a mutation in a homeodomain-leucine zipper I-class homeobox gene. *Proceedings of the National Academy of Sciences USA* 104, 1424–1429.

Konishi, S., Izawa, T., Lin, S.Y., Ebana, K., Fukuta, Y., Sasaki, T. and Yano, M. (2006) An SNP caused loss of seed shattering during rice domestication. *Science* 312, 1392–1396.

Konishi, S., Ebana, K. and Izawa, T. (2008) Inference of the *japonica* rice domestication process from the distribution of six functional nucleotide polymorphisms of domestication-related genes in various landraces and modern cultivars. *Plant Cell Physiology* 49, 1283–1293.

Korban, S. (1986) Interspecific hybridization in *Malus*. *HortScience* 21, 41–48.

Korban, S. and Skirvin, R. (1984) Nomenclature of the cultivated apple. *HortScience* 19, 177–180.

Kosterin, O.E., Zaytseva, O.O., Bogdanova, V.S. and Ambrose, M.J. (2010) New data on three molecular markers from different cellular genomes in Mediterranean accessions reveal new insights into phylogeography of *Pisum sativum* L. subsp. *elatius* (Bieb.) Schmalh. *Genetic Resources and Crop Evolution* 57, 733–739.

Krebs, S. and Hancock, J. (1989) Tetrasomic inheritance of isozyme markers in the highbush blueberry, *Vaccinium corymbosum* L. *Heredity* 63, 11–18.

Krebs, S. and Hancock, J.F. (1991) Embryonic genetic load in the highbush blueberry, *Vaccinium corymbosum* L. (Ericaceae). *American Journal of Botany* 78, 1427–1437.

Kreike, C.M., Van Eck, H.J. and Lebot, V. (2004) Genetic diversity of taro, *Colocasia esculenta* (L.) Schott, in Southeast Asia and the Pacific. *Theoretical and Applied Genetics* 109, 761–768.

Ku, H.-M., Vision, T., Liu, J. and Tanksley, S.D. (2000) Comparing sequenced segments of the tomato and Arabidopsis genomes; large scale duplication followed by selective gene loss creates a network of synteny. *Proceedings of the National Academy of Sciences USA* 97, 9121–9126.

Kumar, O.A., Panda, R.C. and Rao, K.G.R. (1987) Cytogenetic studies of the F₁ hybrids of *Capsicum annuum* with *C. chinense* and *C. baccatum*. *Theoretical and Applied Genetics* 74, 242–246.

Kumar, P.S. and Hymowitz, T. (1989) Where are the diploid (2n = 2x = 20) genome donors of *Glycine* Willd. (Leguminosae, Papilionoideae)? *Euphytica* 40, 221–226.

Kuruvilla, K. and Singh, A. (1981) Karotypic and electrophoretic studies on taro and its origin. *Euphytica* 30, 405–413.

Kwak, M., Kami, J.A. and Gepts, P. (2009) The putative Mesoamerican domestication center of *Phaseolis vulgaris* is located in the Lerma-Santiago Basin of Mexico. *Crop Science* 49, 554–563.

Ladizinsky, G. (1975) On the origin of the broad bean, *Vicia faba* L. *Israel Journal of Botany* 24, 80–88.

Ladizinsky, G. (1979) Seed dispersal in relation to the domestication of Middle East legumes. *Economic Botany* 33, 284–289.

Ladizinsky, G. (1985) Founder effect in crop evolution. *Economic Botany* 39, 191–199.

Ladizinsky, G. (1995) Chickpea, *Cicer arietinum* (Leguminosae – Papilionoideae). In: Smartt, J. and Simmonds, N.W. (eds) *Evolution of Crop Plants*. Longman Scientific & Technical, Harlow, UK, pp. 258–261.

Ladizinsky, G. (1999) Identification of lentil's wild genetic stock. *Genetic Resources and Crop Evolution* 46, 115–118.

Ladizinsky, G. and Adler, A. (1976a) Genetic relationships among annual species of *Cicer*. *Theoretical and Applied Genetics* 48, 197–203.

Ladizinsky, G. and Adler, A. (1976b) The origin of chickpea *Cicer arietinum*. *Euphytica* 88, 181–188.

Ladizinsky, G. and Zohary, D. (1971) Notes on species delimitation, species relationships and polyplody in *Avena*. *Euphytica* 20, 380–395.

Ladizinsky, G., Braun, D., Goshen, D. and Muehlbauer, F.J. (1984) The biological species of the genus *Lens*. *Botanical Gazette* 145, 253–261.

Lagemann, J. (1977) *Traditional African Farming Systems in Eastern Nigeria*. Weltforum Verlag, Munich.

Lagercrantz, U. and Lydiate, D.J. (1996) Comparative genome mapping in *Brassica*. *Genetics* 4, 1903–1910.

Lai, Z., Nakazato, T., Salmaso, M., Burke, J.M., Tang, S., Knapp, S.J. and Rieseberg, L.H. (2005) Extensive chromosomal repatterning and the evolution of sterility barriers in hybrid sunflower species. *Genetics* 171, 291–303.

Lai, Z., Gross, B.L., Zou, Y., Andrews, J. and Rieseberg, L.H. (2006) Microarray analysis reveals differential gene expression in hybrid sunflower species. *Molecular Ecology* 15, 1213–1227.

Lamboy, W.F. and Alpha, C.G. (1998) Using simple sequence repeats (SSRs) for DNA fingerprinting germplasm accessions of grape (*Vitus* L.). *Journal of the American Society for Horticultural Science* 123, 182–188.

Lamppa, G.K. and Bendich, A.J. (1984) Changes in mitochondrial DNA levels during development of pea. *Planta* 162, 463–468.

Langer, R.H.M. and Hill, C.D. (1982) *Agricultural Plants*. Cambridge University Press, New York.

Larter, E.N. (1995) Triticale: Triticosecale spp. (Gramineae – Triticinae). In: Smartt, J. and Simmonds, N.W. (eds) *Evolution of Crop Plants*. Longman Scientific & Technical, Harlow, UK, pp. 181–183.

Lasker, G.W. and Tyzzer, R. (1982) *Physical Anthropology*. Holt, Rinehart and Winston, New York.

Latimer, H. (1958) *A Study of the Breeding Barrier between* Gilia australis *and* Gilia splendens. Claremont University Press, Claremont, California.

Laurie, C.C. Chasalow, S.D., LeDeaux, J.R., McCarroll, R., Bush, D., Hauge, B., Lai, C., Clark, D., Rocheford, T.R. and Dudley, J.W. (2004) The genetic architecture of response to long-term artificial selection for oil concentration in the maize kernel. *Genetics* 168, 2141–2155.

Lauter, N. and Doebley, J. (2002) Genetic variation for phenotypically invariant traits detected in teosinte: implications for the evolution of novel forms. *Genetics* 160, 333–342.

Leakey, M.D. and Lewin, R. (1992) *Origins Reconsidered: In Search of What Makes Us Human*. Anchor Books, New York.

Lebot, V. (2009) *Tropical Root and Tuber Crops: Cassava, Sweet Potato, Yams and Aroids*. CAB International, Wallingford, UK, 413 pp.

Lebot, V. and Aradhya, K.M. (1991) Isozyme variation in taro (*Colocasia esculenta* (L.) Schott.) from Asia and Oceana. *Euphytica* 56, 55–66.

Lee, J.M., Grant, D., Vallejos, C.E. and Shoemaker, R.C. (2001) Genome organization in dicots. II. *Arabidopsis* as a 'bridging species' to resolve genome evolution events among legumes. *Theoretical and Applied Genetics* 103, 765–773.

Lee, R.E. (1968) What hunters do for a living, or how to make out on scarce resources. In: Lee, R.B. and DeVore, I. (eds) *Man the Hunter*. Aldine, Chicago, Illinois, pp. 30–48.

Leggett, J.M. (1992) Classification and speciation in *Avena*. In: Marshall, H.G. and Sorrells, M.E. (eds) *Oat Science and Technology*. Agronomy Monograph 33, ASA and CSSA, Madison, Wisconsin, pp. 29–53.

Leggett, J.M. and Thomas, H. (1995) Oat evolution and cytogenetics. In: Welch, W. (ed.) *The Oat Crop. Production and Utilization*. Chapman and Hall, London, pp. 121–149.

Lentz, D.L., Pohl, M.E.D., Pope, K.O. and Wyatt, A.R. (2001) Prehistoric sunflower (*Helianthus annuus* L.) domestication in Mexico. *Economic Botany* 55, 370–376.

Lentz, D.L., Pohl, M.D., Alvarado, J.L., Tarighat, S. and Bye, R. (2008) Sunflower (*Helianthus annuus* L.) as a pre-Columbian domesticate in Mexico. *Proceedings of the National Academy of Sciences USA* 105, 6232–6239.

Lerceteau-Köhler, E., Guérin, G., Laigret, F. and Denoyes-Rothan, B. (2003) Characterization of mixed disomic and polysomic inheritance in the octoploid strawberry (*Fragaria* x *ananassa*) using AFLP mapping. *Theoretical and Applied Genetics* 107, 619–628.

Levadoux, L. (1946) Study of the flower and sexuality in grapes (in French). *Annales de l'Ecole nationale d'Agriculture de Montpellier N.S.* 27, 1–89.

Levin, D.A. (1983) Polyploidy and novelty in flowering plants. *American Naturalist* 122, 1–25.

Levin, D.A. (2000) *The Origin, Expansion and Demise of Plant Species*. Oxford University Press, New York.

Levin, D.A. and Kerster, H.W. (1967) Natural selelection for reproductive isolation in *Phlox. Evolution* 21, 679–687.

Levin, D.A. and Kerster, H.W. (1969) Density-dependent gene dispersal in *Liatris. American Naturalist* 103, 61–74.

Levin, D.A. and Schmidt, K.P. (1985) Dynamics of a hybrid zone in *Phlox*: an experimental demographic investigation. *American Journal of Botany* 72, 1404–1409.

Levin, D.A., Torres, A.M. and Levy, M. (1979) Alcohol dehyrogenase activity in diploid and autotetraploid *Phlox. Biochemical Genetics* 17, 35–42.

Levin, D.A., Francisco-Ortega, J. and Jansen, R.K. (1995) Hybridization and the extinction of rare plant species. *Conservation Biology* 10, 10–16.

Levinson, C. and Gutman, G.A. (1987) Slipped-strand mispairing: a major mechanism for DNA sequence evolution. *Molecular Biology and Evolution* 4, 203–221.

Levy, M. and Levin, D.A. (1975) Genetic heterozygosity and variation in permanent translocation heterozygotes of the *Oenothera biennis* complex. *Genetics* 79, 493–512.

Levy, M. and Winternheimer, P.L. (1977) Allozyme linkage diseqilibria among chromosome complexes in the permanant translocation heterozygote *Oenothera biennis. Evolution* 31, 465–476.

Levy, M., Steiner, E.E. and Levin, D.A. (1975) Allozyme genetics in permanent translocation heterozygotes of the *Oenothera biennis* complex. *Biochemical Genetics* 13, 487–500.

Lewis, D. (1943) Physiology of incompatibility in plants. III. Autopolyploids. *Journal of Genetics* 45, 171–185.

Lewis, D. (1979) *Sexual Incompatibility in Plants. Studies in Biology no. 110*. Edward Arnold, London.

Lewis, E.B. (1951) Pseudoallelism and gene evolution. *Cold Spring Harbor Symposium of Quantitative Biology* 16, 159–172.

Lewis, H. (1962) Catastrophic selection as a factor in speciation. *Evolution* 16, 257–271.

Lewis, H. (1966) Speciation in flowering plants. *Science* 152, 167–172.

Lewis, W.H. and Lewis, M. (1955) The genus *Clarkia*. *University of California Publications in Botany* 20, 241–392.

Lewontin, R.C. and Kojima, K. (1960) The evolutionary dynamics of complex polymorphisms. *Evolution* 14, 458–472.

Lexer, C., Welch, M.E., Durphy, J.L. and Rieseberg, L.H. (2003a) Natural selection for salt tolerance quantitative trait loci (QTLs) in wild sunflower hybrids: implications for the origin of *Helianthus paradoxus*, a diploid hybrid species. *Molecular Evolution* 12, 1225–1235.

Lexer, C., Welch, M.E., Raymond, O. and Rieseberg, L.H. (2003b) The origin of ecological divergence in *Helianthus paradoxus* (Asteraceae): selection on transgressive characters in a novel hybrid habitat. *Evolution* 57, 1989–2000.

Lexer, C., Lai, Z. and Rieseberg, L.H. (2004) Candidate gene polymorphisms associated with salt tolerance in wild sunflower hybrids: implications for the origin of *Helianthus paradoxus*, a diploid hybrid species. *New Phytologist* 161, 225–233.

Li, C.-D., Rossnagel, B.G. and Scoles, G.J. (2000) Tracing the phylogeny of the hexaploid oat *Avena sativa* with satellite DNAs. *Crop Science* 40, 1755–1763.

Li, C., Zhou, A. and Sang, T. (2005) Genetic analysis of rice domestication syndrome with the wild annual species, *Oryza nivara*. *New Phytologist* 170, 185–194.

Li, C.B., Zhou, A.L. and Sang, T. (2006) Rice domestication by reduced shattering. *Science* 311, 1936–1939.

Li, D.-Z. and Pritchard, H.W. (2009) The science and economics of *ex situ* plant conservation. *Trends in Plant Science* 14, 614–621.

Li, W.-T., Peng, Y.-Y., Wei, Y.-M., Baum, B.R. and Zheng, Y.L. (2009) Relationships among *Avena* species as revealed by consensus chloroplast simple sequence repeat (ccSSR) markers. *Genetic Resources and Crop Evolution* 56, 465–480.

Li, X., Xie, R., Lu, Z. and Zhou, Z. (2010) The origin of cultivated citrus as inferred from internal transcribed spacer and chloroplast DNA sequence and amplified fragment length polymorphism fingerprints. *Journal of the American Society for Horticultural Science* 135, 341–350.

Li, Y., Wu, S.Z. and Cao, Y.S. (1995) Cluster analysis of an international collection of foxtail millet (*Setaria indica* (L.) P. Beauv). *Genetic Resources and Crop Evolution* 45, 279–285.

Li, Y.-H., Li, W., Zhang, L., Chang, R.-Z., Gaut, B.S. and Qiu, L.-J. (2010) Genetic diversity in domesticated soybean (*Glycine max*) and its wild progenitor (*Glycine soya*) for simple sequence repeat and single-nucleotide polymorphism loci. *New Phytologist* 188, 242–253.

Linares, O.F. (2002) African rice (*Oryza glaberrima*): history and future potential. *Proceedings of the National Academy of Sciences USA* 99, 16360–16365.

Linder, C.R., Taha, I., Seiler, G.J., Snow, A.A. and Rieseberg, L.H. (1998) Long-term introgression of crop genes into wild sunflower populations. *Theoretical and Applied Genetics* 96, 339–347.

LingHwa, T. and Morishima, H. (1997) Genetic characterization of weedy rices and the inference on their origins. *Breeding Science* 47, 153–160.

Little, H.A., Grumet, R. and Hancock, J.F. (2009) Modified ethylene signaling as an example of engineering for complex traits: secondary effects and implications for risk assessment. *HortScience* 44, 94–101.

Liu, B., Vega, J.M. and Feldman, M. (1998a) Rapid genomic changes in newly synthesized amphiploids of *Triticum* and *Aegilops*. II. Changes in low copy coding DNA sequences. *Genome* 41, 535–542.

Liu, B., Vega, J.M., Segal, G., Addo, S., Rodova, M. and Feldman, M. (1998b) Rapid genomic changes in newly synthesized amphiploids of *Triticum* and *Aegilops*. I. Changes in low copy non-coding sequences. *Genome* 41, 272–277.

Livingstone, K.D., Lackney, V.K., Blauth, J.R., van Wijk, R. and Jain, M.K. (1999) Genome mapping in *Capsicum* and the evolution of genome structure in the Solanaceae. *Genetics* 152, 1183–1202.

Londo, J.P., Chiang, Y.-C., Hung, K.-H., Chiang, T.-Y. and Schaal, B.A. (2006) Phylogeography of Asian wild rice, *Oryza rufipogan*, reveals multiple independent domestications of cultivated rice, *Oryza sativa*. *Proceedings of the National Academy of Sciences USA* 103, 9578–9583.

Longley, A.E. (1941) Chromosome morphology in maize and its relatives. *Botanical Review* 7, 263–289.

Loomis, W.F. and Gilpin, M.E. (1986) Multigene families and vestigial sequences. *Proceedings of the National Academy of Sciences USA* 83, 2143–2147.

Loskutov, I.G. (2008) On evolutionary pathways of *Avena* species. *Genetic Resources and Crop Evolution* 55, 211–220.

Lu, H., Zhang, J., Liu, K.-B., Wu, N., Li, Y., Zhou, K., Ye, M., Zhang, T., Zhang, H., Yang, X., Shen, L., Xu, D. and Li, Q. (2009) Earliest domestication of common millet (*Panicum miliaceum*) in East Asia extended to 10,000 years ago. *Proceedings of the National Academy of Sciences USA* 106, 7367–7372.

Lukens, L.N. and Doebley, J. (1999) Epistatic and environmental interactions for quantitative trait loci involved in maize evolution. *Genetical Research* 74, 291–302.

Lumaret, R. and Ouazzani, N. (2001) Ancient wild olives in Mediterranean forests. *Nature* 413, 700.

Lumaret, R., Ouazzani, N., Michaud, H., Vivier, G., Deguilloux, M.-F. and Di Giusto, F. (2004) Allozyme variation of oleaster populations (wild olive tree) (*Olea europaea* L.) in the Mediterranean basin. *Heredity* 92, 343–351.

Luo, M.-C., Yang, Z.-L., You, F.M., Kawahara, T., Waines, J.G. and Dvorak, J. (2007) The structure of wild and domesticated emmer wheat populations, gene flow between them, and the site of emmer domestication. *Theoretical and Applied Genetics* 114, 947–959.

Lush, W.M. and Evans, L.T. (1981) The domestication and improvement of cowpeas (*Vigna unguiculata* (L.) Walp.). *Euphytica* 30, 579–587.

Lyman, J.C. and Ellstrand, N.C. (1984) Clonal diversity in *Taraxacum officinale* (Compositae), an apomict. *Heredity* 53, 1–10.

Mabberley, D.J., Jarvis, C.E. and Juniper, B.E. (2001) The name of the apple. *Telopea* 9, 421–430.

MacKey, J. (1954) Neutron and x-ray experiments in wheat and a revision of the speltoid problem. *Hereditas* 40, 65–180.

MacKey, J. (1970) Significance of mating systems for chromosomes and gametes in polyploids. *Hereditas* 66, 165–176.

MacNeish, R.S., Nelken-Terner, A. and Johnson, I.W. (1967) *The Prehistory of the Tehuacán Valley*. University of Texas Press, Austin, Texas.

Maggioni, L., von Bothmer, R., Poulson, G. and Blanca, F. (2010) Origin and domestication of cole crops (*Brassica oleraceae* L.): Linguistic and literary considerations. *Economic Botany* 64, 109–123.

Magoon, M.L., Krishnan, R. and Bai, K.V. (1969) Morphology of pachytene chromosomes and meiosis in *Manihot esculenta*. *Cytologia* 34, 612–624.

Maimberg, R.L. and Mauricio, R. (2005) QTL based evidence for the role of epistasis in evolution. *Genetical Research* 86, 89–95.

Malpa, R., Arnau, G., Noyer, J.L. and Lebot, V. (2005) Genetic diversity of the greater yam (*Dioscorea alata* L.) and relatedness to *D. nummularia* Lam and *D. transversa* Br. as revealed with AFLP markers. *Genetic Resources and Crop Evolution* 52, 919–929.

Malpa, R., Noyer, J.L., Marchand, J.L. and Lebot, V. (2006) Genetic relationship between *Dioscorea alata* L. and *D. nummularia* Lam. As revealed by AFLP markers. In: Motley, T.J., Zerega, N. and Cross, H. (eds) *Darwin's Harvest: New Approaches to the Origins, Evolution and Conservation of Crops*. Columbia University Press, New York, pp. 239–265.

Mangelsdorf, P.C. (1953) *Wheat*. Scientific American, New York.

Mangelsdorf, P.C. (1958) Reconstructing the ancestor of corn. *Proceedings of the American Philosophic Society* 102, 454–463.

Mangelsdorf, P.C. (1974) *Corn: Its Origin, Evolution and Improvement*. Harvard University Press, Cambridge, Massachusetts.

Mangelsdorf, P.C. (1986) The origin of corn. *Scientific American* 255, 80–86.

Mangelsdorf, P.C. and Reeves, R.G. (1939) *The Origin of Indian Corn and its Relatives*. Bulletin 549, Texas Agricultural Experiment Station, College Station, Texas.

Manwell, C. and Baker, C.M. (1970) *Molecular Biology and the Origin of Species*. University of Washington Press, Seattle.

Marble, B.K. (2004) Polyploidy and self-compatibility: is there an association? *New Phytologist* 162, 803–811.

Mariac, C., Jehin, L., Saïdou, A.-A., Thuillet, A.-C., Couderc, M., Sire, P., Jugdé, H., Adam, H., Bezançon, G., Phan, J.-L. and Vigouroux, Y. (2010) Genetic basis of pearl millet adaptation along an environmental gradient investigated by a combination of genome scan and association mapping. *Molecular Ecology* 20, 80–91.

Marsden-Jones, E.M. and Turril, W.B. (1928) Researches on *Silene maritima* and *S. vulgaris*. I. *Kew Bulletin* 1, 1–17.

Marshack, A. (1976) Implications of the Paleolithic symbolic evidence for the origin of language. *American Scientist* 64, 136–145.

Marshall, D. and Allard, R.W. (1970) Isozyme polymorphisms in natural populations of *Avena fatua* and *A. barbata*. *Heredity* 25, 373–382.

Marshall, D.R. and Brown, A.D.H. (1975) Optimum sampling strategies in genetic conservation. In: Frankel, O.H. and Hawkes, J.G. (eds) *Crop Genetic Resources for Today and Tomorrow*. Cambridge University Press, New York, pp. 53–80.

Marshall, E. (2001) Pre-Clovis sites fight for acceptance. *Science* 291, 1730–1732.

Martin, G.B. and Adams, M.W. (1987a) Landraces of *Phaseolus vulgaris* (Fabaceae) in northern Malawi. I. Regional variation. *Economic Botany* 41, 190–203.

Martin, G.B. and Adams, M.W. (1987b) Landraces of *Phaseolus vulgaris* (Fabaceae) in northern Malawi. II. Generation and maintenance of variability. *Economic Botany* 41, 204–215.

Martin, F.W. and Jones, A. (1972) The species of *Ipomoea* closely related to the sweet potato. *Economic Botany* 26, 201–215.

Martinez-Zapater, J.M. and Oliver, J.L. (1984) Genetic analysis of isozyme loci in tetraploid potatoes (*Solanum tuberosum* L.) *Genetics* 108, 669–679.

Masterson, J. (1994) Stomatal size in fossil plants – evidence for polyploidy in the majority of angiosperms. *Science* 264, 421–424.

Mather, K. (1943) Polygenic balance in the canalization of development. *Nature* 151, 68–71.

Mather, K. (1973) *Genetic Structure of Populations*. Chapman & Hall, London.

Mather, K. and Jinks, J.L. (1977) *Introduction to Biometrical Genetics*. Cornell University Press, Ithaca, New York.

Matsuoka, Y. and Nasuda, S. (2004) Durum wheat as a candidate for the unknown female progenitor of bread wheat: an empirical study with a highly fertile F₁ hybrid with *Aegilops tauschii* Coss. *Theoretical and Applied Genetics* 109, 1710–1717.

Matsuoka, Y., Vigouroux, Y., Goodman, M.M., Sanches, J., Buckler, E. and Doebley, J. (2002) A single domestication for maize shown by multilocus microsatellite genotyping. *Proceedings of the National Academy of Sciences USA* 99, 6080–6084.

Matzke, M.A. and Matzke, A.J.M. (1998) Polyploidy and transposons. *Trends in Ecology and Systematics* 13, 241.

Maxted, N., Kell, S., Toledo, Á., Dulloo, E., Heywood, V., Hodgkin, T., Hunter, D., Guarino, L., Jarvis, A. and Ford-Lloyd, B. (2010) A global approach to crop wild relative conservation: securing the gene pool for food and agriculture. *Kew Bulletin* 65, 561–576.

Mayer, M.S. and Bagga, S.K. (2002) The phylogeny of *Lens* (Leguminosae): new insights from ITS sequence analysis. *Plant Systematics and Evolution* 232, 145–154.

Mayer, M.S. and Soltis, P.S. (1994) Chloroplast DNA phylogeny of *Lens* (Leguminosae): origin and diversity of cultivated lentil. *Theoretical and Applied Genetics* 87, 773–781.

Mayr, E. (1942) *Systematics and the Origin of Species*. Columbia University Press, New York.

Mayr, E. (1954) Change of genetic environment and evolution. In: Huxley, J., Hardy, A.C. and Ford, E.B. (eds) *Evolution as a Process*. Allen and Unwin, London, pp. 157–180.

Mayr, E. (1963) *Animal Species and Evolution*. Harvard University Press, Cambridge, Massachusetts.

McDonald, J.A. and Austin, D.F. (1990) Changes and additions in *Ipomoea* section *Batatas* (Convolvulaceae). *Brittonia* 42, 201–215.

McDougal, I., Brown, F.H. and Fleagle, J.E. (2005) Stratigraphic placement and age of modern humans from Kibish, Ethiopia. *Nature* 433, 733–736.

McFadden, E.S. and Sears, E.R. (1946) The origin of *Triticum spelta* and its free-threshing hexaploid relatives. *Journal of Heredity* 37, 81–91.

McGovern, P.E. (2003) *Ancient Wine: the search for the origins of viticulture*. Princeton University Press, Princeton, New Jersey.

McGovern, P.E. and Michel, R.H. (1995) The analytical and archaeological challenge of detecting ancient wine: two case studies from the ancient Near East. In: McGovern, P.E., Fleming, S.J. and Katz, S.H. (eds) *The Origins and Ancient History of Wine*. Gordon and Breach, Amsterdam, pp. 57–67.

McLeod, M.J., Eshbaugh, W.H. and Guttman, S.I. (1979) An electrophoretic study of *Capsicum* (Solanaceae): the purple flowered taxa. *Economic Botany* 106, 326–333.

McLeod, M.J., Eshbaugh, W.H. and Guttman, S.I. (1982) Early evolution of chili peppers (*Capsicum*). *Economic Botany* 36, 361–368.

McNaughton, I.H. and Harper, J.L. (1960) The comparative biology of closely related species living in the same area. I. External breeding barriers between *Papaver* species. *New Phytologist* 59, 15–26.

McNeilly, T. and Antonovics, J. (1968) Evolution in closely adjacent plant populations. IV. Barriers to gene flow. *Heredity* 23, 205–218.

Mehra, K.L. (1963a) Differentiation of cultivated and wild *Eleusine* species. *Phyton* 20, 189–198.

Mehra, K.L. (1963b) Considerations of the African origin of *Eleusine coracana*. *Current Science* 32, 300–301.

Meilleur, B.A. and Hodgkin, T. (2004) *In situ* conservation of crop wild relatives: status and trends. *Biodiversity and Conservation* 13, 663–684.

Menacio-Hautea, D., Fatokum, C.A., Kumar, L., Danesh, D. and Young, N.D. (1993) Comparative genome analysis of mungbean (*Vigna radiata* L. Wilczek) and cowpea (*V. unguiculata*) using RFLP analysis. *Theoretical and Applied Genetics* 86, 797–810.

Mendel, G. (1866) Versuche uber Pflanzen-Hybridan. *Verrhandlugen des Naturforschenden Vereins in Brünn* 4, 1–47.

Mendiburu, A.O. and Peloquin, S.J. (1977) The significance of 2n gametes in potato breeding. *Theoretical and Applied Genetics* 49, 53–61.

Mendoza, H.A. and Haynes, F.L. (1974) Genetic basis of heterosis for yield in the autotetraploid potato. *Theoretical and Applied Genetics* 45, 21–26.

Menkir, A., Goldsbrough, P. and Ejeta, G. (1997) RAPD based assessment of genetic diversity in cultivated races of sorghum. *Crop Science* 37, 564–569.

Merrick, L.C. (1995) Squashes, pumpkins and gourds: *Cucurbita* (Cucurbitaceae). In: Smartt, J. and Simmonds, N.W. (eds) *Evolution of Crop Plants*. Longman Scientific & Technical, Harlow, UK.

Mézard, C., Vignard, J., Drouaud, J. and Mercier, R. (2006) The road to crossovers: plants have their say. *Trends in Genetics* 23, 91–100.

Michaelis, P. (1954) Cytoplasmic inheritance in *Epilobium* and its theoretical significance. *Advances in Genetics* 6, 287–401.

Mignouna, H.D. and Dansi, A. (2003) Yam (*Dioscorea* spp.) domestication by the Nago and Fon ethnic groups in Benin. *Genetic Resources and Crop Evolution* 50, 519–528.

Mignouna, H.D., Mank, R.A., Ellis, T.H.N., Van den Bosch, N., Asiedu, R., Abang, M.M. and Peleman, J. (2002) A genetic linkage map of the water yam (*Dioscorea alata* L.) based on AFLP markers and QTL analysis for anthracnose resistance. *Theoretical and Applied Genetics* 105, 726–735.

Mikami, T., Kishima, Y., Sugiura, M. and Kinoshita, T. (1984) Chloroplast DNA diversity in the cytoplasms of sugarbeet and its related species. *Plant Science Letters* 36, 231–235.

Miki, B., Abdeen, A., Manabe, Y. and MacDonald, P. (2009) Selectable marker genes and unintended changes to the plant transcriptome. *Plant Biotechnology Journal* 7, 211–218.

Miller, J.S. and Venable, D.L. (2000) Polyploidy and the evolution of gender in higher plants. *Science* 289, 2335–2338.

Mitchell-Olds, T. (1996) Genetic constraints on life history evolution – quantitative trait loci influencing growth and flowering in *Arabidopsis thaliana*. *Evolution* 50, 140–145.

Mitton, J.B. and Grant, M.C. (1984) Associations among protein heterozygosity, growth rate and developmental homeostasis. *Annual Review of Ecology and Systematics* 15, 479–499.

Mok, D.W.S. and Peloquin, S.J. (1975) Three mechanisms of 2n pollen formation in diploid potatoes. *Canadian Journal of Genetics and Cytology* 17, 217–225.

Molina, M.D. and Naranjo, C.A. (1987) Cytogenetic studies in the genus *Zea*: 1. Evidence for five as the basic chromosome number. *Theoretical and Applied Genetics* 73, 542–550.

Moore, D.H. and Lewis, H. (1965) The evolution of self-pollination in *Clarkia xantiana*. *Evolution* 19, 104–114.

Moore, P. (1976) How far does pollen travel? *Nature* 260, 280–281.

Morden, C.W., Doebley, J. and Schertz, K.F. (1989) Allozyme variation in Old World races of *Sorghum bicolor*. *American Journal of Botany* 76, 247–255.

Morden, C.W., Doebley, J. and Schertz, K.F. (1990) Allozyme variation among the spontaneous species of *Sorghum* section *Sorghum* (Poaceae). *Theoretical and Applied Genetics* 80, 296–304.

Moretzsohn, M.C., Barbosa, A.V.G., Alves-Freitas, D.M.T., Teixeira, C., Leal-Bertioli, S.C.M., Guimarães, P.M., Pereira, R.W., Lopes, C.R., Cavallari, M.M., Valls, J.F.M., Bertioli, D.J. and Gimenes, M.A. (2009) A linkage map for the B-genome donor of *Arachis* (Fabaceae) and its synteny to the A-genome. *BMC Plant Biology* 9, 40.

Morgan, J. and Richards, A. (1993) *The Book of Apples*. Ebury Press Ltd, London.

Morrell, P.L. and Clegg, M.T. (2007) Genetic evidence for a second domestication of barley (*Hordeum vulgare*) east of the Fertile Crescent. *Proceedings of the National Academy of Sciences USA* 104, 3289–3294.

Mortensen, D.A., Bastiaans, L. and Sattin, M. (2000) The role of ecology in the development of weed management systems: an outlook. *Weed Research* 40, 49–62.

Moscone, E.A., Scaldaferro, M.A., Grabiele, M., Cecchini, N.M., Garcia, Y.S., Jarret, R., Daviña, J.R., Ducasse, D.A., Barboza, G.E. and Ehrendorfer, F. (2007) The evolution of chili peppers (*Capsicum* – Solanaceae): a cytogenetic perspective. *Acta Horticulturae* 745, 137–169.

Motta-Aldana, J.R., Serrano-Serrano, M.L., Hernández-Torres, J., Castillo-Villamizer, G., Debouck, D.G. and Chacón, M.I. (2010) Multiple origins of lima bean landraces in the Americas: evidence from chloroplast and nuclear DNA polymorphisms. *Crop Science* 50, 1773–1787.

Muench, D.G., Slinkard, A.E. and Scoles, G.J. (1991) Determination of genetic variation and taxonomy in lentil (*Lens* Miller) species by chloroplast DNA polymorphism. *Euphytica* 56, 213–218.

Mulcahy, D.L. (1979) The rise of the angiosperms: a genecological factor. *Science* 206, 20–30.

Muños, L.C., Duque, M.C., Debouck, D.G. and Blair, M.W. (2006) Taxonomy of tepary bean and wild relatives as determined by amplified fragment length polymorphism (AFLP) markers. *Crop Science* 46, 1744–1754.

Müntzing, A. (1932) Cyto-genetic investigations on synthetic *Galeopsis tetrahit. Hereditas* 16, 105–164.

Müntzing, A. (1936) The evolutionary significance of autopolyploidy. *Hereditas* 21, 263–378.

Nassar, H.N., Nassar, N.M.A., Vieira, C. and Saraiva, L.S. (1995) Cytogenetic behavior of the interspecific hybrid of *Manihot neusana* Nassar and cassava *Manihot esculenta* Crantz, and its backcross progeny. *Canadian Journal of Plant Science* 75, 675–678.

National Research Council (NRC) (1989) *Lost Crops of the Incas: Little-Known Plants of the Andes with Promise for Worldwide Cultivation.* Report of an Ad Hoc Panel of the Advisory Committee on Technology Innovation Board on Science and Technology for International Development. National Academies Press, Washington, DC.

Nayar, N.M. (2010) The bananas: botany, origin, dispersal. In: Janick, J. (ed.) *Horticultural Reviews* 36, 117–164.

Neale, D.B., Saghai-Maroof, M.A., Allard, R.W. and Jorgensen, R.A. (1988) Chloroplast DNA diversity in populations of wild and cultivated barley. *Genetics* 120, 195–213.

Nee, M. (1990) The domestication of the *Cucurbita* (Cucurbitaceae). *Economic Botany* 44, 56–68.

Negrul, A.M. (1938) Evolucija kuljturnyx from vinograda. *Doklady Akademii nauk SSSR* 8, 585–588.

Nei, M. (1987) *Molecular Evolutionary Genetics.* Columbia University Press, New York.

Nei, M. and Li, W.-H. (1975) Probability of identical monomorphism in related species. *Genetical Research* 26, 31–43.

Nei, M. and Li, W.-H. (1979) Mathematical model for studying genetic variation in terms of restriction endonucleases. *Proceedings of the National Academy of Sciences USA* 76, 5269–5273.

Neumann, K. and Hildebrand, E. (2009) Early bananas in Africa: the state of the art. *Ethnobotany Research and Applications* 7, 353–362.

Newell, C.A. and DeWet, J.M.J. (1973) A cytological survey of *Zea-Tripsicum* hybrids. *Canadian Journal of Genetics and Cytology* 15, 763–778.

Newell, C.J. and Hymowitz, T. (1978) A reappraisal of the genus *Glycine. American Journal of Botany* 65, 168–179.

Newton, K.J. (1983) Genetics of mitochondrial isozymes. In: Tanksley, S.D. and Orton, T.J. (eds) *Isozymes in Plant Genetics and Breeding.* Elsevier, Amsterdam.

Ng, N.Q. (1995) Cowpea, *Vigna unguiculata* (Leguminoseae – Papilionoideae). In: Smartt, J. and Simmonds, N. (eds) *Evolution of Crop Plants.* Longman Scientific & Technical, London, pp. 326–333.

Nichols, R. (2001) Gene trees and species trees are not the same. *Trends in Ecology and Evolution* 16, 358–364.

Nicolosi, E., Deng, Z.N., Gentile, A., La Malfa, S., Continella, G. and Tribulato, E. (2000) Citrus phylogeny and genetic origin of important species as investigated by molecular markers. *Theoretical and Applied Genetics* 100, 1155–1166.

Nikoloudakis, N. and Katsiotis, A. (2008) The origin of the C-genome and cytoplasm of *Avena* polyploids. *Theoretical and Applied Genetics* 117, 273–281.

Nilen, R.A. (1971) *Barley Genetics II.* Pullman, Washington, DC.

Nilsson-Ehle, H. (1909) Kreuzungsuntersuchungen an Hafer and Weizen. *Acta Universitatis Lundensis* 5, 1–122.

Nishiyama, I. and Temamura, T. (1962) Mexican wild form of sweet potato. *Economic Botany* 16, 304–314.

Nishyama, I., Miyasaki, T. and Sakamoto, S. (1975) Evolutionary autopolyploidy of the sweet potato (*Ipomoea batatas* (L.) Lam.) and its progenitors. *Euphytica* 24, 197–208.

Noguti, Y., Oka, H. and Otuka, T. (1940) Studies on the polyploidy of *Nicotiana* induced by the treatment with colchicine. II. Growth rate and chemical analysis of diploid and its autopolyploid in *Nicotiana rustica* and *N. tabacum. Japanese Journal of Botany* 10, 343–364.

Obidiegwu, J., Rodriquez, E., Ene-Obong, E., Loureiro, J., Muoneke, C., Santos, C., Kolesnikova-Allen, M. and Asiedu, R. (2010) Ploidy levels of *Dioscorea alata* L. germplasm determined by flow cytometry. *Genetic Resources and Crop Evolution* 57, 351–356.

O'Brien, P.J. (1972) The sweet potato: its origin and dispersal. *American Anthropology* 74, 342–365.

Ochiai, T., Nguyen, V.X., Tahara, M. and Yoshino, H. (2001) Geographical differentiation of Asian taro, *Colocasia esculenta* (L.) Schott, detected by RAPD and isozyme analysis. *Euphytica* 122, 219–234.

Ochoa, C.M. (1990) *The Potatoes of South America: Bolivia.* Cambridge University Press, New York.

O'Hanlon, P.C., Peakall, R. and Briese, D.T. (1999) Amplified fragment length polymorphism (AFLP) reveals introgression in weedy *Onopordum* thistles: hybridization and invasion. *Molecular Ecology* 8, 1239–1246.

Ohno, S. (1970) *Evolution by Gene Duplication*. Springer-Verlag, New York.

Oka, H.I. (1974) Experimental studies on the origin of cultivated rice. *Genetics* 78, 475–486.

Oka, H.I. (1975) The origin of rice and its adaptive evolution. In: Association of Japanese Agriculture Scientific Societies (ed.) *Rice in Asia*. University of Tokyo Press, Tokyo, pp. 21–34.

Olmo, H.P. (1995) The origin and domestication of the *Vinifera* grape. In: McGovern, P.E. (ed.) *The Origins and Ancient History of Wine*. Gordon and Breach, Amsterdam, pp. 31–43.

Olsen, K.M. (2002) Population history of *Manihot esculenta* (Euphorbiaceae) inferred from nuclear DNA. *Molecular Ecology* 11, 901–911.

Olsen, K.M. (2004) SNPs, SSRs and inferences on cassava's origin. *Plant Molecular Biology* 56, 517–526.

Olsen, K.M. and Schaal, B.A. (1999) Evidence on the origin of cassava: phylogeography of *Manihot esculenta*. *Proceedings of the National Academy of Sciences USA* 96, 5586–5591.

Olsen, K.M. and Schaal, B.A. (2001) Microsatellite variation in cassava (*Manihot esculenta*, Euphorbiaceae) and its wild relatives: further evidence for a southern Amazonian origin of domestication. *American Journal of Botany* 88, 131–142.

Olsen, K.M., Caicedo, A.L., Polato, N., McClung, A., McCouch, S. and Purugganan, M.D. (2006) Selection under domestication: evidence for a sweep in the Rice *Waxy* genomic region. *Genetics* 173, 975–983.

Orabi, J., Backes, G., Wolday, A., Yahyaoui, A. and Jahoor, A. (2007) The Horn of Africa as a centre of barley diversification and a potential domestication site. *Theoretical and Applied Genetics* 114, 1117–1127.

Orgel, L.E. and Crick, F.H.C. (1980) Selfish DNA: the ultimate parasite. *Nature* 284, 604–607.

Orjeda, G., Freyre, R. and Iwanaga, M. (1990) Production of 2n pollen in diploid *Ipomoea trifida* a putative ancestor of sweet potato. *Journal of Heredity* 81, 462–467.

Ortiz, R. (1997) Morphological variation in *Musa* germplasm. *Genetic Resources and Crop Evolution* 44, 393–404.

Ortiz-Garcia, S., Ezcurra, E., Schoel, B., Acevedo, F., Soberón, J. and Snow, A.A. (2005) Absence of detectable transgenes in local landraces of maize in Oaxaca, Mexico (2003–2004). *Proceedings of the National Academy of Sciences USA* 102, 12338–12343.

Otto, S.P. and Whitton, J. (2000) Polyploid incidence and evolution. *Annual Review of Genetics* 34, 401–437.

Ouborg, N.J., Piquot, Y. and Van Groenendael, J.M. (2005) Estimating pollen flow using SSR markers and paternity exclusion: accounting for mistyping. *Molecular Ecology* 14, 3109–3121.

Oumar, I., Mariac, C., Phan, J.L. and Vigouroux, Y. (2008) Phylogeny and origin of pearl millet (*Pennisetum glaucum* [L.] R. Br.) as revealed by microsatellite loci. *Theoretical and Applied Genetics* 117, 489–497.

Ovchinnikova, A., Krylova, E., Gavrilenko, T., Smekalova, T., Zhuk, M., Knapp, S. and Spooner, D.M. (2011) Taxonomy of cultivated potatoes (*Solanum* section *Petota*: Solanaceae). *Botanical Journal of the Linnean Society* 165, 107–155.

Owens, C.L. (2008) Grapes. In: Hancock, J.F. (ed.) *Temperate Fruit Crop Breeding: germplasm to genomics*. Springer Science+Business Media B.V., Dordrecht, the Netherlands.

Ownbey, M. (1950) Natural hybridization and amphiploidy in the genus *Tragopogon*. *American Journal of Botany* 37, 487–499.

Özkan, H., Levy, A.A. and Feldman, M. (2001) Allopolyploidy-induced rapid genome evolution in wheat (*Aegilops-Triticum*) group. *The Plant Cell* 13, 1735–1747.

Özkan, H., Brandolini, A., Schäfer-Pregl, R. and Salamini, F. (2002) AFLP analysis of a collection of tetraploid wheats indicates the origin of emmer and hard wheat domestication in southeast Turkey. *Molecular Biology and Evolution* 19, 1797–1801.

Palaisa, K., Morgante, M., Tingey, S. and Rafalski, A. (2004) Long-range patterns of diversity and linkage disequilibrium surrounding the maize Y1 gene are indicative of an asymmetric selective sweep. *Proceedings of the National Academy of Sciences USA* 101, 9885–9890.

Palmer, J.D. (1985) Comparative organization of chloroplast genomes. *Annual Review of Genetics* 19, 325–354.

Palmer, J.D. (1987) Chloroplast DNA evolution and biosystematic uses of chloroplast DNA variation. *American Naturalist* 130, S6–S29.

Palmer, J.D., Shields, C.R., Cohen, D.B. and Orton, T.J. (1983) Choloroplast DNA evolution and the origin of amphiploid *Brassica* species. *Theoretical and Applied Genetics* 65, 181–189.

Paran, I. and van der Knaap, E. (2007) Genetic and molecular regulation of fruit and plant domestication traits in tomato and pepper. *Journal of Experimental Biology* 58, 3841–3852.

Parfitt, D.E. and Arulsekar, S. (1989) Inheritance and isozyme diversity for GOT and PGM among grape cultivars. *Journal of the American Society of Horticultural Science* 114, 486–491.

Parfitt, D.E., Arulsekar, S. and Ramming, D.W. (1985) Identification of plum x peach hybrids by isozyme analysis. *HortScience* 20, 246–248.

Parisod, C., Alix, K., Just, J., Petit, M., Sarilar, V., Mhiri, C., Ainouche, M., Chalhoub, B. and Grandbastien, M.-A. (2009) Impact of transposable elements on the organization and function of allopolyploid genomes. *New Phytologist* 186, 37–45.

Pasquet, R.S. (1998) Morphiological study of cultivated *Vigna unguiculata* (L.) Walp. Importance of ovule number and definition of cv gr *melanophthalmus. Agronomie* 18, 61–70.

Pasquet, R.S. (1999) Genetic relationships among subspecies of *Vigna unguiculata* (L.) Walp. based on allozyme variation. *Theoretical and Applied Genetics* 98, 1104–1119.

Patel, G.I. and Olmo, H.P. (1955) Cytogenetics of *Vitus.* I. The hybrid *V. vinifera* x *V. rotundifolia. American Journal of Botany* 42, 141–159.

Paterniani, E. (1969) Selection for reproductive isolation between two species of maize, *Zea mays* L. *Evolution* 23, 534–547.

Paterson, A.H., Schertz, K.F., Lin, Y. and Li, Z. (1998) Case history in plant domestication: Sorghum, an example of cereal evolution. In: Paterson, A.H. (ed.) *Molecular Dissection of Complex Traits.* CRC Press, Boca Raton, Florida, pp. 187–196.

Paterson, A.H., Bowers, J.E., Burlow, M.D., Draye, X., Elsik, C.G., Jiang, C.X., Katsar, C.S., Lan, T.-H., Lin, Y.-R. and Wright, R.J. (2000) Comparative genomics of plant chromosomes. *The Plant Cell* 12, 1523–1539.

Peck, S.L., Ellner, S.P. and Gould, F. (1998) A spacially explicit stochastic model demonstrates the feasibility of Wright's shifting balance theory. *Evolution* 52, 1834–1839.

Pedrosa, A., Schweizer, D. and Guerra, M. (2000) Cytological heterozygosity and the hybrid origin of sweet orange (*Citrus sinensis* (L.) Oseck). *Theoretical and Applied Genetics* 100, 361–367.

Pendergast, H.D.V. (1995) *Published Sources of Information on Wild Plant Species.* CAB International, Wallingford, UK.

Perales, H., Brush, S.B. and Qualset, C.O. (2003) Dynamic management of maize landraces in central Mexico. *Economic Botany* 57, 21–34.

Peralta, I.E. and Spooner, D.M. (2007) History, origin and early cultivation of tomato (Solanaceae). In: Razdan, M.K. and Mattoo, A.K. (eds) *Genetic Improvement of Solanaceous Crops.* Vol. 2: *Tomato.* Scientific Publishers, Enfield, New Hampshire, pp. 1–27.

Perl-Treves, R., Zamir, D., Navot, N. and Galun, E. (1985) Phylogeny of *Cucumis* based on isozyme variability and its comparison with plastome phylogeny. *Theoretical and Applied Genetics* 71, 430–436.

Perrier, X., Bakry, F., Carreel, F., Jenny, C., Horry, J.-P., Lebot, V. and Hippolyte, I. (2009) Combining biological approaches to shed light on the evolution of edible apples. *Ethnobotany Research and Applications* 7, 199–216.

Perry, L. and Flannery, K.V. (2007) Precolumbian use of chili peppers in the valley of Oaxaca, Mexico. *Proceedings of the National Academy of Sciences USA* 104, 11905–11909.

Perry, L., Dickau, R., Zarrillo, S., Holst, I., Pearsall, D.M., Piperno, D.R., Berman, M.J., Cooke, R.G., Rademaker, K., Ranere, A.J., Raymond, J.S., Sandweiss, D.H., Scaramelli, F., Tarble, K. and Zeidler, J.A. (2007) Starch fossils and the domestication and dispersal of chili peppers (*Capsicum* ssp. L.) in the Americas. *Science* 315, 986–988.

Perumal, R., Krishnaramanujam, R., Menz, M.A., Katilé, S., Dahlberg, J., Maguill, C.W. and Rooney, W.L. (2007) Genetic diversity among *Sorghum* races and working groups based on AFLPs and SSRs. *Crop Science* 47, 1375–1383.

Pfeiffer, J.E. (1978) *The Emergence of Man.* Harper and Row, New York.

Phillips, L.L. (1966) The cytology and phytogenetics of the diploid species of *Gossypium. American Journal of Botany* 53, 328–335.

Phillips, L.L. (1979) Cotton. In: Simmonds, N.E. (ed.) *Evolution of Crop Plants.* Longman, London.

Phipps, J.B., Robertson, K.R., Rohrer, J.R. and Smith, P.G. (1991) Origins and evolution of subfamily Maloideae (Rosaceae). *Systematic Botany* 16, 303–332.

Pickersgill, B. (1969) The archeological record of chili peppers (*Capsicum* spp.). *American Antiquity* 35, 54–61.

Pickersgill, B. (1989) Cytological and genetical evidence on the domestication and diffusion of crops within the Americas. In: Harris, D. and Hillman, G. (eds) *Farming and Foraging*. Unwin Hyman, London.

Pickersgill, B. (2007) Domestication of plants in the Americas: insights from Mendelian and molecular genetics. *Annals of Botany* 100, 925–940.

Piperno, D.R. (2006) *Phytoliths: a comprehensive guide for archaeologists and paleoecologists*. AltaMira, Lanham, Maryland.

Piperno, D.R. and Dillehay, T.D. (2008) Starch grains on human teeth reveal early broad crop diet in northern Peru. *Proceedings of the National Academy of Sciences USA* 105, 19622–19627.

Piperno, D.R. and Stothert, K.E. (2003) Phytolith evidence for early holocene *Cucurbita* domestication in southwest Ecuador. *Science* 299, 1054–1057.

Piperno, D.R., Ranere, A.J., Holst, I., Iriarte, J. and Dickau, R. (2009) Starch grain and phytolith evidence for early ninth millennium BP maize from the Central Balsas River Valley, Mexico. *Proceedings of the National Academy of Sciences USA* 106, 5019–5024.

Piperno, D.R., Ranere, A.J., Holst, I. and Hansell, P. (2000) Starch grains reveal early root crop horticulture in the Panamanian tropical forest. *Nature* 407, 894–897.

Plucknett, D.L. (1979) Edible aroids. In: Simmonds, N.W. (ed.) *Evolution of Crop Plants*. Longman, London, pp. 10–12.

Plucknett, D.L., Smith, N.J.H., Williams, J.T. and Anishetty, N.M. (1987) *Gene Banks and the World's Food*. Princeton University Press, Princeton, New Jersey.

Pohl, M.E.D., Piperno, D.R., Pope, K.O. and Jones, J.G. (2007) Microfossile evidence from pre-Columbian maize dispersals in the neotropics from San Andrés, Tabasco, Mexico. *Proceedings of the National Academy of Sciences USA* 104, 6870–6875.

Poncet, V., Lamy, F., Enjalbert, J., Joly, H., Sarr, A. and Robert, T. (1998) Genetic analysis of the domestication syndrome in pearl millet (*Pennisetum glaucum* L., Poaceae): inheritance of the major characters. *Heredity* 81, 648–658.

Ponomarenko, W. (1983) History of apple *Malus domestica*. *Journal of Botany USSR* 76, 10–18.

Potter, D., Luby, J.J. and Harrison, R.E. (2000) Phylogenetic relationships among species of *Fragaria* (Rosaceae) inferred from non-coding nuclear and chloroplast DNA sequences. *Systematic Botany* 25, 337–348.

Powell, J.R. (1975) Protein variation in natural populations of animals. *Evolutionary Biology* 8, 79–119.

Powell, W., Baird, E., Duncan, N. and Waugh, R. (1993) Chloroplast DNA variability in old and recently introduced potato cultivars. *Annals of Applied Biology* 123, 403–410.

Prain, D. and Burkill, I.H. (1936) An account of the Genus *Dioscorea* in the East. *Annals of the Royal Botanical Gardens, Calcutta*. Longman, London, UK.

Prakash, S. and Lewontin, R.C. (1968) A molecular approach to the study of genic heterozygosity in natural populations. III. Direct evidence of coadaptation in gene arrangements of *Drosophila*. *Proceedings of the National Academy of Sciences USA* 59, 398–405.

Prakash, S. and Lewontin, R.C. (1971) A molecular approach to the study of geneic heterozygosity. V. Further direct evidence of co-adaptation in inversions of *Drosophila*. *Genetics* 69, 405–408.

Pratt, R.C. and Nabhan, G.P. (1988) Evolution and diversity of *Phaseolus acutifolius* genetic resources. In: Gepts, P. (ed.) *Genetic Resources of Phaseolus Beans*. Kluwer, Dordrecht, the Netherlands, pp. 409–440.

Price, H.J. (1988) DNA content variation among higher plants. *Annals of the Missouri Botanical Garden* 75, 1248–1257.

Price, H.J., Chambers, K.L. and Bachmann, K. (1981) Geographic and ecological distribution of genomic DNA content variation in *Microseris douglasii* (Asteraceae). *Botanical Gazette* 142, 415–426.

Price, H.J., Chambers, K.L., Bachmann, K. and Riggs, J. (1986) Patterns of mean nuclear DNA content in *Microseris douglasii* (Asteraceae) populations. *Botanical Gazette* 142, 496–507.

Prideaux, T. (1973) *Cro-Magnon Man*. Time-Life, New York.

Prince, J.P., Loaiza-Figueroa, F. and Tanksley, S.D. (1992) Restriction fragment length polymorphisms and genetic distance among Mexican accessions of *Capsicum*. *Genome* 38, 224–231.

Prince, J.P., Lackney, V.K., Angeles, C., Blauth, J.R. and Kyle, M.M. (1995) A survey of DNA polymorphism within the genus *Capsicum* and the fingerprinting of pepper cultivars. *Genome* 38, 224–231.

Pringle, H. (1998) The slow birth of agriculture. *Science* 282, 1446–1450.

Proctor, M.C.F. and Yeo, P.F. (1973) *The Pollination of Flowers*. Collins, London.

Provan, J., Powell, W., Dewar, H., Bryan, G., Machray, G.C. and Waugh, R. (1999) An extreme cytoplasmic bottleneck in the modern European cultivated potato (*Solanum tuberosum*) is not reflected in decreased levels of nuclear diversity. *Proceedings of the Royal Society of London, B, Biological Sciences* 266, 633–639.

Pumphrey, M., Bai, J., Laudencia-Chingcuanco, D., Anderson, O. and Gill, B.S. (2009) Nonadditive expression of homoeologous genes is established upon polyploidization in hexaploid wheat. *Genetics* 181, 1147–1157.

Purseglove, J.W. (1972) *Tropical Crops: Monocotyledons*. Longman, London.

Purugganan, M.D. and Fuller, D.Q. (2009) The nature of selection during plant domestication. *Nature* 457, 843–848.

Qu, L., Hancock, J.F. and Whallon, J.H. (1998) Evolution in an autopolyploid group displaying predominantly bivalent pairing at meiosis: genomic similarity of diploid *Vaccinium darrowii* and autotetraploid *V. corymbosum* (Ericaceae). *American Journal of Botany* 85, 698–703.

Quero-Garcia, J., Noyer, J.L., Perrier, X., Marchand, J.L. and Lebot, V. (2004) A germplasm stratification of taro (*Colocasia esculenta*) based on agro-morphological descriptors, validation by AFLP markers. *Euphytica* 137, 387–395.

Quézel, P. (1995) La flore du Bassin Méditerranéen: orgine, misen place, endémisme. *Ecologia Mediterranea* 21, 19–39.

Quiros, C.F. (1982) Tetrasomic inheritance for multiple alleles in alfalfa. *Genetics* 101, 117–127.

Quiros, C.F. and McHale, N. (1985) Genetic analysis of isozyme variation in diploid and tetraploid potatoes. *Genetics* 111, 131–145.

Quist, D. and Capela, I.H. (2001) Transgenic DNA introgressed into traditional landraces in Oaxaca, Mexico. *Nature* 414, 541–543.

Rafalski, A. (2002) Applications of single nucleotide polymorphisms in crop genetics. *Current Opinion in Plant Biology* 5, 94–100.

Rajapakse, S., Nilmalgoda, S.D., Molnar, M., Ballard, R.E., Austin, D.F. and Bohac, J.R. (2004) Phylogenetic relationships in *Ipomoea* series *Batatas* (Convolvulaceae) based on nuclear β-amylase gene sequences. *Molecular Phylogenetics and Evolution* 30, 623–632.

Rajhathy, T. and Thomas, H. (1974) *Cytogenetics of Oats (Avena)*. Miscellaneous Publication 2, Genetic Society of Canada, Ottawa.

Raker, C.M. and Spooner, D.M. (2002) Chilean tetraploid cultivated potato, *Solanum tuberosum*, is distinct from the Andean populations: microsatellite data. *Crop Science* 42, 1451–1458.

Ramachandran, C. and Narayan, R.K.J. (1985) Chromosomal DNA variation in *Cucumis*. *Theoretical and Applied Genetics* 69, 497–502.

Ramsey, J. and Schemske, D.W. (1998) Pathways, mechanisms, and rates of polyploid formation in flowering plants. *Annual Review of Ecology and Systematics* 29, 467–501.

Rana, R.S. and Jain, H.K. (1965) Adaptive role of interchange heterozygosity in the annual *Chrysanthemum*. *Heredity* 20, 21–29.

Randall, D.D., Nelson, C.J. and Asay, K.H. (1977) Ribulose bisphosphate carboxylase: altered genetic expression in tall fescue. *Plant Physiology* 59, 38–41.

Randolph, L.F. (1966) *Iris nelsonii*, a new species of Louisana iris of hybrid origin. *Baileya* 14, 143–169.

Ranere, A.J., Piperno, D.R., Holst, I., Dickau, R. and Itiarte, J. (2009) The cultural and chronological content of early Holocene maize and squash domestication in the central Balsas River Valley, Mexico. *Proceedings of the National Academy of Sciences USA* 106, 5014–5018.

Rao, K.E.P., deWet, J.M.J., Brink, D.E. and Mengesha, M.H. (1987) Infraspecific variation and systematics of cultivated *Setaria italica*, foxtail millet (Poaceae). *Economic Botany* 41, 108–116.

Rapp, R.A., Udall, J.A. and Wendel, J.F. (2009) Genomic expression dominance in allopolyploids. *BMC Biology* 7, 18.

Reed, B.M. (2001) Implementing cryogenic storage of clonally propagated species. *Cryo-Letters* 22, 97–104.

Reed, B.M., Sarasan, V., Kane, M., Bunn, E. and Pence, V.C. (2011) Biodiversity conservation and conservation biotechnology tools. *In Vitro Cell Biology* 47, 1–4.

Rhodes, F.H.T. (1983) Gradualism, punctuated equilibria and the origin of species. *Nature* 305, 269–272.

Rhodes, M.M. and Dempsey, E. (1966) Induction of chromosome doubling at meiosis by the elongate gene in maize. *Genetics* 54, 505–522.

Rick, C.M. (1947) Partial suppression of hair development indirectly affecting fruitfulness and the proportion of cross pollination in a tomato mutant. *American Naturalist* 81, 185–202.

Rick, C.M. (1982) Genetic relationships between self-compatibility and floral traits in the tomato species. *Biologisches Zentralblatt* 101, 185–198.

Rick, C.M. (1995) Tomato: *Lycopersicon esculentum* (Solanaceae). In: Smartt, J. and Simmonds, N.W. (eds) *Evolution of Crop Plants*. Longman, Scientific & Technical, Harlow, UK, pp. 452–457.

Rick, C.M. and Fobes, J.R. (1975) Allozyme variation in the cultivated tomato and closely related species. *Bulletin of the Torrey Botanical Club* 102, 376–384.

Rick, C.M. and Holle, M. (1990) Andean *Lycopersicon esculentum* var *cerasiforme*: genetic variation and its evolutionary significance. *Economic Botany* 43 (Suppl. 3), 69–78.

Rick, C.M., Fobes, J.F. and Holle, M. (1977) Genetic variation in *Lycopersicon pimpinellifolium*: evidence of evolutionary change in mating system. *Plant Systematics and Evolution* 127, 139–170.

Rieseberg, L.H. (1991) Homoploid reticulate evolution in *Helianthus* (Asteraceae): evidence from ribosomal genes. *American Journal of Botany* 78, 1218–1237.

Rieseberg, L.H. (1995) The role of hybridization: old wine in new skins. *American Journal of Botany* 82, 944–953.

Rieseberg, L.H. (2000) Genetic mapping as a tool for studying speciation. In: Soltis, D.E., Soltis, P.S. and Doyle, J.J. (eds) *Molecular Systematics of Plants II. DNA Sequencing*. Kluwer Academic Publishers, Boston, Massachusetts, pp. 459–487.

Rieseberg, L.H. and Blackman, B.K. (2010) Genes in evolution: the control of diversity and speciation. *Annals of Botany* 106, 439–455.

Rieseberg, L. and Burke, J.M. (2008) Molecular evidence and the origin of the domesticated sunflower. *Proceedings of the National Academy of Sciences USA* 105, E46.

Rieseberg, L.H. and Ellstrand, N.C. (1993) What can molecular and morphological markers tell us about plant hybridization? *Critical Reviews in Plant Science* 12, 213–241.

Rieseberg, L.H. and Gerber, D. (1995) Hybridization in the Catalina Island mountain mahogany (*Cerocarpus traskiae*): RAPD evidence. *Conservation Biology* 9, 199–203.

Rieseberg, L.H. and Seiler, G.J. (1990) Molecular evidence and the origin and development of the domesticated sunflower, *Helianthus annuus* (Asteraceae). *Economic Botany* 44, 79–91.

Rieseberg, L.H. and Warner, D.A. (1987) Electrophoretic evidence for hybridization between *Tragopogon mirus* and *T. miscellus* (Compositae). *Sytematic Botany* 12, 281–285.

Rieseberg, L.H. and Willis, J.H. (2007) Plant speciation. *Science* 317, 910–914.

Rieseberg, L.H., Soltis, D.E. and Palmer, J.D. (1988) A molecular examination of introgression between *Helianthus annuus* and *H. bolanderi* (Compositae). *Evolution* 42, 227–238.

Rieseberg, L.H., Beckstrom-Sternberg, S. and Doan, K. (1990) *Helianthus annuus* ssp. *texanus* has chloroplast DNA and nuclear RNA genes of *Helianthus debilis* ssp. *cucumerifolus*. *Proceedings of the National Academy of Sciences USA* 87, 593–597.

Rieseberg, L.H., van Fossen, C. and Desrochers, A. (1995) Hybrid speciation accompanied by genomic reorganization in wild sunflowers. *Nature* 375, 313–316.

Rieseberg, L.H., Sinervo, B., Linder, C.R., Ungerer, M. and Arias, D.M. (1996) Role of gene interactions in hybrid speciation: evidence from ancient and experimental hybrids. *Science* 272, 741–745.

Rieseberg, L.H., Whitton, J. and Gardner, K. (1999) Hybrid zones and genetic architecture of a barrier to gene flow between two sunflower species. *Genetics* 152, 713–727.

Rieseberg, L.H., Baird, S.J.E. and Gardner, K.A. (2000) Hybridization, introgression and linkage evolution. *Plant Molecular Biology* 42, 205–224.

Rieseberg, L.H., Wood, T.E. and Baack, E.J. (2006) The nature of plant species. *Nature* 440, 524–527.

Rieseberg, L.H., Kim, S.-C., Randell, R.A., Whitney, K.D., Gross, B.L., Lexer, C. and Clay, K. (2007) Hybridization and the colonization of novel habitats by annual sunflowers. *Genetica* 129, 149–165.

Riley, H.P. (1938) A character analysis of colonies of *Iris hexagona* var. *giganticaerulea* and natural hybrids. *American Journal of Botany* 25, 727–738.

Riley, R. (1955) The cytogenetics of the differences between some *Secale* species. *Journal of Agricultural Science* 46, 277–283.

Rissler, J. and Mellon, M. (1996) *The Ecological Risks of Engineered Crops*. MIT Press, Cambridge, Massachusetts.

Roa, A.C., Maya, M.M., Duque, M.C., Tohme, J., Allem, A.C. and Bonierbale, M.W. (1997) AFLP analysis of relationships among cassava and other *Manihot* species. *Theoretical and Applied Genetics* 95, 741–750.

Rodríquez, F., Wu, F., Ané, C., Tanksley, S. and Spooner, D.M. (2009) Do potatoes and tomatoes have a single evolutionary history, and what proportion of the genome supports this history? *BMC Evolutionary Biology* 9, 191.

Rodríquez, F., Ghislain, M., Clausen, A.M., Jansky, S.H. and Spooner, D.M. (2010) Hybrid origins of cultivated potatoes. *Theoretical and Applied Genetics* 121, 1187–1198.

Rodriquez, J.M., Berke, T., Engle, L. and Nienhuis, J. (1999) Variation among and within *Capsicum* species revealed by RAPD markers. *Theoretical and Applied Genetics* 99, 147–156.

Rogers, D.J. and Appan, S.G. (1973) *Manihot and Manihoides (Euphorbiaceae)*. Neotropical Monograph No. 13, Hafner Press, New York.

Rohlf, F.J. (1998) *NTSYS-pc Numerical Taxonomy and Multivariate Analysis System, Version 2.0*. Exeter Publishing, Setauket, New York.

Rohlf, F.J. and Schnell, G.D. (1971) An investigation of the isolation-by-distance model. *American Naturalist* 105, 295–324.

Rohweder, H. (1937) Versuch zur Erfassung der mengemmassigen Bedeckung des Dors und Zingst mit polyploiden Pflanzen. *Pflanzen. Planta* 27, 501–545.

Roose, M.L. and Gottlieb, L.D. (1976) Genetic and biochemical consequences of polyploidy in *Tragapogon*. *Evolution* 30, 818–830.

Roose, M.L. and Gottlieb, L.D. (1978) Stability of structural gene number in diploid species with different amounts of nuclear DNA and different chromosome numbers. *Heredity* 40, 159–163.

Roose, M.L. and Gottlieb, L.D. (1980) Biochemical properties and level of expression of alcohol dehydrogenases in the allotetraploid plant *Tragopogon miscellus* and its diploid progenitors. *Biochemical Genetics* 18, 1065–1085.

Roose, M.L., Soost, R.K. and Cameron, J.W. (1995) Citrus (Rutaceae) In: Smartt, J. and Simmonds, N. (eds) *Evolution of Crop Plants*. Longman Scientific and Technical, Harlow, UK, pp. 443–449.

Rossel, G., Kriegner, A. and Zhang, D.P. (2000) From Latin America to Oceania: the historic dispersal of sweet potato re-examined using AFLP. *CIP Program Report 1999–2000*, pp. 315–321.

Roughgarden, J. (1976) Resource partitioning among competing species – a coevolutionary approach. *Theoretical and Population Biology* 9, 388–424.

Rousseau-Gueutin, M., Lerceteau-Köhler, E., Barrot, L., Sargent, D.J., Monford, A., Simpson, D., Arús, P., Guérin, G. and Denoyes-Rothan, B. (2008) Comparative genetic mapping between octoploid and diploid *Fragaria* species reveals a high level of colinearity between their genomes and the essentially disomic behavior of the cultivated octoploid strawberry. *Genetics* 179, 2045–2060.

Rousseau-Gueutin, M., Gaston, A., Aïnouche, A., Aïnouche, M.L., Olbricht, K., Staudt, G., Richard, L. and Denoyes-Rothan, B. (2009) Tracking the evolutionary history of polyploidy in *Fragaria* L. (strawberry): new insights from phylogenetic analysis of low-copy nuclear genes. *Molecular Phylogenetics and Evolution* 51, 515–530.

Rowe, P. and Rosales, F.E. (1996) Bananas and plantains. In: Janick, J. and Moore, J.N. (eds) *Fruit Breeding*. John Wiley & Sons, New York.

Rowewal, S.S., Ramanujam, S. and Mehra, K.L. (1969) Plant type in bengalgrain. *Indian Journal of Genetics* 26, 255–261.

Rowlands, D.C. (1959) A case of mimicry in plants of *Vicia sativa* in lentil crops. *Genetica* 30, 435–446.

Sage, T.L., Bertin, R.I. and Williams, E.G. (1994) Ovarian and other late-acting self-incompatibility systems. In: Williams, E.G., Knox, R.B. and Clark, A.E. (eds) *Genetic Control of Self-incompatibility and Reproductive Development in Plants*. Kluwer Academic, Amsterdam, pp. 116–140.

Sage, T.L., Stumas, F., Cole, W.W. and Barrett, S.C.H. (1999) Differential ovule development following self- and cross-pollination: the basis of self-sterility in *Narcissus triandrus* (Amaryllidaceae). *American Journal of Botany* 86, 855–870.

Saisho, D. and Purugganan, M.D. (2007) Molecular phylogeny of domesticated barley traces expansion of agriculture in the Old World. *Genetics* 177, 1765–1776.

Saitou, N. and Nei, M. (1987) The neighborhood-joining method: a new method for reconstructing phylogenetic trees. *Molecular Biology and Evolution* 4, 406–425.

Salaman, R.N. (1949) *The History and Social Influence of the Potato*. Cambridge University Press, Cambridge.

Salamini, F., Özkan, H., Brandolini, A., Schäfer-Pregl, R. and Martin, W. (2002) Genetics and geography of wild cereal domestication in the Near East. *Nature Reviews* 3, 429–441.

Salamini, F., Heun, M., Brandolini, A., Özkan, H. and Wunder, J. (2004) Comment on 'AFLP data and the origins of domesticated crops'. *Genome* 47, 615–620.

Salick, J., Cellinese, N. and Knapp, S. (1997) Indigenous diversity of cassava: generation, maintenance, use and loss among the Amuesha Peruvian Upper Amazon. *Economic Botany* 51, 6–19.

Salimath, S.S., Deoliveira, A.C., Godwin, I.D. and Bennetzen, J.L. (1995) Assessment of genome origins and genetic diversity in the genus *Eleusine* with DNA markers. *Genome* 38, 757–763.

Salmon, A., Flagel, L., Ying, B. Udall, J.A. and Wendel, J.F. (2010) Homoeologous non-reciprocal recombination in polyploid cotton. *New Phytologist* 186, 123–134.

Sampson, D.R. and Tarumoto, I. (1976) Genetic variances in an eight parent half dialle of oats. *Canadian Journal of Genetics and Cytology* 18, 419–427.

Sanderson, J.A. and Hulbert, E.O. (1955) Sunlight as a source of radiation. In: Hollaender, A. (ed.) *Radiation Biology II. Ultraviolet and Related Radiations*. McGraw-Hill, New York.

SanMiguel, P. and Bennetzen, J.L. (1998) Evidence that a recent increase in maize genome size was caused by the massive amplification of intergene retrotranspositions. *Annals of Botany* 82, 37–44.

Sang, T. and Ge, S. (2007a) The puzzle of rice domestication. *Journal of Integrative Plant Biology* 49, 760–768.

Sang, T. and Ge, S. (2007b) Genetics and phylogenetics of rice domestication. *Current Opinion in Genetics and Development* 17, 533–538.

Sanjur, O.I., Piperno, D.R., Andres, T.C. and Wessel-Beaver, L. (2002) Phylogenetic relationships among domesticated and wild species of *Cucurbita* (Cucurbitaceae) inferred from a mitochondrial gene: implications for crop plant evolution and areas of origin. *Proceedings of the National Academy of Sciences USA* 99, 535–540.

Sasaki, K. (1986) Development and type of East Asian agriculture. In: Hanihara, K. (ed.) *The Origin of Japanese*. Shogakukan, Tokyo, pp. 86–105.

Sauer, C.O. (1952) *Agricultural Origins and Dispersals*. MIT Press, Cambridge, Massachusetts.

Sauer, J.D. (1967) The grain amaranths and their relatives – a revised taxonomic and geographic survey. *Annals of the Missouri Botanical Garden* 54, 103–137.

Sauer, J.D. (1993) *Historical Geography of Crop Plants: A Select Roster*. CRC Press, Boca Raton, Florida.

Sax, K. (1931) The origin and relationships of the Pomoideae. *Journal of the Arnold Arboretum* 12, 3–22.

Sax, K. (1933) The origin of the Pomoideae. *Proceedings of the American Society of Horticultural Science* 30, 147–150.

Scandalios, J.G. (1969) Genetic control of multiple molecular forms of enzymes in plants: a review. *Biochemical Genetics* 3, 37–79.

Scascitelli, M., Whitney, K.D., Randell, R.A., King, M., Buerkle, C.A. and Rieseberg, L.H. (2010) Genome scan of hybridizing sunflowers from Texas (*Helianthus annuus* and *H. debilis*) reveals asymmetric patterns of introgression and small islands of genomic differentiation. *Molecular Ecology* 19, 521–541.

Schaal, B.A. (1975) Population structure and local differentiation in *Liatris cylindracea*. *American Naturalist* 109, 511–528.

Schaal, B.A. and Olsen, K.M. (2000) Gene genealogies and population variation in plants. *Proceedings of the National Academy of Sciences USA* 97, 7024–7029.

Schäfer, H.I. (1973) Zur Taxonomie der *Vicia narbonensis*-Gruppe. *Kultupflanze* 21, 211–273.

Schemske, D.W. (2000) Understanding the origin of species. *Evolution* 54, 1069–1073.

Schemske, D.W. and Bierzychudek, P. (2001) Perspective: evolution of flower color in the desert annual *Linanthus parryae*: Wright revisited. *Evolution* 55, 1269–1282.

Schemske, D.W. and Bradshaw, H.D., Jr (1999) Pollinator preference and the evolution of floral traits in monkeyflowers (*Mimulus*). *Proceedings of the National Academy of Sciences USA* 96, 11910–11915.

Schilling, E. and Heiser, C. (1981) Infragenetic classification of *Helianthus* (Compositae). *Taxon* 30, 393–403.

Schlötterer, C. (2004) The evolution of markers – just a matter of fashion? *Nature Reviews Genetics* 5, 63–69.

Schmidt, T. and Heslop-Harrison, J.S. (1998) Genomes, genes and junk: the large scale organization of plant chromosomes. *Trends in Plant Science* 3, 195–199.

Schmit, V. and Debouck, D.G. (1991) Observations on the origin of *Phaseolus polyanthus* Greenman. *Economic Botany* 45, 345–364.

Schumann, C.M. and Hancock, J.F. (1990) Paternal inheritance of plastids in *Medicago sativa*. *Theoretical and Applied Genetics* 78, 863–866.

Schwarzbach, A.E. and Rieseberg, L.H. (2002) Likely multiple origins of a diploid hybrid sunflower species. *Molecular Ecology* 11, 1703–1715.

Scorza, R. and Okie, W. (1990) Peaches (*Prunus*). In: Moore, J.N. and Ballington, J.R. (eds) *Genetic Resources of Temperate Fruit and Nut Crops.* International Society of Horticultural Science, Wageningen, the Netherlands.

Scott, N.S. and Possingham, J.V. (1981) Chloroplast DNA in expanding spinach leaves. *Journal of Experimental Botany* 31, 1082–1092.

Sears, B. (1980) Elimination of plastids during spermatogenesis and fertilization in the plant kingdom. *Plasmid* 4, 233–255.

Sears, E.R. (1944) Cytogenetic studies with polyploidy species of wheat. II. Additional chromosome aberrations in *Triticum vulgare. Genetics* 29, 232–246.

Seavey, S.R. and Bawa, K.S. (1986) Late-acting self incompatibility in angiosperms. *Botanical Review* 52, 195–219.

Seavey, S.R. and Carter, S.K. (1994) Self-fertility in *Epilobium obcordatum* (Onagraceae). *American Journal of Botany* 81, 331–338.

Sefc, K.M., Lopes, M.S., Lefort, F., Botta, R. and Roubelakis-Angelakis, H. (2000) Microsatellite variability in grapevine cultivars from different European regions and evaluation of assignment testing to assess the geographic origin of cultivars. *Theoretical and Applied Genetics* 100, 498–505.

Seijo, G., Lavia, G.I., Fernández, A., Krapovickas, A., Ducasse, D.A., Bertioli, D.J. and Moscone, E.A. (2007) Genomic relationships between the cultivated peanut (*Arachis hypogaea*, Leguminosae) and its close relatives revealed by double GISH. *American Journal of Botany* 94, 1963–1971.

Selvi, A., Nair, N.V., Noyer, J.L., Singh, N.K., Balasundaram, N., Bansal, K.C., Koundal, K.R. and Mohapatra, T. (2006) AFLP analysis of the phenetic organization and genetic diversity in the sugarcane complex, *Saccharum* and *Erianthus. Genetic Resources and Crop Evolution* 53, 831–842.

Serre, D., Langaney, A., Chech, M., Teschler-Nicola, M., Paunovic, M., Mennecier, P., Hofreiter, M., Possnert, G. and Pääo, S. (2004) No evidence of Neanderthal mtDNA contribution to early modern humans. *PLoS Biology* 2, 313–317.

Settler, T.L., Schrader, L.E. and Bingham, E.T. (1978) Carbon dioxide exchange rates, transpiration and leaf characteristics in genetically equivalent ploidy levels in alfalfa. *Crop Science* 18, 327–332.

Shaked, H., Kashkush, K., Özkan, H., Feldman, M. and Levy, A.A. (2001) Sequence elimination and cytocine methylation are rapid and reproducible responses of the genome to wide hybridization and allopolyploidy in wheat. *Plant Cell* 13, 1749–1759.

Shands, H.L. (1990) Plant genetic resources and conservation: the role of the gene bank in delivering useful genetic materials to the research scientist. *Journal of Heredity* 81, 7–10.

Shii, C.T., Mok, M.C. and Mok, D.W.S. (1981) Developmental controls of morphological mutants of *Phaseolus vulgaris* L: differential expression of mutant loci in plant organs. *Developmental Genetics* 2, 279–290.

Shiotani, I. and Kawase, T. (1989) Genomic structure of the sweet potato and haploids in *Ipomoea trifida. Japanese Journal of Breeding* 39, 57–66.

Shipman, P. (1988) What does it take to be a meateater? *Discover*, September, p. 44.

Simmonds, N.W. (1962) *The Evolution of Bananas.* Tropical Science Series, Longmans, London.

Simmonds, N.W. (1985) *Principles of Crop Improvement.* Longman, London.

Simmonds, N.W. (1995a) Bananas: *Musa* (Musaceae). In: Smartt, J. and Simmonds, N.W. (eds) *Evolution of Crop Plants.* Longman Scientific & Technical, Harlow, UK, pp. 370–375.

Simmonds, N.W. (1995b) Potatoes: *Solanum tuberosum* (Solanaceae). In: Smartt, J. and Simmonds, N.W. (eds) *Evolution of Crop Plants.* Longman Scientific & Technical, Harlow, UK, pp. 466–471.

Simmonds, N.W. (1995c) Food crops; 500 years of travel. In: Duncan, R.R. (ed.) *International Germplasm Transfer: Past and Present. Special Publication No. 23,* CSSA, Madison, Wisconsin, pp. 31–46.

Simons, K.J., Fellers, J.P., Trick, H.N., Zhang, Z., Tai, Y.-S., Gill, B.S. and Faris, J.D. (2006) Molecular characterization of the major wheat domestication gene Q. *Genetics* 172, 547–555.

Simpson, G.G. (1961) *Principles of Animal Taxonomy.* Columbia University Press, New York.

Sims, L.E. and Price, H.J. (1985) Nuclear DNA content variation in *Helianthus* (Asteraceae). *American Journal of Botany* 72, 1213–1219.

Singh, A., Devarumath, R.M., RamaRao, S., Soghn, V.P. and Raina, S.N. (2008) Assessment of genetic diversity, and phylogenetic relationships based on ribosomal DNA repeat unit length variation and internal transcribed spacer (ITS) sequences in chickpea (*Cicer arietinum*) cultivars and its wild species. *Genetic Resources and Crop Evolution* 55, 65–79.

Singh, K.B. and Ocampo, B. (1993) Interspecific hybridization in annual *Cicer* species. *Journal of Genetics and Breeding* 47, 199–204.

Singh, K.B. and Ocampo, B. (1997) Exploitation of wild species for yield improvement in chickpea. *Theoretical and Applied Genetics* 95, 418–423.

Singh, K.B., Malhorta, R.S., Halila, H., Knights, E.J. and Verma, M.M. (1994) Current status and future strategy in breeding chickpea for resistance to biotic and abiotic stresses. *Euphytica* 73, 137–149.

Sladkin, M. (1985) Gene flow in natural populations. *Annual Review of Ecology and Systematics* 16, 393–430.

Small, R.L. and Wendel, J.F. (1998) The mitochondrial genome of allopolyploid cotton (*Gossypium* L.). *Journal of Heredity* 90, 251–253.

Smartt, J. (1984) Gene pools in grain legumes. *Economic Botany* 38, 24–35.

Smartt, J. (1999) *Grain legumes: Evolution and Genetic Resources.* Cambridge University Press, Cambridge.

Smartt, J., Gregory, W.C. and Gregory, M.P. (1978) The genomes of *Arachis hypogea* I. Cytogical studies of putative genome donors. *Euphytica* 27, 665–675.

Smith, B.D. (1989) Origins of agriculture in Eastern North America. *Science* 246, 1566–1571.

Smith, B.D. (1997) The initial domestication of *Curcurbita pepo* in the Americas 10,000 years ago. *Science* 276, 932–934.

Smith, B.D. (1998) *The Emergence of Agriculture.* Scientific American Library, New York.

Smith, B.D. (2001) Documenting plant domestication: the consilience of biological and archeological approaches. *Proceedings of the National Academy of Sciences USA* 98, 1324–1326.

Smith, B.D. (2006) Eastern North America as an independent center of plant domestication. *Proceedings of the National Academy of Sciences USA* 103, 12223–12228.

Smith, D.C. (1946) *Sedum pulchellum*: a physiological and morphological comparison of diploid, tetraploid and hexaploid races. *Bulletin of the Torrey Botanical Club* 73, 495–541.

Smith, F.H. and Clarkson, Q.D. (1956) Cytological studies of interspecific hybridization in *Iris*, subsection Californicae. *American Journal of Botany* 43, 582–588.

Smith, L.B. and King, G.J. (2000) The distribution of BoCAL-a alleles in *Brassica oleracea* is consistent with a genetic model for curd development and domestication of the cauliflower. *Molecular Breeding* 6, 603–613.

Smýkal, P., Kenicer, G., Flavell, A.J., Corander, J., Kosterin, O., Redden, R.J., Ford, R., Coyne, C.J., Maxted, N., Ambrose, M.J. and Ellis, N.T.H. (2010) Phylogeny, phylogeography and genetic diversity in the *Pisum* genus. *Plant Genetic Resources, Characterization and Utilization* 8, 1–15.

Snow, A.A. and Palma, P.M. (1997) Commercialization of transgenic plants: potential ecological risks. *Bioscience* 47, 86–96.

Snow, A., Andow, D.A., Gepts, P.E., Hallerman, E.M., Power, A., Tiedje, J.M. and Wolfenbarger, L.L. (2003) Genetically engineered organisms and the environment: current status and recommendations. *Ecological Applications* 15, 377–404.

Snow, R. (1960) Chromosomal differentiation in *Clarkia dudleyana. American Journal of Botany* 47, 302–309.

Snowden, J.D. (1936) *The Cultivated Races of Sorghum.* Allard and Son, London.

Snowdon, R.J., Friedrich, T., Friedt, W. and Kohler, W. (2002) Identifying the chromosomes of the A- and C-genome diploid *Brassica* species *B. rapa* (syn. *campestris*) and *B. oleracea* in their amphidiploid *B. napus. Theoretical and Applied Genetics* 104, 533–538.

Sokal, R. R. and Rohlf, F.J. (1995) *Biometry: The Principles and Practice of Statistics in Biological Research,* 3rd edition. W.H. Freeman and Co, San Francisco, California.

Sokal, R.R., Oden, N.L. and Wilson, C. (1991) Genetic evidence for the spread of agriculture in Europe by demic diffusion. *Nature* 351, 143–145.

Solbrig, O.T. and Simpson, B.B. (1977) A garden experiment on competition between biotypes of the common dandelion (*Taraxacum officinale*). *Journal of Ecology* 65, 427–430.

Soleri, D., Cleveland, A. and Cuevas, F.A. (2006) Transgenic crops and crop varietal diversity: the case of maize in Mexico. *BioScience* 56, 503–513.

Soltis, D.E. (1984) Autopolyploidy in *Tolmiea menziesii* (Saxifragaceae). *American Journal of Botany* 71, 1171–1174.

Soltis, D.E. and Soltis, P.S. (1988) Electrophoretic evidence for tetrasomic segregation in *Tolmiea menziesii* (Saxifragaceae). *Heredity* 60, 375–382.

Soltis, D.E. and Soltis, P.S. (1989a) Allopolyploid speciation in *Tragopogon*: insights from chloroplast DNA. *American Journal of Botany* 76, 1119–1124.

Soltis, D.E. and Soltis, P.S. (1989b) Tetrasomic inheritance in *Heuchera micrantha* (Saxifragaceae). *Journal of Heredity* 80, 123–126.

Soltis, D.E. and Soltis, P.S. (1993) Molecular data and the dynamic nature of polyplody. *Critical Reviews in Plant Sciences* 12, 243–273.

Soltis, D.E. and Soltis, P.S. (1995) The dynamic nature of polyploidy genomes. *Proceedings of the National Academy of Sciences USA* 92, 8089–8091.

Soltis, D.E. and Soltis, P.S. (1999) Polyploidy: recurrent formation and genome evolution. *Trends in Ecology and Systematics* 14, 348–352.

Soltis, D.E., Soltis, P.S., Schemske, D.W., Hancock, J.F., Thompson, J.N., Husband, B.C. and Judd, W.D. (2007) Autopolyploidy in angiosperms: have we grossly underestimated the number of species? *Taxon* 56, 13–30.

Soltis, P.S. and Soltis, D.E. (2000) The role of genetic and genomic attributes in the success of polyploids. *Proceedings of the National Academy of Sciences USA* 97, 7051–7057.

Soltis, P.S. and Soltis, D.E. (2009) The role of hybridization in plant speciation. *Annual Review of Plant Biology* 60, 561–588.

Song, K.M., Osborn, T.C. and Williams, P.H. (1990) *Brassica* taxonomy based on nuclear restriction fragment length polymorphisms (RFLPs). 3. Genome relationships in *Brassica* and related genera and the origin of *B. oleraceae* and *B. rapa* (syn. *campestris*). *Theoretical and Applied Genetics* 79, 497–506.

Song, K.M., Lu, P., Tang, K. and Osborn, T.C. (1995) Rapid genome change in synthetic polyploids of *Brassica* and its implications for polyploidy evolution. *Proceedings of the National Academy of Sciences USA* 92, 7719–7723.

Souza Machado, V., Bandeen, J.D., Taylor, W.D. and Lavigne, P. (1977) Atrazine resistant biotypes of common ragweed and birds rape. *Research Report of the Canadian Weed Commission (East Section)* 22, 305.

Spencer, L.J. and Snow, A.A. (2001) Fecundity of transgenic wild-crop hybrids of *Cucurbita pepo* (Cucurbitaceae): implications for crop-to-weed gene flow. *Heredity* 86, 694–702.

Spooner, D.M. and Hijmans, R.J. (2001) Potato systematics and germplasm collecting, 1989–2000. *American Journal of Potato Research* 78, 237–268.

Spooner, D.M., Bryan, G.L., van den Berg, R.G. and del Rio, A. (2003) Species concepts and relationships in wild and cultivated potatoes. *Acta Hoticulturae* 619, 63–75.

Spooner, D.M., Peralta, I.E. and Knapp, S. (2005a) Comparison of AFLPs with other markers for phylogenetic inference in wild tomatoes (*Solanum* L. section *Lycopersicon* (Mill.) Wettst.). *Taxon* 54, 43–61.

Spooner, D.M., McLean, K., Ramsay, G., Waugh, R. and Bryan, G.J. (2005b) A single domestication for potato based on mulilocus amplified fragment length polymorphism genotyping. *Proceedings of the National Academy of Sciences USA* 102, 14694–14699.

Spooner, D.M., Núñez, J., Trujillo, G., del Rosario Herrera, M., Guzmán, F. and Ghislain, M. (2007) Extensive simple sequence repeat genotyping of potato landraces supports a major reevaluation of their gene pool structure and classification. *Proceedings of the National Academy of Sciences USA* 104, 19398–19403.

Sporne, K.R. (1971) *The Mysterious Origin of Flowering Plants.* Oxford University Press, London.

Sreekumari, M.T. and Mathew, P.M. (1991) Karomorphology of five morphotypes of taro (*Colocasia esculenta* (L.) Schott.). *Cytologia* 56, 215–218.

Srisuwan, S., Sihachakr, D. and Siljak-Yakovlev, S. (2006) The origin and evolution of sweet potato (*Ipomoea batatas* Lam.) and its wild relatives through cytogentic approaches. *Plant Science* 171, 424–433.

Stadler, L.J. (1942) Some observations on gene variability and spontaneous mutation. *Spragg Memorial Lectures* (Michigan State University) 3, 3–15.

Stapf, O. and Hubbard, C.E. (1934) *Pennisetum.* In: Prain, D. (ed.) *Flora of Tropical Africa 9.* Reeve Brothers, London.

Staudt, G. (1984) Cytological evidence of double restitution in *Fragaria. Plant Systematics and Evolution* 146, 171–179.

Stebbins, G.L. (1947) Types of polyploids: their classification and significance. *Advances in Genetics* 1, 403–429.

Stebbins, G.L. (1950) *Variation and Evolution in Plants.* Columbia University Press, New York.

Stebbins, G.L. (1956) Cytogenetics and evolution of the grass family. *American Journal of Botany* 43, 890–905.

Stebbins, G.L. (1957) The hybrid origin of microspecies in the *Elymus glaucus* complex. *Cytologia* 36, 336–340.

Stebbins, G.L. (1959) Genes, chromosomes and evolution. In: Turrill, W. (ed.) *Vistas in Botany.* Pergamon Press, Elmsford, New York.

Stebbins, G.L. (1971) *Chromosomal Evolution in Higher Plants*. Arnold, London.

Stebbins, G.L. (1972) Research on the evolution of higher plants: problems and prospects. *Canadian Journal of Genetics and Cytology* 14, 453–462.

Stebbins, G.L. (1974) *Flowering Plants*. Harvard University Press, Cambridge, Massachusetts.

Stebbins, G.L. (1980) Polyploidy in plants: unsolved problems and prospects. In: Lewis, W.H. (ed.) *Polyploidy – Biological Relevance*. Plenum, New York, pp. 495–520.

Stephens, S.G. (1946) The genetics of 'corky'. I. The New World alleles and their possible role as an interspecific isolating mechanism. *Journal of Genetics* 47, 150–161.

Stephens, S.G. (1951) Possible significance of duplication in evolution. *Advances in Genetics* 4, 247–265.

Stevenson, G.C. (1965) *Genetics and Breeding of Sugarcane*. Longman, London.

Story, D.A. (1985) Adaptive strategies of archaic cultures of the West Gulf Coastal Plain. In: Ford, R.I. (ed.) *Prehistoric Food Production in North America*, Museum of Anthropology, University of Michigan, Ann Arbor, Michigan.

Strasburg, J.L., Scotti-Saintagne, C., Scotti, I., Lai, Z. and Rieseberg, L.H. (2009) Genomic patterns of adaptive divergence between chromosomally differentiated sunflower species. *Molecular Biology and Evolution* 26, 1341–1355.

Strauss, S.H. and Libby, W.J. (1987) Allozyme heterosis in *Radiata* pine is poorly explained by overdominance. *American Naturalist* 130, 879–890.

Stubbe, W. (1960) Untersuchungen zur Genetischen Analyses des Plastoms von *Oenothera*. *Zeitschrift für Botanisch* 48, 191–218.

Stubbe, W. (1964) The role of the plastome in evolution of the genus *Oenothera*. *Genetics* 35, 28–33.

Stuber, C.W. and Goodman, M.M. (1983) Inheritance, intracellular localization and genetic variation of phosphogluco-mutase enzymes in maize (*Zea mays* L.). *Biochemical Genetics* 21, 667–689.

Sudré, C.P., Conçalves, L.S.A., Rodriques, R., do Amaral Júnior, A.T., Riva-Souza, E.M. and dos S. Bento, C. (2010) Genetic variability in domesticated *Capsicum* spp as assessed by morphological and agronomic data. *Genetics and Molecular Research* 9, 283–294.

Sun, G., Dilcher, D.L., Zheng, S. and Zhou, Z. (1998) In search of the first flower: a Jurassic angiosperm, *Archaefructus*, from northeast China. *Science* 282, 1692–1694.

Swingle, W.T. (1967) The botany of citrus and its wild relatives. In: Reuther, W., Webber, H.J. and Bachelor, L.D. (eds) *The Citrus Industry*. Division of Agricultural Science, University of California, Berkeley, California, pp. 1–39.

Tai, G.C.C. (1976) Estimation of general and specific combining abilities in potato. *Canadian Journal of Cytogenetics and Cytology* 18, 463–470.

Takahashi, S., Furukawa, T., Asano, T., Terajima, Y., Shimada, H., Sugimoto, A. and Kadowaki, K. (2005) Very close relationship of the chloroplast genomes among *Saccharum* species. *Theoretical and Applied Genetics* 110, 1523–1529.

Takayama, S. and Isogai, A. (2005) Self-incompatibility in plants. *Annual Review of Plant Biology* 56, 467–489.

Takhtajan, A. (1969) *Flowering Plants – Origin and Dispersal*. Oliver and Boyd, Edinburgh, UK.

Talbert, L.E., Doebley, J.F., Larson, S. and Chandler, V.L. (1990) *Tripsicum andersonii* is a natural hybrid involving *Zea* and *Tripsicum*: molecular evidence. *American Journal of Botany* 77, 722–726.

Talbert, L.E., Blake, N.K., Storie, E.W. and Lavin, M. (1995) Variability in wheat based on low-copy DNA sequence comparisons. *Genome* 38, 951–957.

Talbert, L.E., Smith, L.Y. and Blake, K.K. (1998) More than one origin of hexaploid wheat is indicated by sequence comparison of low-copy DNA. *Genome* 41, 402–407.

Tanaka, T. (1954) *Species Problems in Citrus*. Japanese Society for the Promotion of Science, Tokyo.

Tang, T., Lu, J., Huang, J., He, J., McCouch, S.R., Shen, Y., Kai, Z., Purugganan, M.D., Shi, S. and Wu, C.-I. (2006) Genomic variation in rice: Genesis of highly polymorphic linkage blocks during domestication. *PLo5 Genet* 2(11): e199.doi:10.1371/journal.pgen.0020199

Tanksley, S.D. (1993) Mapping polygenes. *Annual Review of Genetics* 27, 205–233.

Tanksley, S.D. and McCouch, S.R. (1997) Seed banks and molecular maps: unlocking the genetic potential from the wild. *Science* 277, 1063–1066.

Tanksley, S.D., Ganal, M.W., Price, J.P., de Vicente, M.C., Bonierbale, M.W., Broun, P., Fulton, T.M., Giovannoni, J.J., Grandillo, S. and Martin, G.B. (1992) High density molecular linkage maps of the tomato and potato genomes. *Genetics* 132, 1141–1160.

Tanno, K.I. and Willcox, G. (2006a) How fast was wild wheat domesticated? *Science* 311, 1886.

Tanno, K-I. and Willcox, G. (2006b) The origins of cultivation of *Cicer arietinum* L. and *Vicia faba* L.: early finds from Tell el-Kerkh, northwest Syria, late 10th millennium BP. *Vegetation History and Archaeobotany* 15, 197–204.

Tate, J.A., Joshi, P., Soltis, K.A., Soltis, P.S. and Soltis, D.E. (2009) On the road to diploidization? Homoeolog loss in independently formed populations of the allotetraploid *Tragopogon miscellus* (Asteraceae). *BMC Plant Biology* 9, 80.

Tattersall, I. (1998) *Becoming Human*. Harcourt, Brace & Company, Orlando, Florida.

Tattersall, I. (2009) Human origins: out of Africa. *Proceedings of the National Academy of Sciences USA* 106, 16018–16021.

Templeton, A.R. (1981) Mechanisms of speciation – a population genetics approach. *Annual Review of Ecology* 12, 23–41.

Tenaillon, M.I., Sawkins, M.C., Long, A.D., Gaut, R.L., Doebley, J.F. and Gaut, B.S. (2001) Patterns of DNA sequence polymorphism along chromosome 1 of maize (*Zea mays* ssp. *mays* L.). *Proceedings of the National Academy of Sciences USA* 98, 9161–9166.

Teruachi, R., Chikaleke, V.A., Thottappilly, G. and Hahn, S.K. (1992) Origin and phylogeny of Guinea yams as revealed by RFLP analysis of chloroplast DNA and nuclear ribosomal DNA. *Theoretical and Applied Genetics* 83, 743–751.

Thompson, K.F. (1979) Cabbages, kales, etc.: *Brassica oleracea* (Cruciferae). In: Simmonds, N.E. (ed.) *Evolution of Crop Plants*. Longman, London, pp. 49–52.

Tian, F., Stevens, N.M. and Buckler, E.S. (2009) Tracking footprints of maize domestication and evidence for a massive selection sweep on chromosome 10. *Proceedings of the National Academy of Sciences USA* 106, 9979–9986.

Timko, M.P. and Vasconcelos, A.C. (1981) Photosynthetic activity and chloroplast membrane polypeptides in euploid cells of *Ricinus*. *Physiologia Plantarum* 52, 192–196.

Ting, Y.C. (1985) Meiosis and fertility of anther derived maize plants. *Maydica* 30, 161–169.

Ting, Y.C. and Kehr, A.E. (1953) Meiotic studies in the sweet potato. *Journal of Heredity* 44, 207–211.

Townsend, C.E. and Remmenga, E.E. (1968) Inbreeding in tetraploid alsike clover, *Trifolium hybridum*. *Crop Science* 8, 213–217.

Turesson, G. (1925) The plant species in relation to habitat and climate. *Hereditas* 6, 147–236.

U, N. (1935) Genome analysis of *Brassica* with special reference to experimental formation of *B. napus*. *Journal of Heredity* 73, 335–339.

Udall, J.A., Quijada, P.A. and Osborn, T.C. (2005) Detection of chromosomal rearrangements derived from homeologous recombination in four mapping populations of L. *Brassica napus*. *Genetics* 169, 967–979.

Ugent, D., Pozorski, S. and Pozorski, T. (1982) Prehistoric remains of sweet potato from the Casma Valley, Peru. *Phytologia* 49, 401–415.

Ugent, D., Pozorski, S. and Pozorski, T. (1986) Archeological manioc (*Manihot*) from coastal Peru. *Economic Botany* 40, 78–102.

Vaillancourt, R.E. and Weeden, N.F. (1992) Chloroplast DNA polymorphism suggests a Nigerian center of domestication for the cowpea, *Vigna unguiculata*, Leguminosae. *American Journal of Botany* 79, 1194–1199.

Valenzuela, H.R. and DeFrank, J. (1995) Agroecology of tropical underground crops for small scale agriculture. *Critical Reviews in Plant Sciences* 14, 213–238.

van den Berg, R.G., Miller, J.T., Ugarte, M.L., Kardolus, J.P., Villand, J., Niehuis, J. and Spooner, D.M. (1998) Collapse of morphological species in the wild potato *Solanum brevicaulae* complex (Solanaceae: Sect. Petota). *American Journal of Botany* 85, 92–109.

Van Oss, H., Aron, Y. and Ladizinsky, G. (1997) Chloroplast DNA variation and evolution in the genus *Lens* Mill. *Theoretical and Applied Genetics* 94, 452–457.

Van Slageren, M.W. (1994) *Wild Wheats: A monograph of Aegilops L. and Amblyopyrum (Jaub. & Spaach) Eig (Poaceae)*. Papers 1994, Wageningen Agricultural University, Wageningen, the Netherlands.

Van Zeist, W. and Bakker-Heeres, J.A.H. (1985) Archeological studies in the Levant 1. Neolithic sites in the Damascus Basin: Aswad, Ghoraifé, Ramad. *Palaeohistoria* 24, 165–256.

Vander Kloet, S.P. and Lyrene, P.M. (1987) Self-incompatibility in diploid, tetraploid and hexaploid *Vaccinium corymbosum*. *Canadian Journal of Botany* 65, 660–665.

Vanderplank, J.E. (1978) *Disease Resistance in Plants*. Academic Press, New York.

Vaughan, D.A. (1994) *The Wild Relatives of Rice*. International Rice Research Institute, Manila, the Philippines.

Vavilov, N.I. (1926) *Studies on the Origins of Cultivated Plants*. Institute of Applied Botany and Plant Breeding, Leningrad.

Vavilov, N.I. (1949–1950) *The Origin, Variation, Immunity and Breeding of Cultivated Crops*. Chronica Botanica, Waltham, Massachusetts.

Velasco, R., Zharkikh, A., Affourtit, J. *et al.* (83 others) (2010) The genome of the domesticated apple (*Malus* × *domestica* Borkh). *Nature Genetics* 42, 833–841.

Vershinin, A.V., Allnutt, T.R., Knox, M.R., Ambrose, M.J. and Ellis, N.T.H. (2003) Transposable elements reveal the impact of introgression, rather than transposition in *Pisum* diversity, evolution, and domestication. *Molecular Biology and Evolution* 20, 2067–2075.

Vetukhiv, M. (1956) Fecundity of hybrids between geographic populations of *Drosophila pseudoobscura*. *Evolution* 10, 139–146.

Viard, F., Bernard, J. and Desplanque, B. (2002) Crop-weed interactions in the *Beta vulgaris* complex at a local scale: alleleic diversity and gene flow within sugar beet fields. *Theoretical and Applied Genetics* 104, 688–697.

Vickery, R.K. (1953) An experimental study of the races of the *Mimulus guttalis* complex. *Proceedings of the 7th International Botanical Congress*, Stockholm, p. 272.

Vigouroux, Y., McMullen, M., Hittinger, C.T., Houchins, K., Schulz, L., Kresovich, S., Matsuoka, Y. and Doebley, J. (2002) Identifying genes of agronomic importance in maize by screening microsatellites for evidence of selection during domestication. *Proceedings of the National Academy of Sciences USA*. 99, 9650–9655.

Vijendra Das, L.D. (1970) Chromosome associations in diploid and autotetraploid *Zea mays* L. *Cytologia* 35, 259–261.

Von Bothmer, R., Yen, C. and Yang, J. (1990) Does wild, six-row barley, *Hordeum agriocrithon* really exist? *FAO/IBPGR Plant Genetic Resource Newsletter* 77, 17–19.

Vorsa, N. and Bingham, E.T. (1979) Cytology of 2n pollen production in diploid alfalfa, *Medicago sativa*. *Canadian Journal of Genetics and Cytology* 21, 525–530.

Waddington, C.H. (1953) Genetic assimilation of an aquired characteristic. *Evolution* 7, 118–126.

Waddington, C.H. (1957) *The Strategy of Genes*. Allen & Unwin, London.

Wahl, I. and Segal, A. (1986) Evolution of host–parasite balance in natural indigenous populations of wild barley and oats in Israel. In: Barigozzi, C. (ed.) *Origin and Domestication of Wild Plants*. Elsevier, Amsterdam, pp. 129–142.

Wahlund, S. (1928) Zussammersetung von Populationen und Korrelation – sercheinungen von Standpunkt der Verebungslehre aus betrachtet. *Hereditas* 1, 165–206.

Walbot, V. and Cullis, C.A. (1985) Rapid genome change in higher plants. *Annual Review of Plant Physiology* 36, 367–396.

Walker, A.R., Lee, E. and Robinson, S.P. (2006) Two new grape cultivars, bud sports of Cabernet Sauvignon bearing pale-coloured berries, are the result of deletion of two regulatory genes of the berry color locus. *Plant Molecular Biology* 62, 623–635.

Wallace, B. (1968) *Topics in Population Genetics: Coadaptation*. Norton, New York.

Wallace, B. (1970) *Genetic Load*. Prentice Hall, Englewood Cliffs, New Jersey.

Walsh, B.M. and Hoot, S.B. (2001) Phylogenetic relationships of *Capsicum* (Solanaceae) using DNA sequences from two noncoding regions: the chloroplast *atpB-rbcL* spacer region and nuclear *waxy* introns. *International Journal of Plant Science* 162, 1409–1418.

Walters, J.L. (1952) Heteromorphic chromosome pairs in *Paeonia californica*. *American Journal of Botany* 39, 145–151.

Wang, G.-Z., Miyashita, N.T. and Tsunewaki, K. (1997) Plastom analysis of *Triticum* (wheat) and *Aegilops*: PCR-single-strand conformational polymorphism (PCR-SSCP) analysis of organellar DNAs. *Proceedings of the National Academy of Sciences USA* 94, 14570–14577.

Wang, J., Tian, L., Madlung, A., Lee, H.-S., Chen, M., Lee, J.J., Watson, B., Kagochi, T., Coumai, L. and Chen, Z.J. (2004) Stochastic and epigenetic changes of gene expression in *Arabidopsis* polyploids. *Genetics* 167, 1961–1973.

Wang, J., Tian, L., Lee, H.-S., Wei, N.E., Jiang, H., Watson, B., Madlung, A., Osborn, T.C., Doerge, R.W., Comai, L. and Chen, Z.J. (2006) Genomewide nonadditive gene regulation in *Arabidopsis* allotetraploids. *Genetics* 172, 507–517.

Wang, K.-J., Li, X.-H. and Li, F.-S. (2008) Phenotypic diversity of the big seed type subcollection of wild soybean (*Glycine soja* Sieb. et Zucc.) in China. *Genetic Resources and Crop Evolution*. 55, 1335–1346.

Wang, K.-J., Li, X.-H., Zhang, J.-J., Chen, H., Zhang, Z.-L. and Yu, G.-D. (2010) Natural introgression from cultivated soybean (*Glycine max*) into wild soybean (*Glycine soya*) with the implications for the origin of populations of semi-wild type and for biosafety of wild species in China. *Genetic Resources and Crop Evolution* 57, 747–761.

Wang, R.-L., Wendell, J. and Dekker, J. (1995) Weedy adaptation in *Setaria* ssp.: I. Isozyme analysis of the genetic diversity and population genetic structure in *S. viridis*. *American Journal of Botany* 82, 308–317.

Wang, X., Shi, X., Hao, B., Ge, S. and Luo, J. (2005) Duplication and DNA segmental loss in the rice genome: implications for diploidization. *New Phytologist* 165, 937–946.

Warwick, S.I. and Black, L. (1980) Uniparental inheritance of atrizine resistance in *Chenopodium album*. *Canadian Journal of Plant Science* 60, 751–753.

Washburn, S.L. and Moore, R. (1980) *Ape into Human: a Study of Human Evolution*. Little and Brown, Boston, Massachusetts.

Watkins, R. (1995) Apple and pear. In: Smartt, J. and Simmonds, N.W. (eds) *Evolution of Crop Plants*. Longman, London, pp. 418–422.

Watrud, L.S., Lee, E.H., Fairbrother, A., Burdick, C., Reichman, J.R., Bollman, M., Strom, M., King, G. and Van de Water, P.K. (2004) Evidence for landscape-level, pollen-mediated gene flow from genetically modified creeping bentgrass with *CP4 EPSPS* as a marker. *Proceedings of the National Academy of Sciences USA* 101, 14533–14538.

Way, R., Aldwinckle, H., Lamb, R., Rejman, A., Sansavini, S., Shen, T., Watkins, R., Westwood, M. and Yoshida, Y. (1991) Apples. In: Moore, J. and Ballington, J. (eds) *Genetic Resources in Temperate Fruit and Nut Crops*. International Society of Horticultural Science, Wageningen, the Netherlands, pp. 1–62.

Weatherwax, P. (1954) *Indian Corn in Old America*. Academic Press, New York.

Weeden, N.F. (1983) Evolution of plant isozymes. In: Tanksley, S.D. and Orton, T.J. (eds) *Isozymes in Plant Genetics and Breeding*. Elsevier Science Publishers, Amsterdam, pp. 175–208.

Weeden, N.F. (2007) Genetic changes accompanying the domestication of *Pisum sativum*: is there a common genetic basis to the domestication syndrome for legumes? *Annals of Botany* 100, 1017–1025.

Weeden, N. and Gottlieb, L.D. (1980) Isolation of cytoplasmic enzymes from pollen. *Plant Physiology* 66, 400–403.

Weeden, N. and Lamb, R. (1987) Genetics and linkage analysis of 19 isozyme loci in apple. *Journal of the American Society of Horticultural Sciences* 112, 865–872.

Weeden, N.F. and Robinson, R.W. (1990) Isozyme studies in *Cucurbita*. In: Bates, D.M., Robinson, R.W. and Jeffrey, C. (eds) *Biology and Utilization of the Cucurbitaceae*. Cornell University Press, Ithaca, New York, pp. 51–59.

Weeden, N.F., Reisch, B.I. and Martins, M.-H.E. (1988) Genetic analysis of isozyme polymorphism in grape. *Journal of the American Society for Horticultural Science* 113, 765–769.

Weeden, N.F., Muehlbauer, F.G. and Ladizinsky, G. (1992) Extensive conservation of linkage relationships between pea and lentil genetic maps. *Journal of Heredity* 83, 123–129.

Weinberg, W. (1908) On the demonstration of heredity in man (translated by S.H. Boyan) (1963) *Papers in Human Genetics*. Presentations Hall, Englewoods Cliffs, New Jersey.

Weinreich, D.M. and Chao, L. (2005) Rapid evolutionary escape by large populations from local fitness peaks is likely in nature. *Evolution* 59, 1175–1182.

Weir, B.S., Allard, R.W. and Kahler, A.L. (1972) Analysis of complex allozyme polymorphisms in a barley population. *Genetics* 72, 505–523.

Weiss, E., Kislev, M.E. and Hartmann, A. (2006) Autonomous cultivation before domestication. *Science* 312, 1606–1610.

Wendel, J.F. (1989) New World tetraploid cottons contain old world cytoplasm. *Proceedings of the National Academy of Sciences USA* 86, 4132–4136.

Wendel, J.F. (2000) Genome evolution in polyploids. *Plant Molecular Biology* 42, 225–249.

Wendel, J.F. and Albert, V.A. (1992) Phylogenetics of the cotton genus (*Gossypium*) – character state weighted parisimony analysis of chloroplast restriction site data and its systematic and biogeographic implications. *Systematic Botany* 17, 115–143.

Wendel, J.F. and Percy, R.G. (1990) Allozyme diversity and introgression in the Galapagos Islands endemic *Gossypium darwinii* and its relationship to continental *G. barbadense*. *Biochemical Ecology and Sytematics* 18, 517–528.

Wendel, J.F., Cronn, R.C., Johnston, J.S. and Price, H.J. (2002) Feast and famine in plant genomes. *Genetica* 114, 37–47.

Wendel, J.F., Brubaker, C.L. and Seelanan, T. (2010) The origin and evolution of *Gossypium*. In: Stewert, J.M., Oosterhuis, D., Heitholt, J.J. and Mauney, J.R. (eds) *Physiology of Cotton*. Springer Science+Business Media B.V., the Netherlands, pp. 1–18.

Whitaker, T.W. (1944) The inheritance of certain characters in a cross of two American species of *Lactuca*. *Bulletin of the Torrey Botanical Club* 71, 347–355.

White, M.J.D. (1978) *Modes of Speciation*. Freeman, San Francisco, California.

Whitehouse, M.L.K. (1950) Multiple-allelomorph incompatibility of pollen and style in the evolution of the angiosperms. *Annals of Botany* 14, 199–216.

Whitman, T.G., Morrow, P.A. and Potts, B.M. (1991) Conservation of hybrid plants. *Science* 254, 779–780.

Wijnheijmer, E.H.M., Brandenburg, W.A. and TerBorg, S.J. (1989) Interactions between wild and cultivated carrots (*Daucus carota* L.) in the Netherlands. *Euphytica* 40, 147–154.

Wilkes, H.G. (1977) Hybridization of maize and teosinte in Mexico and Guatemala and the improvement of maize. *Economic Botany* 31, 254–293.

Willcox, G., Fornite, S. and Herveux, L. (2008) Early Holocene cultivation before domestication in northern Syria. *Vegetation History and Archaeobotany* 17, 313–325.

Williams, E.G., Clarke, A.E. and Knox, R.B. (eds) (1994) *Genetic Control of Self-incompatibility and Reproductive Development in Flowering Plants*. Advances in Cellular and Molecular Biology of Plants, Vol. 2. Kluwer Academic, Dordrecht, the Netherlands.

Wills, D.M. and Burke, J.M. (2006) Chloroplast DNA variation confirms a single origin of domesticated sunflower (*Helianthus annuus* L.). *Journal of Heredity* 2006, 403–408.

Wilson, A.C., Carlson, S.S. and White, T.J. (1975) Biochemical evolution. *Annual Review of Biochemistry* 46, 573–639.

Wilson, C.D. and Heiser, D.B. (1979) The origin and evolutionary relationships of 'Huauzontle' (*Chenopodium nutalliae* Safford), a domesticated chenopod of Mexico. *American Journal of Botany* 66, 198–206.

Wilson, H.D. (1981) Genetic variation among South American populations of tetraploid *Chenopodium* sect. *Chenopodium* subsect. *Cellulata*. *Systematic Botany* 6, 380–398.

Wilson, H.D., Barber, S.C. and Walker, T.W. (1982) Loss of duplicate gene expression in tetraploid *Chenopodium*. *Biochemical Systematics and Ecology* 11, 7–13.

Wilson, R.E. and Rice, L.L. (1968) Allelopathy as expressed by *Helianthus annuus* and its role in old field succession. *Bulletin of the Torrey Botanical Club* 95, 432–448.

Wittwer, S., Youtai, Y., Han, S. and Lianzheng, W. (1987) *Feeding a Billion. Frontiers of Chinese Agriculture*. Michigan State University Press, East Lansing, Michigan.

Wolf, D.E., Takebayashi, N. and Rieseberg, L.H. (2001) Predicting the risk of extinction through hybridization. *Conservation Biology* 5, 1039–1053.

Wolfenbarger, L.L. and Grumet, R. (2003) Executive summary. In: Wolfenbarger, L.L. (ed.) *Proceedings of a Workshop on: Criteria for Field Testing of Plants with Engineered Regulatory, Metabolic, and Signaling Pathways, 3–4 June 2002*. Information Systems for Biotechnology, Blacksburg, Virginia, pp. 5–12.

Wong, C., Argent, G., Set, O., Lee, S.K. and Gan, Y.Y. (2002) Assessment of the validity of the sections of *Musa* (Musaceae) using AFLPs. *Annals of Botany* 90, 231–238.

Wood, T.E., Takebayashi, N., Barker, M.S., Mayrose, I., Greenspoon, P.B. and Rieseberg, L.H. (2009) The frequency of polyploidy speciation in vascular plants. *Proceedings of the National Academy of Sciences USA* 106, 13875–13879.

Workman, P.L. (1969) The analysis of simple genetic polymorphism. *Human Biology* 41, 97–114.

Wright, H.E. (1968) Natural environment of early food production north of Mesopotamia. *Science* 161, 334–339.

Wright, S. (1922) Coefficients of inbreeding and relationship. *American Naturalist* 56, 330–338.

Wright, S. (1931) Evolution in Mendelian populations. *Genetics* 16, 97–159.

Wright, S. (1943) Isolation by distance. *Genetics* 28, 114–138.

Wright, S. (1969) *Evolution and the Genetics of Populations. Vol. 2. The Theory of Gene Frequencies*. University of Chicago Press, Chicago, Illinois.

Wright, S.I., Bi, I.V., Schroeder, S.G., Yamasaki, M., Doebley, J.F., McMullen, M.D. and Gaut, B.S. (2005) The effects of artificial selection on the maize genome. *Science* 308, 1310–1314.

Wu, C.-I. (2001) The genic view of the process of speciation. *Journal of Evolutionary Biology* 14, 851–865.

Xiao, J., Li, J., Grandillo, S., Ahn, S.N., Yuan, L., Tanksley, S.D. and McCouch, S.R. (1998) Identification of trait-improving quantitative trait loci alleles from a wild rice relative, *Oryza rufipogan*. *Genetics* 150, 899–909.

Xiong, L.Z., Liu, K.D., Dai, X.K., Xu, C.G. and Zang, Q. (1999) Identification of genetic factors controlling domestication traits of rice using an F_2 population of a cross between *Oryza sativa* and *O. rufipogan*. *Theoretical and Applied Genetics* 98, 243–251.

Yabuno, T. (1966) Biosystematic study of the genus *Echinochloa*. *Japanese Journal of Botany* 19, 277–323.

Yamamoto, T., Lin, H., Sasaki, T. and Yano, M. (2000) Identification of heading date quantitative trait locus *Hd6* and characterization of its epistatic interactions with *Hd2* in rice using advanced backcross progeny. *Genetics* 154, 885–891.

Yamane, H., Tao, R., Sugiura, A., Hauck, N.R. and Iezzoni, A.F. (2001) Identification and characterization of S-RNases in tetraploid sour cherry (*Prunus cerasus* L.) *Journal of the American Society for Horticultural Sciences* 126, 661–667.

Yang, B., Thorogood, D., Armstead, I. and Barth, S. (2008) How far are we from unravelling self-incompatibility in grasses? *New Phytologist* 178, 740–753.

Yang, B., Thorogood, D., Armstead, I. and Barth, S. (2009) Identification of genes expressed during the self-incompatibility response in perennial ryegrass (*Lolium perenne* L.). *Plant Molecular Biology* 70, 709–723.

Yatabe, Y., Kane, N.C., Scott-Saintagne, C. and Rieseberg, L.H. (2007) Rampant gene exchange across a strong reproductive barrier between the annual sunflowers, *Helianthus annuus* and *H. petiolaris*. *Genetics* 175, 1883–1893.

Yen, D.E. (1982) Sweet potato in historical perspective. In: Villareal, R. and Griggs, T. (eds) *Sweet Potato – Proceedings 1st International Symposium*. AVRDC, Tainan, Taiwan.

Yu, S.B., Li, J.X., Gao, Y.J., Li, X.H., Zhang, Q. and Maroff, M.A.S. (1997) Importance of epistasis as the genetic basis of heterosis in an elite rice hybrid. *Proceedings of the National Academy of Sciences USA* 94, 9226–9231.

Yule, G.L. (1902) Mendel's laws and their probable relationship to intra-racial heredity. *New Phytologist* 1, 193–207.

Zhang, D., Cervantes, J., Huamán, Z., Carey, E. and Ghislain, M. (2000) Assessing genetic diversity of sweet potato (*Ipomoea batatas* (L.) Lam.) cultivars from tropical America using AFLP. *Genetic Research and Crop Evolution* 47, 659–665.

Zhang F., Zhai, H.-Q., Paterson, A.H., Xu, J.-L., Gao, Y.-M., et al. (2011) Dissecting Genetic Networks Underlying Complex Phenotypes: The Theoretical Framework. *PLoS ONE* 6(1), e14541. doi:10.1371/journal.pone.0014541

Zhang, L.-B., Zhu, Q., Wu, Z.-Q., Ross-Ibarra, J., Gaut, B.S., Ge, S. and Sang, T. (2009) Selection on grain shattering genes and rates of rice domestication. *New Phytologist* 184, 708–720.

Zhang, W.B., Zhang, J.R. and Hu, X.L. (1993) Distribution and diversity of *Malus* germplasm resources in Yunnan, China. *HortScience* 28, 978–980.

Zhao, Q., Thuillet, A.-C., Uhlmann, N.K., Weber, A., Rafalski, J.A., Allen, S.M., Tingey, S. and Doebley, J. (2008) The role of regulatory genes during maize domestication: evidence from nucleotide polymorphism and gene expression. *Genetics* 178, 2133–2143.

Zheng, X.-M. and Ge, S. (2010) Ecological divergence in the presence of gene flow in two closely related *Oryza species* (*Oryza rufipogan* and *O. nivara*). *Molecular Ecology* 19, 2439–2454.

Zhou, Z.Q. (1999) The apple genetic resources in China: the wild species and their distributions, informative characteristics and utilisation. *Genetic Resources and Crop Evolution* 46, 599–609.

Zhu, Q. and Ge, S. (2005) Phylogenetic relationships among A-genome species of the genus *Oryza* revealed by intron sequences of four nuclear genes. *New Phytologist* 167, 249–265.

Zhukovskii, P.M. (1970) Spontaneous and experimental introgression, its role in evolution and breeding. *Soviet Genetics* 6, 449–453.

Zijlstra, S., Purimahua, C. and Lindhout, P. (1991) Pollen tube growth in interspecific crosses between *Capsicum* species. *HortScience* 26, 585–587.

Zohary, D. (1965) Colonizer species in the wheat group. In: Baker, H.G. and Stebbins, G.L. (eds) *The Genetics of Colonizing Species*. Academic Press, New York.

Zohary, D. (1973) The origin of cultivated cereals and pulses in the Near East. In: Wahrman, J. and Lewis, K.R. (eds) *Chromosomes Today*, Vol. 4. John Wiley & Sons, New York.

Zohary, D. (1986) The origin and spread of agriculture in the old world. In: Barigozzi, C. (ed.) *The Origin and Domestication of Cultivated Plants*. Elsevier, Amsterdam, pp. 3–20.

Zohary, D. (1989) Pulse domestication and cereal domestication: how different are they? *Economic Botany* 43, 31–34.

Zohary, D. (1995a) Lentil: *Lens culinaris* (Leguminoseae – Papilionoideae). In: Smartt, J. and Simmonds, N.W. (eds) *Evolution of Crop Plants*. Longman Scientific & Technical, Harlow, UK, pp. 271–274.

Zohary, D. (1995b) Olive: *Olea europaea* (Oleaceae). In: Smartt, J. and Simmonds, N.W. (eds) *Evolution of Crop Plants*. Longman Scientific & Technical, Harlow, UK, pp. 379–382.

Zohary, D. and Feldman, M. (1962) Hybridization between amphidiploids and the evolution of polyploids in the wheat (*Aegilops-Triticum*) Group. *Evolution* 16, 44–61.

Zohary, D. and Hopf, M. (1973) Domestication of pulses in the Old World. *Science* 182, 887–894.

Zohary, D. and Hopf, M. (2000) *Domestication of Plants in the Old World*, 3rd edn. Clarendon Press, Oxford.

Zohary, D. and Spiegel-Roy, P. (1975) Beginnings of fruit growing in the Old World. *Science* 187, 319–327.

Zohary, M. (1972) *Flora Palaestina*. Israel Academy of Sciences and Humanities, Jerusalem.

Index